MAJOR STRUCTURAL ZONES AND FAULTS OF THE NORTHERN APPALACHIANS

Edited by
P. St-Julien and
J. Béland

1982

P. St-Julien
Département de Géologie,
Université Laval
Québec, P.Q. G1K 7P4

J. Béland
Département de Géologie
Université de Montreal
Montréal, P.Q. H3C 3J7

International Standard Book Number: 0-919216-20-X

Geological Association of Canada
Department of Earth Sciences
University of Waterloo
Waterloo, Ontario N2L 3G1

Printed by: Johanns Graphics Limited, Waterloo, Ontario

CONTENTS

PREFACE

The papers of this Special Paper were presented at a symposium held during the joint meeting of the Geological Association of Canada and the Mineralogical Association of Canada at Quebec City in May 1979. At that time the symposium was entitled "Major Lineaments in the Northern Appalachians"; this has now been changed to "Major Structural Zones and Faults of the Northern Appalachians" so it will reflect better the content of the papers that were presented and are included herein.

After having received an agreement with the Technical Programme Committee chaired by J.Y. Chagnon to hold the symposium and an indication from the Association's Editorial Committee, at the time chaired by W.G.E. Caldwell, that the papers could be published as a Special Paper, the researchers well known for their work in the northern Appalachians were asked to contribute to the symposium. All authors were invited to contribute to this volume but not all contributions presented at the symposium are included in this volume.

The editors wish to express their sincere thanks to the authors for their co-operation and the courtesy and patience they have shown during the preparation of this volume. We also wish to acknowledge the contributions of the following persons who generously gave their time in a variety of tasks: A.L. Albee, A. Baer, R. Béland, G. Boone, J.R. Butler, W.M. Cady, W.R. Church, K.L. Currie, A.A. Drake, T. Feininger, G.W. Fisher, W.K. Fyson, D.S. Harwood, A.D. Hatcher, Jr., H. Helmstaedt, A.M. Hussey, A. Hynes, P.B. Kink, R.S. Naylor, R.B. Newman, P.H. Osberg, W.E.A. Phillips, R.A. Price, D.W. Rankin, P.E. Schenk, J.W. Skehan, P. Stringer, and I. Zietz.

Our gratitude also goes to John Rodgers and William H. Poole who kindly gave their support to this enterprise at the very beginning.

Finally we would like to thank the Editorial Committee of Geological Association of Canada, especially W.G.E. Caldwell, J.W. Kramers and M.E. Czerneda, for their valuable assistance. The Geological Association of Canada acknowledges the generous support of the Natural Sciences and Engineering Research Council of Canada and the Geological Survey of Canada for providing grants to offset part of the publication costs of this volume.

P. St-Julien and J. Béland, Editors

MAJOR STRUCTURAL ZONES AND FAULTS OF THE NORTHERN APPALACHIANS

INTRODUCTION

The Northern Appalachians, comprised mainly of Paleozoic rocks, have been divided into longitudinal tectono-stratigraphic domains. Traditional subdivisions into anticlinoria and synclinoria have been used until recently by Cady (1960), E-An-Zen (1968) and Rodgers (1970). In general the anticlinoria expose Cambro-ordovician rocks and the synclinoria are covered by Silurian and Devonian formations.

More recently Williams (1978, 1979) from work done in Newfoundland and Williams et al. (1972, 1974) have recognized in the Appalachian Orogen five tectono-stratigraphic zones based essentially on features particular to the rocks of Middle Ordovician age and older. From west to east these zones have been named: Humber, Dunnage, Gander, Avalon and Meguma. Most of the characteristics particular to each zone persist along the whole length of the orogen.

Hall and Robinson (this volume), in a synthesis of the geology of New England, have also recognized five tectono-stratigraphic zones which can be correlated with those previously established by Williams. Their Zones I and IIA are equivalents of the Humber Zone; their Zones IIB and III correspond to the Dunnage Zone and their Zones IV and V to the Gander and Avalon Zones respectively. They also suggest that the limit between their Zones II and III could be a Taconian suture; this limit is nearly along the extension of the Baie Verte-Brompton Line considered by Williams and St-Julien (this volume) to represent the surface trace of an ancient continent-ocean interface destroyed by the Taconian Orogeny.

On the basis of lithologic distinctions and of the ranges of age, basement in New England has been divided into four domains by Osberg (1978). Basement A, of radiometric age of 1100 Ma, is exposed primarily along the east side of the western margin. Basement B, with a minimum isotopic age of 950 Ma, is exposed in western Maine. Basement C is exposed in small areas in west-central Massachusetts, east-central Maine, southeastern New Hampshire and in central Massachusetts. Radiometric age of 575 Ma is reported for these rocks in west central Massachusetts. Osberg includes the stratified core rocks of the Oliverian gneisses, in this basement C domain. Basement D is exposed in coastal Maine in eastern Massachusetts and in Rhode Island. Radiometric ages fall in the range 540 to 600 Ma. Lyons et al. (this volume) have recognized plates and zones in east central New England. Their zone G in the west with a Grenvillian basement correlates with the Humber zone. Their plate I to the east, in large part covered by the Siluro-Devonian of the Connecticut Valley-Gaspé synclinorium, is thought to be probably a remnant of the Dunnage Zone. Further to the east follows their plate A with an Avalonian basement extending from the west side of the Bronson Hill anticlinorium to a point beyond the present seacoast; they view the anticlinorium as the result of a paleosubduction. The Chain Lakes Massif in northern Maine and southwestern Quebec with rocks 1600 Ma old and its presumed extension to the north is their plate C and is equivalent to basement B of Osberg (1978).

The tectono-stratigraphic zones in the northern Appalachians are in general bounded by major faults. One of these is the Baie Verte-Brompton Line (Williams and St-Julien, this volume) which separates the Humber and Dunnage Zones. Another is the Dover-Hermitage Bay Fault described by Kennedy *et al.* (this volume) which divides the Gander and Avalon Zones. A third fault is the Minas Geofracture discussed by Keppie (this volume) which separates the Avalon and Meguma Zones. As pointed out by Keppie, this last fault has had a long and complexe history that lasted throughout the Acadian and Hercynian Orogenies. With its sister fault, the Chinecto Bay Fault, it is closely related to folding and major dextral transcurrent movements in that part of the northern Appalachians.

Within the Gander zone in central western Maine, Boudette (this volume) recognizes an ophiolitic assemblage that rests on the southern edge of the Chain Lakes Precambrian Massif. Small isolated bodies of steatite and serpentinite along strike are thought to represent remnants of this same ophiolite caught in a mélange and the whole zone is interpreted as the surface expression of a suture between two microplates within the Gander Zone.

Since the advent of the theory of plate tectonics several models of the evolution of the northern Appalachians (Dewey, 1969; Bird and Dewey, 1970; Stevens, 1970; Williams *et al.*, 1972; St-Julien and Hubert, 1975; Osberg, 1978) have been proposed. All models imply the formation and the destruction of the Iapetus Ocean from the Late Precambrian to the Early Paleozoic and all are consistent with the zonation of the Early Paleozoic proposed by Williams (1978) particularly, the Humber, Dunnage and Gander Zones.

CAMBRO ORDOVICIAN

The *Humber Zone* represents the formation and the destruction of an Atlantic type continental margin, that is, the ancient margin of the North American continent (Rodgers, 1968; Williams and Stevens, 1974; St-Julien and Hubert 1975; Williams, 1979). The Cambrian and Ordovician in this zone rest with an angular unconformity on the Grenvillian crystalline basement; the eastern limit is the Baie Verte-Brompton Line with its ophiolites.

In the western sector, the rocks resting unconformably on the Grenvillian basement are mostly platformal carbonates at the base, overlain directly by Middle Ordovician turbidites, followed by easterly derived olistostromes on which came to rest allochthonous slices of psammetic and pelitic rocks originally deposited at the continental slope and rise and later thrust onto the platform (Bird and Dewey 1970; Williams and Stevens, 1974; St-Julien and Hubert, 1975). In Newfoundland, sheets of ophiolites crown the allochthonous stack. In general the allochthonous rocks towards the east and southeast are progressively more deformed and are covered by nappes of schists and locally of Grenvillian basement rocks.

The *Dunnage Zone* is what is left of the Iapetus Ocean crust (Upadhyay *et al.*, 1971; Smitheringale, 1972; Kay, 1975; Norman and Strong, 1975; Kidd, 1977), and of the calc-alkaline volcanic rocks (Kean and Strong, 1975) and the sedimentary rocks (St-Julien and Hubert, 1975; Williams, 1979) that were deposited on it. This zone reaches its maximum width in northeastern Newfoundland but is reduced to a fault line in the southwestern part of the island. Haworth and Miller (this volume) have shown that the magnetic and gravimetric anomalies at Notre Dame Bay in Newfoundland indicate that the oceanic crust extends here offshore in a northeasterly trending synclinal structure. From other regional geophysical informations (Haworth *et al.*, 1978) they believe that this synclinal structure is a secondary feature in an overall southeasterly dipping crust indicating that subduction was towards the southeast.

The Dunnage Zone according to Williams (1978, 1979) re-appears on the mainland in northern New Brunswick and southeastern Quebec. In northern New Brunswick it is represented by the Fournier Group ophiolitic complex (Rast *et al.*, 1976; Pajari *et al.*, 1977; Ruitenberg *et al.*, 1977).

In Newfoundland the southeastern limit of the Dunnage Zone is at the Gander River Belt mafic and ultramafic complex (Williams, 1979) and in New Brunswick it coincides with the Rocky Brook-Millstream Fault (Rast and Stringer, 1974; Rast *et al.*, 1976). In Quebec, Williams (1978) has set the limit of the Dunnage Zone at the southeastern edge of the ophiolitic complex but ought to be included also, the olistostromal pebbly mudstone and other rocks of the St-Daniel Formation (Williams and St-Julien, this volume) the turbidite sequence of the St-Victor synclinorium and the calc-alkaline volcanic rocks of the Ascot and Weedon Formations (St-Julien and Hubert, 1975). Thus in Quebec, a complete section of the Dunnage Zone begins with an ophiolitic complex followed in successive order by red and green pelagic shales, calc-alkaline volcanic and pyroclastic rocks, an assemblage of shale and oliostostromal pebbly mudstone (St-Daniel Formation), an alternance of felsic tuffs, pelagic shales and chert (Beauceville Formation) and finally a sequence of turbidites (St-Victor Formation). The Ascot and Weedon calc-alkaline volcanic rocks are considered to be southeastern lateral equivalents of the Beauceville and St-Victor Formations (St-Julien and Hubert, 1975).

Doolan *et al.* (this volume) explains the disappearance of the ophiolitic complexes and calc-alkaline volcanic sequences of the Dunnage Zone to the south towards Vermont by a diachronous impingement of irregular margins during a continent-arc collision. To the northeast (northeast of Thetford Mines in Quebec) a similar truncation is attributed to transcurrent faults formed during the closure of the Iapetus Ocean against the Chain Lakes Massif plate. In this regard it ought to be mentioned that within the Dunnage Zone, deformation and metamorphism are relatively less pronounced than in either of the adjacent zones (Humber and Gander).

The *Gander Zone* comprises pre-Middle Ordovician metamorphic rocks that have been subjected to multiple deformations. It includes, in Newfoundland, the Gander Group, in New-Brunswick, the lower part of the Tetagouche Group and in New England, metamorphic rock correlatives of the Lower Tetagouche (Williams, 1978, 1979). The southeastern limit of the Gander Zone is, in Newfoundland, the Dover-Hermitage Bay Fault (Blackwood and Kennedy, 1975; Blackwood and O'Driscoll, 1976; Kennedy *et al.*, this volume) and in New Brunswick, the Lubec-Belle Isle Fault (Rast and Dickson, this volume; Kennedy *et al.*, this volume). The Clinton-Newbury Fault in Massachusetts to the southwest occupies a similar position (Kennedy *et al.*, this volume).

The thick terrigenous sequence that characterizes the Gander Zone was according to Williams (1964) deposited at a continental margin which according to Pajari *et al.* (1977) and Williams and Doolan (1978) was of Andean type. Thus the Gander Zone resulted from the construction and the destruction of the Iapetus Ocean southeastern margin (Kennedy, 1976). In Newfoundland the most ancient deformation is assigned to the Lower Ordovician because a Caradocian conglomerate with tectonized clasts rests with angular unconformity on ultramafic rocks (Kennedy, 1976). Similarly in New-Brunswick within the Tetagouche Group a coeval unconformity separates the lower clastic rocks from the upper volcanic assemblage (Pajari *et al.*, 1977).

The *Avalon Zone* is constituted mainly of Late Precambrian volcanic and sedimentary rocks which in Newfoundland are relatively mildly deformed compared to the adjacent Gander Zone to the north. Minor local unconformities are observed in places whereas, elsewhere, the Late Precambrian passes up conformably into the Lower Paleozoic. The lithology of the Cambrian indicates that a platform existed there as the Iapetus Ocean was opening to the northwest (Williams, 1979). To the southwest, in Cape Breton Island, southern New-Brunswick and southeastern New England similar Late Precambrian rocks point to an extension of the Avalon Zone (Lyons *et al.*, this volume; Hall and Robinson, this volume).

The *Meguma Zone* is restricted to southwestern Nova Scotia and is separated from the Avalon Zone by the east-west transcurrent Minas Geofracture (Keppie, this volume). It is constituted by the Meguma Group which is a greywacke sequence overlain by a shale bearing in its upper levels a Lower Ordovician graptolite fauna (Schenk, 1970, 1976). Conformably above the Meguma Group is an assemblage of undated volcanic and sedimentary rocks which begins with a tillite believed to represent a Middle Ordovician glacial episode (Schenk 1972; Lane, 1976) and still higher in the stratigraphic sequence, are Devonian sedimentary rocks.

Sedimentological features of the Meguma Group point to a southeasterly source in what must have been a metamorphic terrane which because of the large amount of sediments derived is thought to have had continental dimensions (Schenk 1970, 1971, 1976). Thus the Meguma Zone is interpreted as a remnant of the African continent left behind glued to the North American continent as the present Atlantic Ocean began to open in Mesozoic time (Schenk 1971, 1975; Keppie, 1977a).

SILURIAN AND DEVONIAN

The Siluro-Devonian rocks in the northern Appalachians reveal a paleogeographic and tectonic evolution in great contrast with that indicated by the Cambro-Ordovician. They are mainly marine and continental deposits resting with angular unconformity on a Cambro-Ordovician basement thus pointing to a near complete destruction of the continental margins of the Iapetus Ocean (Williams, 1979). In the Quebec Appalachians the Siluro-Devonian strata of the Connecticut Valley-Gaspé synclinorium rest with angular unconformity on folded Cambro-Ordovician formations and the synclinorium axis cuts obliquely across the Baie Verte-Brompton Line separating the Humber and Dunnage Zones (Béland, 1974; St-Julien and Hubert, 1975; Williams and St-Julien, 1978; Williams and St-Julien, this volume). Similarly, at White Bay in northern Newfoundland, Silurian volcanic and sedimentary rocks rest with angular unconformity on the Humber Zone allochthons (Williams, 1977, 1979). In the Dunnage Zone also, non-fossiliferous strata correlated with the Silurian overlie unconformably Lower Ordovician ophiolites and island arc volcanic sequences (Schroeter, 1973; Neale *et al.*, 1975; Kidd, 1977). Furthermore, the lack of Siluro-Devonian ophiolites and the absence of extensive continental margins, pelagic or semi-pelagic marine sediments, ophiolitic mélanges and calc-alkaline volcanic suites strongly indicate that the Iapetus Ocean had disappeared prior to the Siluro-Devonian accumulation. So it is thought by several authors (Williams, 1979; Williams and St-Julien, this volume; Fyffe, this volume) that the history of the Silurian and Devonian systems in the northern Appalachians is that of sedimentary basins formed on the destroyed margins and some relics of the Iapetus Ocean. Other authors (Dewey, 1969; Bird and Dewey, 1970; McKerrow and Ziegler, 1971, 1972; Schenk 1971; Dewey and Kidd, 1974; St-Julien and Hubert, 1975; McKerrow and Cocks, 1977, 1978; Keppie, 1977a, 1977b; Osberg, 1978) however, have expressed the view that the Acadian Orogeny coincided with the final closure of the Iapetus Ocean in Late Silurian-Early Devonian. The positions of possible Siluro-Devonian sutures have been proposed by McKerrow and Cocks (1977, 1978) and Hall and Robinson (this volume).

CARBONIFEROUS

With the Carboniferous here are some Late Devonian strata which in terms of tectono-stratigraphic units belong to the same sequence. Most of this sequence is located in a large basin bordered on the north and south by what may be referred to as the New Brunswick and Meguma basements respectively. Two smaller basins: Boston and Narragansett occur in the states of Massachussetts and Rhode Island respectively.

The strata involved rest with angular unconformity on the metamorphic Precambrian or the Early Paleozoic and they comprise mostly continental sandstones, conglomerates, siltstones and shales with minor limestones and evaporites; red beds abound. Their environment of sedimentation, vast arid land surfaces with but brief marine incursions, changed little.

The larger Fundy Basin has been interpreted as a fault-bounded rift valley with the faults remaining active during the whole episode of sedimentation (Belt, 1965, 1968). Folds are ascribed to block movements and diapiric surges of salt beds. Other authors (Ruitenberg and McCutcheon, this volume; Ruitenberg *et al.*, 1973; Rast and Grant 1973; Rast *et al.*, 1976), however, have attributed to an Hercynian orogenic phase a marked deformation observed in Carboniferous rocks in northern New-Brunswick. Similar narrow zones of highly deformed rocks are also encountered in relation to the Minas Geofracture (Keppie, this volume) in Nova Scotia and in Carboniferous rocks of Newfoundland.

TRIASSIC

The triassic in the northern Appalachians is restricted to two basins in part bounded by normal faults. They are the Connecticut basin in the states of Connecticut and Massachusetts and the Minas basin in Nova Scotia. Both contain essentially nondeformed red beds and basalts.

OROGENIES

The *northern Appalachians* have been affected by several orogenies; the principal ones being: the Avalonian, the Taconian, the Acadian and the Hercynian (Alleghenian). Less well established orogenic phases are indicated during the Late Precambrian and the Siluro-Devonian.

The Paleozoic rocks in general rest with angular unconformity on a Precambrian basement. This contact however is locally difficult to trace in as much as the assignment of certain units to the Precambrian or the Paleozoic is not always unequivocal. Besides, in some area metasedimentary rocks considered to be Precambrian overlie with angular unconformity other Precambrian rocks of higher metamorphic grade (granulite facies).

The *Avalonian Orogeny* restricted to the Avalon Zone is Late Precambrian. In Newfoundland it is revealed by an angular unconformity and plutonic intrusions. In New Brunswick and Nova Scotia it involved folding, metamorphism and mylonitization in steeply dipping zones. In this regard Rast and Dickson (this volume) have proposed that at the opening of the Iapetus Ocean, in Late Precambrian, the separation of the North American and Avalonian plates was achieved by strike slip faulting as indicated by the major Pocologan mylonite zone. Later, other movements led to further opening, deformation and metamorphism.

The *Taconian Orogeny*, as is generally admitted, took place during the Ordovician and resulted from a subduction followed by a collision. Several hypotheses however have been proposed as to how many zones of subduction were involved, in what direction were they dipping and what geological bodies came into collision. For some authors (Williams, 1979; Fyffe, this volume; Hall and Robinson, this volume) the Taconian Orogeny resulted from the closing of the Iapetus Ocean and the destruction of the ancient continental margin of North America (continent-continent collision). For Hall and Robinson it involved the suturing of the Avalonian basement onto the North American Grenvillian basement. For others (Bird and Dewey, 1980; Dewey and Kidd, 1974; St-Julien and Hubert, 1975; Osberg, 1978; Hatch, this volume) the collision was between the continent and islands arcs.

The tectono-stratigraphic zone most affected by the Taconian episode is the Humber Zone although the Dunnage and the Gander Zones were also involved (Williams, 1979). In the Humber Zone thrusts and nappes, marked polyphase deformation and metamorphism can be ascribed to this orogeny. Pre-Middle Ordovician polyphased deformation and high grade metamorphism are also recognized in the Gander Zone (Kennedy and McGonigal, 1972). Kennedy (1976), however, has held the view that deformation and metamorphism in the Gander Zone did take place in the Late Precambrian.

The effect of the Taconian Orogeny in the Dunnage Zone as compared to what is observed in the bordering Humber and Gander Zones is much less pronounced.

The *Acadian Orogeny* which took place during the Devonian is obviously the main diastrophic event recorded in the northern Appalachians. Practically all rocks of the orogen are strongly affected by it and it is held responsible for the structuration into the major anticlinoria and synclinoria that now characterize the mountain belt. Upright tight folding accompanied by a mild metamorphism is generalized in the northeast and northwest whereas recumbent folding and high grade metamorphism prevail in the southeast. Scores of late to post-orogenic granitic plutons intrude the Siluro-Devonian strata (Williams *et al.*, 1972; Rast and Lutes, 1979).

Most authors relate the Acadian events to the final closure of the Iapetus Ocean in a continent-continent collision (Dewey 1969; McKerrow and Ziegler 1971, 1972; Schenk, 1971; McKerrow and Cocks 1977, 1978; Keppie 1977a, 1977b; Osberg, 1978; Hatch, this volume). According to Osberg (1978) and to Hatch (this volume), during both the Taconian and the Acadian orogenies subduction was effected with a southeasterly dip under the Avalonian plate advancing to the northwest. Williams (1979) however holds the view that a continent-continent collision had taken place during the Taconian Orogeny and that there is little evidence of the persistence of a continental margin in Siluro-Devonian time and little indication of a Devonian suture. Hall and Robinson (this volume) in southern New England locate the Acadian suture in a Siluro-Devonian sedimentary trough between two zones of Avalonian or correlative basement without any clear evidence that an intervening oceanic crust ever existed in this basin. Fyffe (this volume) considers that the Matapedia Basin represents a remnant seaway that persisted after the closure of the Early Paleozoic Ocean during the Taconian Orogeny. He holds that the Fredericton trough evolved from a graben structure formed in what he assumes was in Silurian times a continuous landmass uniting the present Miramichi Highlands to these of the Caledonian Range of southern New Brunswick. In his opinion since no Siluro-Devonian trench environment can be here recognized, plate tectonics models that attempt to relate volcanism of this age to subduction are erroneous. He also considers that the southward verging Taconian folds in the Miramichi highlands are compatible with the closing of an Ordovician Ocean in northern New Brunswick and that the upright Acadian folds and high angle reverse faults can be attributed to further continental convergence.

The *Hercynian (Alleghenian) Orogeny* has affected the southeastern sector of the northern Appalachians. In southern New Brunswick and Nova Scotia it has produced folds and thrust faults (Fyson, 1967; Currie, 1977; Rast, 1980) and in eastern New England, the intrusion, over a distance of some 120 km of two mica granites of the Concord plutonic series, dated 327 to 375 Ma (Early Pennsylvanian to Early Permian) and accompanied by metamorphism, has been attributed to this orogeny (Lyons *et al.*, this volume). According to Ruitenberg and McCutcheon (this volume) the structural trends resulting from both the Acadian and the Hercynian Orogenies were largely controlled by dextral movements along the Cobequid-Chedebucto Fault (Minas Geofracture) that separates the Avalon and the Meguma Zones.

To conclude this introductory presentation, it can be said that of the twelve contributions included in this volume, seven deal with the tectonostratigraphic zonation in the northern Appalachians or discuss related problems, and five are concerned with major faults many of which are boundaries to the structural zones. Hall and Robinson and Lyons *et al.* propose zonations of respectively southern and east central New England. Hatch expresses the view that in western New England the Taconian and the Acadian Orogenies can be tied to the subduction of an oceanic plate that descended southeasterly under the Avalon Continental Plate as the latter advanced to the northwest. Doolan *et al.* explains the lack of continuity of the Dunnage Zone in Quebec-Vermont by a diachronous impingement of irregular adjacent margins in a continent-arc collision on one end and a truncation by transcurrent faults at the other end, all of this taking place during closure of the Iapetus Ocean against the Chain Lakes Massif Plate. Fyffe analyses the effects of the Taconian and Acadian Orogenies in the Gander and Dunnage Zones in central and northern New Brunswick. Ruitenberg and McCutcheon attempt to relate the structural trends within the Avalon Zone in southern New Brunswick to the Acadian and Hercynian (Alleghenian) Orogenies. Haworth and Miller use geophysical data to establish that the Dunnage Zone extends offshore in Newfoundland in a synclinal structure which they consider is a secondary feature in an overall southeasterly dipping oceanic plate.

In regard to the contributions on faults, Williams and St-Julien stress the significance of the Baie Verte-Brompton Line separating the Humber and Dunnage Zones. It is in their view the surface trace of an ancient continent-ocean interface that was but all destroyed by the Taconian Orogeny. Kennedy *et al.* imply that the Dover-Hermitage Bay Fault that separates the Gander and the Avalon Zones is essentially a dip-slip fault that played its role essentially during the Late Precambrian. Boudette sees in an ophiolitic complex at the southern edge of the Chain Lakes Massif and in the isolated small bodies of ultramafic rocks dispersed along the extension of this complex, a suture within the Gander Zone. From the major mylonite zone observed at the northwest margin of the Avalon Zone in southern New Brunswick, Rast and Dickson deduce that in Late Precambrian the Iapetus Ocean began to open up by a strike-slip fault that led to an eventual separation of the North American and Avalonian Plates. Finally, according to Keppie the major fault zone (Minas Geofracture) that separates the Meguma and Avalon Zones had a complex history of successive movements that lasted through the Acadian and Hercynian (Alleghenian) orogenies involving dextral movements accompanied by folding.

REFERENCES

Béland, J., 1974, La tectonique des Appalaches du Québec: Geosci. Canada, v. 1, no. 4, p. 26-32.

Belt, E.S., 1965, Stratigraphy and paleogeography of the Mabou Group and related middle Carboniferous facies, Nova Scotia, Canada: Geol. Soc. America Bull., v. 76, p. 777-801.

―――――――――, 1968, Carboniferous continental sedimentation, Atlantic Provinces, Canada: in Klein, G.V., ed., Symposium on continental sedimentation: Geological Society of America Spec. Paper 106, p. 127-176.

Bird, J.M. and Dewey, J.F., 1970, Lithosphere plate-continental margin tectonics and the evolution of the Appalachian Orogen: Geol. Soc. America Bull., v. 81, p. 1031-1060.

Blackwood, R.F. and Kennedy, M.J., 1975, The Dover Fault: western boundary of the Avalon Zone in northeastern Newfoundland: Canadian Jour. Earth Sci., v. 12, p. 320-325.

Blackwood, R.F. and O'Driscoll, C.E., 1976, The Gander-Avalon Zone boundary in southeastern Newfoundland: Canadian Jour. Earth Sci., v. 13, p. 1155-1159.

Cady, W.M., 1960, Stratigraphic and geotectonic relationships in northern Vermont and southern Quebec: Geol. Soc. America Bull., v. 71, p. 531-576.

Currie, K.L. 1977, A note on post-Mississippian thrust faulting in northwestern Cape Breton Island: Canadian Jour. Earth Sci., v. 14, p. 2937-2941.

Dewey, J.F., 1969, The evolution of the Caledonian/Appalachian Orogen: Nature, v. 222, p. 124-128.

Dewey, J.F. and Kidd, W.S.F., 1974, Continental collision in the Appalachian/Caledonian orogenic belt: variations related to complete and incomplete suturing: Geology, v. 2, p. 543-546.

Fyson, W.K., 1967, Gravity sliding and cross folding in Carboniferous rocks, Nova Scotia: American Jour. Sci., v. 265, p. 1-11.

Haworth, R.T., Lefort, T.P. and Miller, H.G., 1978, Geophysical evidence for an east dipping Appalachian subduction zone beneath Newfoundland: Geology, v. 6, p. 522-526.

Kay, M., 1975, Campbellton sequence, manganiferous beds adjoining the Dunnage Mélange, northeastern Newfoundland: Geol. Soc. America Bull., v. 86, p. 105-108.

Kean, B.F. and Strong, D.F., 1975, Geochemical evolution of an Ordovician island arc of the Central Newfoundland Appalachians: American Jour. Sci., v. 275, p. 97-118.

Kennedy, M.J., 1976, Southeastern margin of the Northeastern Appalachians: Late Precambrian orogeny on a continental margin: Geol. Soc. America Bull., v. 87, p. 1317-1325.

Kennedy, M.J. and McGonigal, M., 1972, The Gander Lake and Davidsville Groups of northeastern Newfoundland: new data and geotectonic implication: Canadian Jour. Earth Sci., v. 9, p. 452-459.

Keppie, J.D., 1977a, Tectonics of Southern Nova Scotia: Nova Scotia Dept. Mines Paper 77-1, 34 p.

_____, 1977b, Plate tectonic interpretation of Paleozoic World Maps: Nova Scotia Dept. Mines Paper 77-3, 45 p.

Kidd, W.S.F., 1977, The Baie Verte Lineament, Newfoundland: Ophiolite complex floor and mafic volcanic fill of a small Ordovician marginal basin: in Talwani, M. and Pitman, W.C., eds., Islands arcs, deep sea trenches and back-arc basins: American Geophys. Union, Maurice Erwing Series 1, p. 407-418.

Lane, T.E., 1976, Stratigraphy of the White Rock Formation: Maritime Sediments, v. 12, p. 87-106.

McKerrow, W.S. and Cocks, L.R.M., 1977, The location of the Iapetus Ocean suture in Newfoundland: Canadian Jour. Earth Sci., v. 14, p. 488-489.

_____, 1978, A Lower Paleozoic trench-fill sequence, New World Island, Newfoundland: Geol. Soc. America Bull., v. 89, p. 1121-1132.

McKerrow, W.S. and Ziegler, A.M., 1971, The Lower Silurian paleogeography of New Brunswick and adjacent areas: Jour. Geol., v. 79, p. 635-646.

_____, 1972, Silurian paleogeographic development of the Proto-Atlantic Ocean: Proceeding 24th International Geological Congress, Montreal, Section 6, p. 4-10.

Neale, E.R.W., Kean, B.F. and Upadhyay, H.D., 1975, Post-ophiolite unconformity, Tilt Cove-Betts Cove area, Newfoundland: Canadian Jour. Earth Sci., v. 12, p. 880-886.

Norman, R.E. and Strong, D.F., 1975, The geology and geochemistry of ophiolitic rocks exposed at Mings Bight, Newfoundland: Canadian Jour. Earth Sci., v. 12, p. 777-797.

Osberg, P.H., 1978, Synthesis of the geology of the northeastern Appalachians: in IGCP Project 27, Caledonian-Appalachian orogen of the North Atlantic Region: Geol. Survey Canada Paper 78-13, p. 137-147.

Pajari, G.E., Jr., Rast, N. and Stringer, P., 1977, Paleozoic volcanicity along the Bathurst-Dalhousie Geotraverse, New Brunswick, and its relations to structure: in Baragar, W.R.A., Coleman, L.C. and Hall, J.M., eds., Volcanic Regimes in Canada: Geol. Assoc. Canada Spec. Paper 16, p. 111-124.

Rast, N., 1980, The Avalonian plate in the Northern Appalachians and Caledonides: in Wones, D.R., ed., The Caledonides of the U.S.A.: Virginia Polytechnic Instit. Geological Sciences, Memoir 2, p. 63-66.

Rast, N. and Grant, R., 1973, Transatlantic correlation of the Variscan-Appalachian Orogeny: American Jour. Sci., v. 273, p. 572-579.

Rast, N. and Lutes, G.G., 1979, The metamorphic aureole of the Pokiok-Skifflake granite, southern New Brunswick: Geol. Survey Canada, Paper 79-1A, p. 267-271.

Rast, N. and Stringer, P., 1974, Recent advances and the interpretation of geologic structure of New Brunswick: Geosci. Canada, v. 1, no. 4, p. 15-25.

Rast, N., Stringer, P. and Burke, K.B.S., 1976, Profiles across the northern Appalachians of Maritime Canada: in Drake, C.L., ed., Geodynamics: Progress and Projects: American Geophys. Union, p. 193-202.

Rodgers, John, 1968, The eastern edge of the North American continent during the Cambrian and Early Ordovician: in E-an Zen *et al.*, eds., Studies of Appalachian Geology: Northern and Maritime: New York, John Wiley and Sons, p. 141-150.

——————, 1970, The tectonics of the Appalachians: Wiley Interscience, New York, 271 p.

Ruitenberg, A.A., Fyffe, L.R. and McCutcheon, S.R., 1977, Evolution of pre-Carboniferous tectonostratigraphic zones in the New Brunswick Appalachians: Geosci. Canada, v. 4, p. 171-181.

Ruitenberg, A.A., Venugopal, D.V. and Giles, P.S., 1973, Fundy Cataclastic Zone, New Brunswick: evidence for post-Acadian penetrative deformation: Geol. Soc. America Bull., v. 84, p. 3029-3044.

St-Julien, P. and Hubert, C., 1975, Evolution of the Taconian Orogen in the Quebec Appalachians: American Jour. Sci., v. 275A, p. 337-362.

St-Julien, P., Hubert, C. and Williams, H., 1976, The Baie Verte-Brompton Line and its possible tectonic significance in the northern Appalachians (abst.): Geol. Soc. America, Abstracts with Programs, v. 8, p. 259-260.

Schenk, P.E., 1970, Regional variation of the flysch-like Meguma Group (Lower Paleozoic) of Nova Scotia compared to recent sedimentation of the Scotian Shelf: in Lajoie, J., ed., Flysch Sedimentology in North America: Geol. Assoc. Canada Spec. Paper 7, p. 127-153.

——————, 1971, Southeastern Atlantic Canada, northwestern Africa, and continental drift: Canadian Jour. Earth Sci., v. 8, p. 1218-1251.

——————, 1972, Possible Late Ordovician glaciation of Nova Scotia: Canadian Jour. Earth Sci., v. 9, p. 95-107.

——————, 1975, Paleozoic evolution of African Nova Scotia-polar and deep to equatorial and continental: IXth International Congress of Sedimentology, Nice, Theme I, p. 181-186.

——————, 1976, A regional synthesis (of Nova Scotia geology): Maritime Sediments, v. 12, p. 17-24.

Schroeter, T.M., 1973, Geology of the Nippers Harbour area, Newfoundland: University of Western Ontario, London, Ontario, M.Sc. Thesis, 88 p.

Smitheringale, W.G., 1972, Low potash Lushs Bight tholeiites: ancient oceanic crust in Newfoundland: Canadian Jour. Earth Sci., v. 9, p. 574-588.

Stevens, R.K., 1970, Cambro-Ordovician flysch sedimentation and tectonics in west Newfoundland and their possible bearing on a Proto-Atlantic Ocean: in Lajoie, J., ed., Flysch Sedimentology in North America: Geol. Assoc. Canada, Spec. Paper 7, p. 165-177.

Upadhyay, H.D., Dewey, J.F. and Neale, E.R.W., 1971, The Betts Cove ophiolite complex, Newfoundland: Appalachian oceanic crust and mantle: Geol. Assoc. Canada Proceedings, v. 24, p. 27-34.

Williams, H., 1964, The Appalachians in Northeastern Newfoundland – a two-sided symmetrical system: American Jour. Sci., v. 262, p. 1137-1158.

——————, 1977, The Coney Head Complex: another Taconic allochthon in West Newfoundland: American Jour. Science, v. 277, p. 1279-1295.

——————, 1978, Tectonic-lithofacies map of the Appalachian orogen: Dept. Geol., Memorial University of Newfoundland, St. Johns, Map No. 1, 1:1 000 000.

——————, 1979, Appalachian Orogen in Canada: Canadian Jour. Earth Sci., v. 16, p. 792-807.

Williams, H. and Doolan, B.L., 1978, Margins and vestiges of Iapetus (abst.): Geol. Soc. America-Geol. Assoc. Canada, Abstracts with Programs, v. 10, No. 7, p. 517.

Williams, H., Kennedy, M.J. and Neale, E.R.W., 1972, The Appalachian structural province: *in* Price, R.A., and Douglas, R.J.W., eds., Variations in Tectonic Styles in Canada: Geol. Assoc. Canada Spec. Paper 11, p. 181-261.

_____, 1974, The northwestward termination of the Appalachian orogen: *in* Nairn, A.E.M., and Stehli, F.G., eds., The Ocean Basins and Margins: v. 2, New York Plenum, p. 79-123.

Williams, H. and St. Julien, P., 1978, The Baie Verte-Brompton Line in Newfoundland and regional correlations in the Canadian Appalachians: Geol. Survey of Canada, Paper 78-1A, p. 225-229.

Williams, H. and Stevens, R.K., 1974, The ancient continental margin of eastern North America: *in* Burk, C.A., and Drake, C.L., eds., The geology of continental margins: Springer-Verlag, New York, p. 781-796.

Zen, E-an, 1968, Nature of the Ordovician orogeny in the Taconic area: *in* Zen, E-an *et al.*, eds, Studies of Appalachian Geology, Northern and Maritime: Wiley Interscience, New York, p. 129-140.

P. St-Julien and J. Béland

STRUCTURAL ZONES

Major Structural Zones and Faults of the Northern Appalachians, edited by
P. St-Julien and J. Béland, Geological Association of Canada Special Paper 24, 1982

STRATIGRAPHIC-TECTONIC SUBDIVISIONS OF SOUTHERN NEW ENGLAND

Leo M. Hall and Peter Robinson
Department of Geology and Geography, University of Massachusetts
Amherst, Massachusetts 01003

ABSTRACT

Pre-Carboniferous rocks of southern New England and southeastern New York divide into five stratigraphic-tectonic zones based mainly on the ages and types of Precambrian, Cambrian, and Ordovician rocks. From west to east they are: I) Manhattan-Williamstown Zone with Grenvillian and some Late Precambrian basement and unconformable cover of Cambrian-Ordovician clastics and carbonates; II) Greenwich-Rowe Zone with Precambrian and Late Precambrian(?) continental basement (IIA) or Late Precambrian-Early Paleozoic oceanic crust (IIB) overlain by Cambrian-Ordovician clastic and volcanic cover; III) Haddam-Warwick Zone of Late Precambrian gneisses and quartzose sedimentary rocks along with other gneisses of uncertain age, possibly Ordovician, overlain with apparent unconformity by Middle Ordovician volcanics and black shales; IV) Milford-Nashoba Zone with basement rocks of Late Precambrian and uncertain age as in Zone III, but overlain by a probable Middle Ordovician section dominated by oxidized shales with lesser black shales and volcanics; and V) Dedham Zone with unmetamorphosed to weakly metamorphosed Late Precambrian volcanic, clastic, and plutonic rocks overlain unconformably by Cambrian clastic strata of the Baltic faunal realm.

The eastern limit of exposed Cambrian-Ordovician carbonate section essentially defines the boundary between I and II. This section is structurally overlain by thrust sheets of Cambrian-Ordovician clastics and some volcanics transported from Zone II into Zone I during Taconian deformation. Intense metamorphism seems to have accompanied late stages of Taconian deformation.

Zone II is divided into two parts separated by a thrust fault referred to as Cameron's Line. The western part, Zone IIA, contains Lower Cambrian clastics resting unconformably on North American Grenvillian basement, and is the probable source of thrust sheets that lie to the west. The eastern part, Zone IIB, is characterized by a Cambro-Ordovician clastic-volcanic section resting on what are interpreted as fragmented oceanic crustal rocks, predominantly mafic and ultramafic rocks. In southwestern Connecticut Zone IIA is not exposed and the basal mafic-ultramafic rocks of Zone IIB were thrust against Zone I during Taconian convergence.

Zone III basement is characterized by Late Precambrian gneisses and quartzose sedimentary rocks along with layered gneisses and gneisses of plutonic derivation that may be Precambrian in part and Ordovician in part. Zone III Late Precambrian rocks appear to have undergone a pre-Acadian, probably pre-Middle Ordovician metamorphism to sillimanite-orthoclase grade. Middle Ordovician volcanics, similar to those in Zone II, and black shales, identical to those in Zones II and I, overlie the Zone III basement rocks with apparent unconformity, indicating that these three zones were unified by Middle Ordovician. The boundary between pre-Middle Ordovician rocks of Zones II and III is covered nearly everywhere, but is inferred from the types of basement, and the apparent lack of Cambrian cover in III as compared to II. The boundary between Zone II and Zone III may thus be a Taconian suture. Unlike Zone I, or the western part of Zone II, there is no evidence for significant late Taconian regional metamorphism in the eastern part of Zone II or in Zone III.

Basement of Zone IV is nearly identical to that of Zone III, although the Late Precambrian rocks in IV are more extensive and better dated. The gneisses of uncertain or mixed age are concentrated in the western part of the zone and include the Late Precambrian-Ordovician Massabesic Gneiss complex. The probable Middle Ordovician cover (Nashoba, etc.) is in part a more oxidized shale facies than in II and III, and apparently underwent intense pre-Silurian metamorphism up to sillimanite – K-feldspar grade. In addition to these differences, Zones III and IV are distinguished because the boundary between them is believed to have been the locus for Acadian convergent tectonics and metamorphism with west-directed nappes subsequently backfolded toward the east, so that structurally high rocks in the east moved down beneath basement rocks of III. Subsequently this basement was activated by gravity to form gneiss domes. Acadian tectonics are thus similar to and on the same scale as in the Pennine Zone of the Alps and seem to be in the style expected for continental collision, though evidence of post-Taconian oceanic crust and mantle seems totally lacking.

The older rocks of Zone V are barely metamorphosed and include Late Precambrian granites intrusive into volcanic and sedimentary rocks. These rocks are unconformably overlain by clastic rocks that contain fossils of the Baltic faunal realm. Late Ordovician alkali granite plutons intrude these rocks and the entire assemblage is locally overlain by unmetamorphosed fossiliferous Late Silurian-Early Devonian volcanics and extensively overlain by Carboniferous clastic rocks. Except for Carboniferous metamorphism in southern Rhode Island, Zone V seems to have escaped any important effects of Paleozoic regional metamorphism. Faults that may be of major displacement and are of uncertain age mark the Zone IV-V boundary in many places. A metamorphic gradient occurs in Precambrian rocks in the vicinity of this boundary and the boundary may be thought of as a locally faulted metamorphic gradient.

RÉSUMÉ

Les roches pré-carbonifères du sud de la Nouvelle Angleterre et du sud-est de l'état de New York sont regroupées en cinq zones stratigraphiques et tectoniques d'après les âges et les types de roches précambriennes, cambriennes et ordoviciennes qu'on y rencontre. D'ouest en est, ces zones sont: I) la zone de Manhattan-Williamstown de soubassement grenvillien et partiellement Précambrien supérieur et une couverture discordante de roches clastiques et carbonatées cambro-ordoviciennes; II) la zone de Greenwich-Rowe de soubassement continental précambrien et Précambrien supérieur (?) (IIA) ou une croûte océanique du Précambrien supérieur-Paléozoïque inférieur (IIB) et une couverture de roches clastiques et volcaniques cambro-ordoviciennes; III) la zone de Haddam-Warwick constituée de gneiss et de roches sédimentaires quartzeuses du Précambrien supérieur et d'autres gneiss d'âge incertain possiblement ordoviciens, le tout recouvert probablement en discordance par des roches volcaniques et des shales noirs de l'Ordovicien moyen; IV) la zone de Milford-Nashoba avec son soubassement de roches du Précambrien supérieur ou d'âge incertain, comme dans la zone III, et une couverture de roches probablement de l'Ordovicien moyen où prédominent des shales oxydés et en moindre proportion des shales noirs et des roches volcaniques; V) la zone de Dedham constituée de roches volcaniques, clastiques et plutoniques non ou faiblement métamorphisées et d'âge Précambrien supérieur, recouvertes en discordance par des couches d'origine clastique portant une faune d'affinité baltique.

La limite orientale d'affleurement de la séquence carbonatée cambro-ordovicienne démarque essentiellement les zones I et II. Cette séquence est recouverte de nappes de chevauchement constituées surtout de roches clastiques et, en proportion moindre, de roches volcaniques, le tout d'âge cambro-ordovicien. Le transport s'est fait de la zone II vers la zone I au cours de la déformation taconienne dont les phases tardives ont été caractérisées par un métamorphisme prononcé.

La zone II est séparée en deux parties par une faille de chevauchement appelée Ligne de Cameron; la partie ouest (zone IIA) contient des roches clastiques du Cambrien inférieur reposant en discordance sur le soubassement grenvillien nord-américain; les nappes transportées vers l'ouest trouvent probablement leur origine dans cette zone. La partie est, zone IIB, comporte une séquence clastique et volcanique reposant sur ce qu'on croit être des fragments de croûte océanique de compositions surtout basique et ultrabasique. Dans le sud-ouest du Connecticut la zone IIA n'affleure pas et les roches basiques et ultrabasiques de la zone IIB ont chevauché la zone I lors de la convergence taconienne.

Les roches caractérisant la zone III sont des gneiss et des sédiments quartzeux du Précambrien supérieur auxquels s'ajoutent des gneiss lités et d'autres gneiss de dérivation plutonique du Précambrien et en partie de l'Ordovicien. Les roches du Précambrien de la zone III semblent avoir subi, probablement avant l'Ordovicien moyen, un métamorphisme de grade sillimanite-orthose. Des roches volcaniques de l'Ordovicien moyen semblables à celles de la zone II et des shales noirs identiques à ceux des zones I et II recouvrent apparemment en discordance le soubassement de la zone III indiquant par là que ces trois zones étaient juxtaposées à l'Ordovicien moyen.

La limite entre les roches de l'Ordovicien moyen des zones II et III est presque partout cachée mais on peut la tracer à partir des types de soubassement et l'absence apparente de couverture cambrienne dans la zone III alors que celle-ci est présente dans la zone II. La limite des zones II et III pourrait donc représenter une suture taconienne. Contrairement à ce qu'on observe dans la zone I et le secteur occidental de la zone II, la partie orientale de la zone II et la zone III ne montrent aucun signe probant de métamorphisme régional qu'on pourrait relier au taconien tardif.

Le soubassement de la zone IV est à peu près identique à celui de la zone III quoique les roches du Précambrien supérieur en IV sont plus étendues et aussi mieux datées. Les gneiss d'âges mixtes ou incertains sont concentrés dans le secteur ouest de la zone et ceci inclut le complexe gneissique de Massabesic du Précambrien supérieur-Ordovicien. La couverture, probablement d'âge Ordovicien moyen (Nashoba, etc), est en partie un facies de shale plus oxydé que celui des zones II et III et le premier a apparemment subi, avant le Silurien, un métamorphisme prononcé s'élevant jusqu'au grade sillimanite-feldspath potassique. En plus de ces différences, les zones III et IV se distinguent parce que leur limite commune pourrait représenter le lieu du tectonisme et du métamorphisme de la convergence acadienne suivis par un rétrocharriage vers l'est des nappes originellement transportées vers l'ouest, de sorte que les roches tectoniquement au-dessus ont été enfouies sous le soubassement de la zone III. Subséquemment le soubassement aurait par gravité formé des domes de gneiss. Ainsi la tectonique acadienne aurait des traits communs et à la même échelle que la zone pennique des Alpes et rappellerait le style propre à une collision continentale; cependant les indices d'une croûte océanique et d'un manteau post-taconien manquent totalement.

Les roches les plus anciennes de la zone V sont à peine métamorphisées et comprennent des granites du Précambrien supérieur recoupant des roches volcaniques et sédimentaires.

Celles-ci sont recouvertes en discordance par des roches clastiques comportant une faune d'affinité baltique. Ces dernières sont envahies par des plutons de granite alcalin de l'Ordovicien supérieur et sur cet ensemble reposent par endroits des roches volcaniques, fossilifères, non métamorphisées, du Silurien supérieur-Dévonien inférieur, le tout recouvert abondamment par des roches clastiques du Carbonifère. Mis à part le métamorphisme carbonifère dans le sud du Rhode Island, la zone V ne montre aucun effet significatif du métamorphisme régional paléozoïque. Des failles possiblement soumises à des déplacements importants mais d'âges incertains marquent à plusieurs endroits la limite commune des zones IV et V. Un gradient de métamorphisme dans des roches précambriennes est visible au voisinage de cette limite; on pourrait l'interpréter comme un gradient métamorphique localement faillé.

INTRODUCTION

The purpose of this paper is to present a broad classification of the metamorphic-plutonic terrane of southern New England (Figs. 1, 2, and 3) that will be of general use, particularly for workers from outside the region. In doing this we draw on some forty collective years of experience in the region and particularly on recent experiences preparing a compilation for a Penrose Conference in 1972 and preparing contributions for new bedrock maps of Massachusetts and Connecticut, and for the Caledonide Orogen Project of the IGCP (International Geological Correlation Programme). The ideas and map patterns presented depend heavily on work of many colleagues, part of which is not generally available although Naylor (1975) and Osberg (1978) have summarized basement age provinces in the region. This is our personal assessment of the state of knowledge in 1979 and will prove controversial to many, if not most, workers. Because we are dealing with generalization on a broad scale, we have not attempted to provide detailed documentation of available evidence, although some details are formally shown here for the first time. It is our hope that the general framework presented will provide a useful context for discussions of tectonic history and a stimulus for more extensive research particularly in field and structural geology, geochemistry and geochronology.

The five stratigraphic – tectonic subdivisions (I-V) used here (Fig. 4), were originated as part of our contribution to the Tectonic Lithofacies Map of the Appalachian Orogen prepared by Williams (1978). We had in mind his subdivisions A through I for the Canadian Maritime Appalachians (Williams *et al.*, 1972), but did not wish to attempt the same subdivision in New England. Subsequently we realized the importance of two different terranes within Zone II, hence IIA and IIB. From our present outlook it appears certain that Zone I corresponds to Zones A and B in Newfoundland, Zone IIA to Zone C, and Zone IIB to Zone D. III relates to E and F mainly in being intermediate geographically. Zone IV can be related, but with little confidence, to Zone G. Zone V corresponds with certainty to Zone H. In terms of the broader zones outlined by Williams (1978), I and IIA correspond closely to the Humber Zone, IIB and III roughly to the Dunnage Zone, IV roughly to the Gander Zone, and V closely to the Avalon Zone.

To retain a geologically familiar context we have organized our discussion around broad age subdivisions. This is followed by definition of the boundaries of the stratigraphic – tectonic zones, and finally by some speculations on the tectonic significance of the boundaries.

STRATIGRAPHY

Precambrian and Late Precambrian

Grenville basement is present in Zones I and IIA and consists of metamorphosed sedimentary, volcanic, and plutonic igneous rocks. The Precambrian age of these basement rocks is reasonably well documented. A zircon date of 980 Ma has been determined from the Fordham Gneiss (Grauert and Hall, 1973) and zircons from gneisses in the Hudson Highlands have yielded an 1150 Ma age (Tilton and Davis, 1959). Where data are available, radiometric ages given in this paper have been converted to the decay constants recommended by the IUGS Subcommission of Geochronology (Steiger and Jäger, 1977). Rb-Sr whole-rock studies on rocks from the Hudson Highlands indicate ages from 895 Ma to 1145 Ma (Mose and Helenek, 1976). Basement gneisses in the Housatonic Highlands are unconformably overlain by Cambrian clastics and zircon data from rocks in the core of the Berkshires yield 1022 Ma to 1055 Ma ages (Ratcliffe and Zartman, 1976). Rb-Sr work on muscovite from core rocks in the Green Mountain anticlinorium indicates an age of 964 Ma (Faul et al., 1963). Basement cores of gneiss domes along the east flank of the anticlinorium are unconformably overlain by Late Precambrian to Cambrian basal clastics. Part of the core of the Waterbury Dome in Connecticut may be Grenville basement although an Rb-Sr whole-rock isochron determined on rocks from there indicates an apparent age of 455 ±50 Ma (Clark and Kulp, 1968). No known Grenville basement occurs in southern New England outside of Zones I and IIA.

Rock units included in the Late Precambrian gneisses in Zone I are the Yonkers Gneiss (560 ±30 Ma; Long, 1969) and Pound Ridge Granite (583 ±19 Ma; Mose and Hayes, 1975). The Yonkers Gneiss is a pink granitic gneiss that contains biotite and a dark bluish-green amphibole with a 2V of 0° to 20° identified optically as hastingsite. The Dry Hill Gneiss, in the Pelham Dome in Massachusetts, has been dated at 560 ±30 Ma (Naylor et al., 1973), and is lithically similar to the Yonkers Gneiss. It contains an amphibole with essentially the same optical properties described above, on which a microprobe analysis indicates a high ferric iron component but a normal amount of Na_2O. The age and lithic similarities of these rocks implies some relationship between them but the nature of this relationship is not understood. Perhaps they represent igneous rocks that were emplaced or erupted coevally in different places possibly during extensional tectonics related to the opening of Iapetus.

Late Precambrian rocks are exposed in Zone III in the cores of the Pelham, Stony Creek, Clinton, and Lyme domes (Fig. 1). The dominant rock of the Pelham dome is the Dry Hill Gneiss believed to be of volcanic derivation (Ashenden, 1973). It is a pink microcline-oligoclase gneiss with green biotite, iron-rich hornblende with 2V close to 0°, magnetite, allanite, and garnet, and thus very closely resembles the Yonkers Gneiss of Zone I. An essentially concordant zircon age of 560 ±30 Ma was obtained from this rock (Naylor et al., 1973). Closely associated with Dry Hill Gneiss is Poplar Mountain Gneiss, a feldspathic rock of sedimentary derivation with brown biotite, and a variety of quartzites: vitreous, actinolitic, or micaceous. At the south end of the Pelham dome the Poplar Mountain Gneiss grades into the Mount Mineral Formation, predominantly garnet-mica schist, quartzite, and amphibolite (Robinson et al., 1973). Mount Mineral Formation schists contain clear evidence of pre-

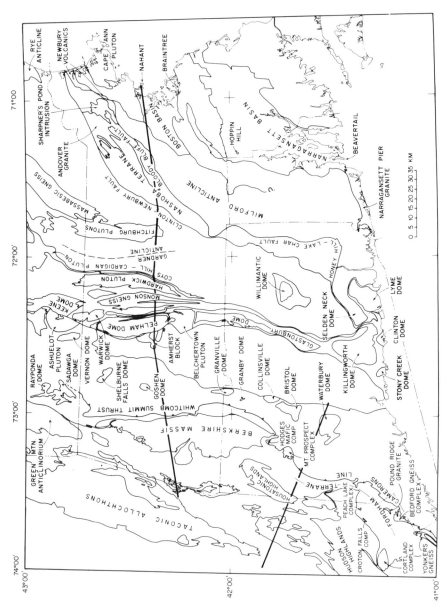

Figure 1. Locality map for area of geologic map in Figure 2. Shows locations of sections in Figure 5.

Acadian, possibly Precambrian, metamorphism to sillimanite-orthoclase grade in the form of relict coarse sillimanite, high temperature orthoclase megacrysts and high Mg garnets. Relict sillimanite is preserved as inclusions in relict orthoclase megacrysts overprinted by an Acadian kyanite-muscovite schist matrix (Robinson, *et al.*, 1975). The Mount Mineral Formation also contains four lenses of metamorphosed harzburgite. The lack of obvious unconformity between Late Precambrian and younger rocks in the core of the Pelham dome is probably mainly a result of intense Acadian folding. The "dome" is in fact an up-arched stack of Acadian recumbent folds.

Hills and Dasch (1972) obtained a Late Precambrian Rb-Sr whole rock age of 603 ±78 Ma for the granite in the core of the Stony Creek dome which cuts quartzites similar to those in the Pelham dome. Similar though undated rocks occur in the Clinton and Lyme Domes, although relations in these domes are complicated by intense metamorphism of probable Permian age (Lundgren, 1966).

Late Precambrian rocks of Zone IV are exposed in coastal southeastern Connecticut, in the Willimantic dome, and the Milford anticlinorium of eastern Massachusetts and Rhode Island. There is also a Precambrian component in the Massabesic Gneiss Complex (Aleinikoff, 1978) although it is classed here with other gneisses of uncertain or mixed age. The Plainfield Formation of Connecticut, in particular, closely resembles the metamorphosed sedimentary rocks of the Pelham dome. The Sterling Gneiss, a pink microcline gneiss, locally with Fe-rich hornblende, that is closely similar to the Dry Hill and Yonkers Gneisses has, in part, been interpreted as intrusive. Dated rocks in the Milford anticlinorium include the Milford Granite yielding 630 Ma zircons (Zartman and Naylor, 1972), and the Northbridge Granite Gneiss yielding an Rb-Sr whole-rock age of 569 Ma (Fairbairn *et al.*, 1967).

The Late Precambrian rocks of coastal Connecticut have clearly been involved in intense Acadian deformation and probably Alleghenian deformation as well. It is not presently clear if the rocks of the Milford anticlinorium had already become gneisses in the Precambrian, or if the main metamorphic effect on them was during the intense Ordovician metamorphism now believed to have occurred in the adjacent Nashoba and related units to the west. The unmetamorphosed state of the adjacent Late Silurian-Early Devonian Newbury Volcanics is against any intense Acadian metamorphism in this region.

The Late Precambrian of Zone V resembles the Avalon Zone of southeastern Newfoundland but has a greater abundance of plutonic rocks. The oldest rocks are a variety of weakly metamorphosed sedimentary and volcanic rocks, mainly inclusions and roof pendants in Dedham granodiorite and related intrusions. These little-metamorphosed plutons yield a variety of Late Precambrian isotopic ages (Rb-Sr 514 ±17 Ma for Dedham at Hoppin Hill; 591 ±28 Ma for Westwood Granite, Fairbairn *et al.*, 1967), and are also overlain unconformably by fossiliferous Lower Cambrian strata. Recent isotopic data on the Mattapan Volcanics (Kaye and Zartman, 1980) strongly suggest a Late Precambrian to Cambrian rather than a Late Paleozoic age for the long controversial conglomerates and shales of the Boston Basin.

Cambrian, Lower Ordovician, and Uncertain Age

Zone I has autochthonous and allochthonous Cambrian and Lower Ordovician cover rocks. Autochthonous rocks include basal clastics (Lowerre Quartzite, Poughquag Quartzite, Dalton Formation, and Cheshire Quartzite) that unconforma-

bly overlie basement and are succeeded by the Cambrian-Lower Ordovician carbonate bank deposits (Inwood, Wappingers, and Stockbridge). Allochthonous rocks,
which were originally deposited in Zone IIA, were transported onto the Zone I
autochthonous rocks during Taconian deformation. Clastic sedimentary rocks and
some volcanics constitute the allochthonous rocks and are commonly referred to as
the Taconic Sequence (Zen, 1967) or Eugeosynclinal (allochthonous) Sequence
(Fisher *et al.*, 1970). In addition we include Members B and C of the Manhattan
Schist (Hall, 1968) and much of the Waramaug Formation in western Connecticut
(Rodgers *et al.*, 1959) with allochthonous rocks in Zone I. Thus the allochthonous
rocks are an eastern facies of the autochthonous rocks in Zone I. They were originally deposited in Zone IIA on Grenville basement along with other clastics and some
volcanics. The Hoosac Formation in Connecticut and Massachusetts is equivalent to
some of the allochthonous section but remains in Zone IIA. Even here the Hoosac is
in thrust contact with Grenville basement in a number of places (Norton, 1976). The
eastern facies of Lowerre Quartzite in western Connecticut and southeastern New
York (Hall, 1976) also represents Zone IIA, but is not extensive enough to be shown
on Figure 1.

Zone IIB contains clastic and some volcanic strata of Cambrian and Lower
Ordovician age: the Rowe and Moretown Formations in Massachusetts (Hatch and
Stanley, 1973), much of the Hartland terrane of Connecticut (Gates and Martin, 1976)
as well as the Taine Mountain Formation (Stanley, 1964), Orange and Trap Falls
Formations (Crowley, 1968), Derby Hill and Fairfield Formations (Dietrich, 1968),
and the schist and granulite and felsic gneisses of the Norwalk, Connecticut region
(Kroll, 1977). Apparently oceanic crust formed the substrate upon which rocks in
Zone IIB were deposited. Mafic and ultramafic rocks within the zone are evidence of
this. Precambrian and/or Late Precambrian continental crustal rocks are known in all
zones except Zone IIB and, except in Zone V, all have been intensely metamorphosed. Gneisses of uncertain age, Lower Ordovician (?), are in Zones III and IV.
Thus there are two distinct basement terranes, the western basement (North American or Grenvillian) and the eastern basement (Avalonian and gneisses of uncertain
age) and the strata of Zone IIB were deposited on oceanic crust that lay between
them.

The only reasonably documented Cambrian or Lower Ordovician strata in Zone
III occur at the western edge where strata presumed to be Lower Ordovician and
equivalent to the Moretown Formation (Taine Mountain Formation) rest against
gneisses of the Waterbury and Bristol domes. There is an analogous occurrence to
this in the case of the Albee Formation on the west flank of the Whitefield dome in
northwestern New Hampshire (Billings, 1937).

Elsewhere in Zone III, physically between documented or suspected Late Precambrian and Middle Ordovician rocks, there is a variety of gneisses including
layered quartz-plagioclase gneisses and amphibolites of volcanic derivation. The
question of the origin and age of these gneisses represents one of the major unsolved
problems of New England geology. The fact that we have chosen to group these with
Cambrian and Lower Ordovician strata in other zones does *not* imply that we are
convinced they are of that age, but only that they occupy a similar *physical position*
to Cambrian and Lower Ordovician strata elsewhere.

The rocks in this grouping in New Hampshire were included in the Oliverian Plutonic Series by Billings (1956) and assigned to the Lower Devonian on the basis of rare localities where they were thought to intrude Silurian-Devonian strata. In considering similar rocks in southeastern Connecticut, Rosenfeld and Eaton (1958) noted the concept of gneiss domes developed by Eskola (1949) and suggested that gneisses in the cores of the domes might be reactivated basement as old as Precambrian. This conception was further supported by Robinson (1963) who noted the wide variety of gneisses in different domes in the Orange, Massachusetts area, all overlain by essentially the same coherent sequence of Middle Ordovician strata. He also noted that a Devonian age of the dome gneisses is apparently ruled out by the fact that the long recognized unconformity at the base of Silurian strata bevels across Middle Ordovician units and locally onto the dome gneisses.

An important contribution to this problem was made by Naylor (1968, 1969) in his study of the Mascoma dome in New Hampshire where he recognized two major rock types in the core of the dome, stratified felsic gneiss of probable volcanic derivation and massive gneiss of plutonic derivation. Field relations strongly indicate that the massive gneiss intrudes the layered gneiss. Naylor tentatively suggested that the stratified gneiss is part of the Middle Ordovician Ammonoosuc Volcanic sequence and is intruded by the massive gneiss, and he strongly indicated that the massive gneiss is unconformably overlain by a feldspathic sandstone and quartz conglomerate of the basal Silurian. Detailed U-Pb isotopic work by Naylor on zircons yielded an average of 440 ±25 Ma for four specimens of massive gneiss and the same for one specimen of the layered gneiss. On this basis Naylor suggested that the stratified gneiss may represent the felsic lower part of an Ordovician volcanic sequence that continued into the overlying Ammonoosuc Volcanics and that the massive gneiss represents a subvolcanic pluton younger or contemporaneous with the Ammonoosuc. He indicated there was no evidence that the volcanic rocks represented by the stratified gneiss were deposited on sialic crust, leaving the implication they were deposited on oceanic crust.

Since Naylor's work in New Hampshire there have been several further contributions to this problem: 1) The establishment by Ashenden (1973) of a stratigraphic sequence for the northern portion of the Pelham dome with a lower sequence of probable Late Precambrian rocks and an upper sequence of plagioclase gneisses and amphibolites of uncertain age which he named Fourmile Gneiss. Zircons from the Dry Hill Gneiss of the lower sequence are Late Precambrian (Naylor et al., 1973). The Fourmile Gneiss remains undated. 2) Detailed mapping by Hall (unpublished) in the Shelburne Falls dome supports Emerson's 1898 assignment of these rocks to "Monson Gneiss" and favors more specific correlation with the nearby Fourmile Gneiss. Reconnaissance with R.S. Stanley, N.L. Hatch, and John Rodgers to the cores of the Goshen, Granville, Collinsville, Bristol and Waterbury domes suggested a similar correlation for rocks in these domes. Although part of the core of the Waterbury dome had been suggested as Grenville basement, an Rb-Sr whole rock isochron from there indicates an apparent age of 455 ±50 Ma (Clark and Kulp, 1968). 3) Robinson (1977) has discovered a local lens of quartz-pebble conglomerate and quartzite at the base of the Ammonoosuc Volcanics where it rests on Monson Gneiss favoring an unconformity at this position. 4) R.E. Zartman (pers. commun., 1979) has obtained a preliminary Pb-Pb zircon age of ~440 Ma on a sample of Monson Gneiss.

Thus, at present it is uncertain whether this widespread series of gneisses is a "basement" of rocks metamorphosed before deposition of the Middle Ordovician strata or whether it is merely a pile of Ordovician volcanics. In either case it is in contact with Late Precambrian (Avalonian) rocks, is clearly a completely different facies from Cambrian-Lower Ordovician clastic rocks further west, or from Cambrian rock further east in Zone V, and hence is very significant to paleogeographic reconstructions.

In the portion of Zone IV in southeastern Connecticut (Goldsmith, 1976) the stratigraphic sequence appears similar to that of the Pelham dome, with probable Late Precambrian quartzite-bearing gneisses overlain by a group of layered feldspathic gneisses, in ascending order, Mamacoke Formation, New London Gneiss and Monson Gneiss (as locally defined). Our recent examination of these rocks with Goldsmith convinces us that a finer stratigraphic subdivision has been achieved in southeastern Connecticut than in central Connecticut or Massachusetts, and that Monson Gneiss in the latter areas contains equivalents of at least two, if not all three, of the southeastern Connecticut units. These units appear to pinch out eastward and

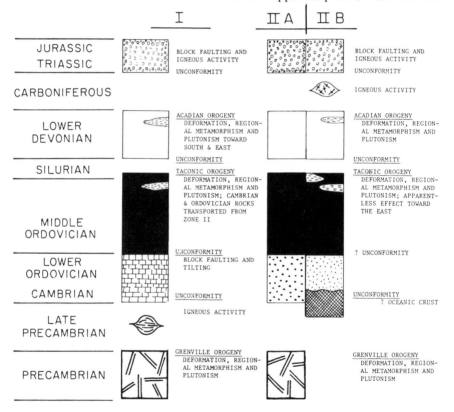

Figure 3. Correlation chart of columnar sections for stratigraphic-tectonic zones of southern New England. Patterns are the same as those on Figure 2 and Figure 5. Patterns for Ordovician-Silurian-Devonian plutonic rocks are as follows: Oriented single dashes – foliated

northeastward along the Lake Char fault zone although clearly represented in part just outside the core of the Willimantic dome. Eastward into Rhode Island we have shown (Fig. 2) in this general category a rather large patch of plutonic rocks of uncertain age which laps eastward into the western edge of Zone V.

The Massabesic Gneiss in the northwest part of Zone IV has also been placed with gneisses of uncertain or mixed age. In this case Aleinikoff (1978) has demonstrated two components, a group of layered gneisses of probable volcanic derivation yielding Late Precambrian zircon ages that indicate they are at least 650 Ma old and gneisses of more homogeneous plutonic aspect yielding Ordovician ages on the order of 480 Ma.

Unmetamorphosed or weakly metamorphosed Cambrian clastic and calcareous strata of the Baltic faunal realm are well represented in Zone V (Theokritoff, 1968; Skehan *et al.*, 1977). They occur either unconformably overlying Late Precambrian plutonic or volcanic rocks as at Hoppin Hill or Beavertail, or in contact with intrusions of probable Ordovician age as at Braintree, Weymouth (near Braintree), and Nahant. Fossils include Lower Cambrian at Hoppin Hill, Weymouth and Nahant,

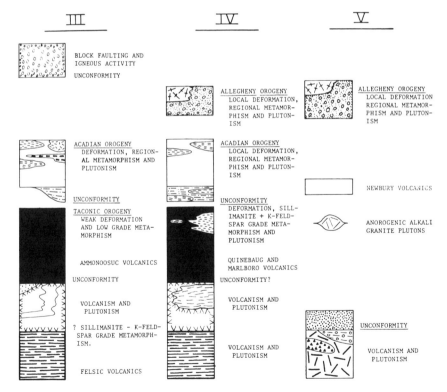

granite; random single dashes – non-foliated granite; heavy single dashes – Coys Hill – Cardigan and Ashuelot plutons of porphyritic granite; paired dashes – granodiorite, monzodiorite, tonalite, gabbro.

and Middle Cambrian at Braintree and Beavertail. The Cambrian sections at Weymouth and Hoppin Hill can be correlated, suggesting a stable environment of sedimentation with only slight facies changes. The faunas and general character of the rocks are identical to Cambrian rocks unconformably resting on Late Precambrian in southeastern Newfoundland and near St. John, New Brunswick.

Middle and Late Ordovician

Black shales and volcanics with local calcareous rocks are common and widespread across Zones I, II, III, and IV where they rest unconformably on older rocks in many places. In Zone I black shale (Walloomsac, Normanskill, and Manhattan A) is the most abundant rock type with a basal limestone commonly present (Balmville Limestone and the marble or limestone members of Manhattan A and Walloomsac). Middle Ordovician rocks in Zone II consist of black shales and volcanics, along with some coarser clastics and volcaniclastics, that are included in the Hawley and Cobble Mountain Formations in Massachusetts and Connecticut (Hatch and Stanley, 1973) and the Prospect Gneiss (Crowley, 1968), mafic gneisses (Kroll, 1977), and the Harrison Gneiss (Hall, 1976) north of Long Island Sound in Connecticut and New York (Plate I). Ultramafic rocks, mainly serpentinite, occur locally in the Cambrian-Ordovician strata of Zone II.

In Zone III Middle Ordovician strata rest directly on a variety of dome gneisses except in the extreme west where they locally overlie Moretown-equivalent strata. The dominant rock type is sulfidic mica schist derived from euxenic shale (Partridge Formation and equivalents), although non-sulphidic schists are important at the top of the sections around the domes west of the Connecticut River (Fig. 2). Rocks of volcanic derivation including amphibolites and felsic gneisses are abundantly interspersed with the sulphidic schists in places, particularly near the base. These volcanics locally make up as much as 50 per cent of the section, as for example, in the section near the Monson Gneiss (Figs. 2 and 3). Pods of ultramafic rocks, variously metamorphosed, are abundant in places in these Middle Ordovician strata.

A distinctive section of volcanics, Ammonoosuc Volcanics, rests on the gneisses of the Glastonbury, Monson, Warwick, and Keene domes (Figs. 2 and 3). The lower portion of the Ammonoosuc is dominated by amphibolites whereas the upper portion is dominantly felsic gneisses. A variety of primary features including pillows and conglomerates are recognizable and attest to the volcanic nature of these rocks. We think the Ammonoosuc is represented by a similar, although very thin section of rocks, on the Collinsville dome (Figs. 1, and 2). The Ammonoosuc Volcanics do not interfinger with the overlying sulphidic schists and their included volcanics, but do pinch out along strike between gneisses and overlying schists.

Generally the Middle Ordovician schists exposed in the various isoclinal anticlines in Zone III east of the Bronson Hill anticlinorium in central Massachusetts contain fewer volcanic rocks than further west, but amphibolites are fairly abundant in the wider zones close to the Connecticut state line.

The age and correlation of rocks in Zone IV that are here assigned to the Middle Ordovician are uncertain and in dispute. These rocks include a basal zone of metamorphosed mafic and felsic volcanics, Marlboro Formation in Massachusetts and Quinebaug Formation in Connecticut, that might correlate with the Ammonoosuc Volcanics, although parts of the Quinebaug have also been compared to the Monson Gneiss. The volcanics are overlain by generally thick units of pelitic schists and gneisses, the Nashoba Formation in Massachusetts and the Tatnic Hill Formation in Connecticut. These consist in part of sulphidic schists and amphibolites (including ultramafic blocks) identical to the Partridge Formation of Zone III, but the dominant rock types are gray-weathering sillimanite- and magnetite-bearing schist and gneiss that were apparently deposited as clastic sediments of a more oxidized facies than the Partridge Formation. Also different in Zone IV is the evidence of intense pre-Silurian regional metamorphism of the Nashoba Formation (and probably also Tatnic Hill Formation) recently derived by R.E. Zartman, A.F. Shride, and R.M. Goldsmith (pers. commun., 1979). This is based on a new radiometric age of ~430 Ma on the Sharpners Pond Intrusion, a completely undeformed, unmetamorphosed unit that crosscuts the Andover Granite. The Andover Granite is a highly deformed migmatitic gneiss believed to have formed during regional metamorphism of the Nashoba Formation. The tectonic setting and significance of this metamorphism has yet to be determined. It clearly did not extend east into Zone V where there could have been no intense pre-Carboniferous Paleozoic metamorphism. Indeed, the locus of most intense metamorphism of this phase, up to sillimanite-orthoclase grade, seems to have been along the centre of the Nashoba and Tatnic Hill Formations, with lower grade rocks to the east toward the Marlboro and Quinebaug Formations.

Ordovician strata seem totally absent from Zone V, but a series of anorogenic alkaline to peralkaline granitic plutons have yielded Middle to Late Ordovician radiometric ages (Zartman and Marvin, 1971). These include the well known Cape Ann and Quincy plutons, previously thought to be post-Acadian pre-Pennsylvanian on the basis of contact relations and lack of deformation.

Silurian

In the region considered (Fig. 2) Silurian rocks are known in Zones II, III, IV, and V although Silurian is also present in the region west of the map area. The Silurian of Zone II and the western part of Zone III is of extreme importance to the geology of the region but it is too thin to be shown on Figure 2. We adopt the view that both the Gile Mountain Formation and Waits River Formation are Devonian, not partly or entirely Silurian as is favored by some (Boucot, 1968; Thompson and Rosenfeld, 1979). In Zone II the Silurian is represented by the Shaw Mountain and Russell Mountain Formations which consist of local lenses of quartz pebble conglomerate, quartzite, calc-silicate rock, and amphibolite distributed along the unconformity between Ordovician and Devonian strata. Silurian rocks in western Zone III consist of the Clough Quartzite, quartz-pebble conglomerate, quartzite, and calc-silicate rock, that contains Llandovery fossils in northern Massachusetts and southwestern New Hampshire; and the Fitch Formation, calc-silicate granulites and minor mica schists containing Ludlow fossils in northwestern New Hampshire. The rock types and their variations and lens-like distribution strongly suggest an irregular shoreline-like environment with localized stream, beach, and marine sedimentation and deep chemical weathering of locally exposed source areas along a relatively stable margin of Silurian North America.

In the eastern part of Zone III and in Zone IV the Silurian thickens dramatically into a continuous section dominated by feldspathic calcareous granulite, sulfidic schist, and quartzite (Paxton Formation and equivalents). This eastward thickening corresponds to a better exposed thickening documented in northwestern Maine (Moench and Boudette, 1970) and marks the west margin of a Silurian trough traceable from southeastern Connecticut and through central Maine to Fredericton, New Brunswick. The eastern limit of exposure of this basin facies in Zone IV is along the Clinton-Newbury fault zone in Massachusetts and the top of the Tatnic Hill Formation in Connecticut. It is not yet known if any of these rocks along the eastern limit in southern New England contain sedimentary facies appropriate to the east margin of the trough, although it has been suggested that the feldspathic granulite had a partial source in the Late Silurian-Early Devonian "coastal volcanic belt" to the east (McKerrow and Ziegler, 1972).

As compared to eastern Maine (Gates, 1978), the "coastal volcanic belt" has very limited exposure in southeastern New England, consisting of two fault-bounded wedges of fossilferous Newbury Volcanics, uppermost Silurian and lowermost Devonian (Shride, 1976), at the west edge of Zone V. Unlike the Silurian trough facies immediately west of the Clinton-Newbury fault, the Newbury Volcanics are neither folded nor significantly metamorphosed, although they are overturned toward the southeast, apparently as a result of severe faulting. The "coastal volcanic belt" has been proposed by McKerrow and Ziegler (1972) as a volcanic chain above a southeast-dipping subduction zone on the southeast consuming margin of the Fredericton trough. Moench and Gates (1976) and Gates (1978) found this unsatisfactory on the basis of the bimodal basalt-rhyolite suite in the Eastport, Maine area. However, the abundant andesites of the Newbury Volcanics (Shride, 1976; Martha Godchaux, pers. commun., 1979) do not seem against it. We find Cameron and Naylor's (1976) coupling of the Newbury Volcanics with the Middle to Late Ordovician alkaline granites curious and improbable. We think relations to various granitic to dioritic rocks in Zone IV yielding 430 Ma ages (R.E. Zartman and R.M. Goldsmith, pers. commun., 1979) need exploration.

Devonian

Devonian rocks are present in all five zones and overlie Silurian or rest unconformably on pre-Silurian rocks. In Zone I there are clean clastics and carbonates in the Hudson Valley represented mainly by the Helderberg and Ulster Groups (Fisher *et al.*, 1970). No Silurian or Devonian strata are known between the Hudson Valley and the Taconian unconformity in western Massachusetts and Connecticut. There has apparently been slip along this unconformity and it has been interpreted as a décollement surface in western Massachusetts (Hatch and Stanley, 1976).

The stratigraphic problems and facies relations of Lower Devonian strata in Zones II, III, and IV particularly in the Connecticut Valley-Gaspé synclinorium, are too complex to be discussed here. Throughout, the dominant rock type is graphitic, non-sulfidic, quartzose schist. Another major rock type particularly abundant west of the Connecticut River is schist with interbedded quartzose marble (Waits River Formation and portions of Gile Mountain Formation). Minor but locally important rock types include calc-silicate granulite, amphibolite or greenstone derived from mafic volcanics, or gneiss derived from felsic volcanics. Eastern Zone III and Zone IV are dominated by gray-weathering graphitic schist of the Littleton Formation, closely similar to the Seboomook Formation of Maine. The depositional model for the Seboomook as a northwestward-spreading deltaic wedge derived from active uplifts to the southeast (Hall *et al.*, 1976) seems compatible with relations in southeastern New England.

Zone III, as well as parts of Zones I, II, and IV, is dominated by Devonian plutons ranging from gabbro to granite. These include early stratiform sheets involved in all phases of Acadian deformation and locally yielding ages of 405 to 395 Ma (Lyons and Livingston, 1977) such as the Coys Hill-Cardigan pluton, Ashuelot pluton, Hardwick pluton, and parts of the Fitchburg pluton. Other plutons such as the Belchertown and Prescott, clearly crosscutting early folds but themselves deformed and metamorphosed during the *dome stage* of deformation, yield ages around 380 Ma (Ashwal *et al.*, 1979; Naylor, 1970). A few weakly foliated granites appear to be late- or post-tectonic, and might even be post-Devonian as was found for a pluton of Concord granite in New Hampshire (Lyons and Livingston, 1977).

The only Devonian rocks in Zone V are part of the Newbury Volcanics (see under Silurian) and one or two dated small alkali granite plutons too small to be shown on Figure 1.

Carboniferous

It is not our purpose here to review Carboniferous stratigraphy, structure or thermal events. Carboniferous or probable Carboniferous rocks are exposed in the Narragansett basin and attached Norfolk basin in southeastern Massachusetts, and in three fault-bounded wedges west of the Clinton-Newbury fault in central Massachusetts. Carboniferous to Permian plutons include the Narragansett Pier Granite in Rhode Island, the Pinewood Adamellite in south-central Connecticut and the Milford and related granites associated with the Massabesic Gneiss in southern New Hampshire. Late- or post-Carboniferous metamorphism locally reaches sillimanite grade in southern Rhode Island and up to garnet grade near Worcester, Massachusetts. Carboniferous metamorphism along the coast of southeastern Connecticut, locally up to sillimanite-orthoclase grade (Lundgren, 1966), appears to anneal away the cataclastic effects of the Honey Hill fault. A pattern of ~250 Ma K-Ar ages on micas covers about two-thirds of southern New England (Zartman *et al.*, 1970) and brings into question the age of the metamorphic minerals in this region. In the case of the Belchertown pluton a K-Ar age on metamorphic hornblende produced during the *gneiss dome stage* gives a minimum age of 361 Ma for this stage (Ashwal *et al.*, 1979).

CHARACTERIZATION OF ZONE BOUNDARIES

Zones I and IIA

Basement rocks are the same in both zones and represent "North American" continental crust. The east margin of Zone I as shown on Figure 4 is drawn along the east edge of presently exposed carbonate bank deposits and is extrapolated through the Precambrian in the Berkshires and the southern end of the Green Mountains

(Figs. 1 and 2). There is a problem with this, however, because carbonates undoubtedly extend beneath the surface and thus farther east than this line. The facies relationship between the carbonate bank, Zone I and the clastics of Zone IIA, is implied but no direct evidence of this is available in southern New England. Rodgers (1968) has cited evidence for the eastern edge of the carbonate bank in northern Vermont and Pennsylvania. Presumably the transition zone in southern New England is covered by Zone II rocks that have been thrust over this boundary (Fig. 5). Such may be the case along the east flank of the basement anticlinoria in Massachusetts where much of the Hoosac Formation has been thrust westward and into contact with basement (Norton, 1976) and in Vermont where the Sherman Marble (Skehan, 1961) may be an eastern extension of the carbonate bank. We suggest that the Sherman Marble and associated older rocks may be unconformably overlain by (?)Middle Ordovician rocks, the Heartwellville, and then covered by the Hoosac Formation along a thrust fault. In a paleogeographic reconstruction, however, one would consider the east edge of the carbonate bank facies as the Zone I-IIA boundary and it would thus be represented by a facies change.

Zones I-IIA, and IIB

This boundary is a thrust fault referred to as Cameron's Line in southeastern New York and Connecticut, and the Whitcomb Summit thrust in Massachusetts (Fig. 1). There is no known basement in Zone IIB compared to the "North American" basement of Zones I and IIA. The Cambrian and Lower Ordovician strata in Zone IIB are similar to those in IIA with a few more volcanics. Mafic and some ultramafic rocks occur associated with Cambrian through Middle Ordovician rocks in Zone IIB and are particularly common in the vicinity of Cameron's Line (Figs. 1 and 2) along the west edge of Zone IIB in Connecticut.

The Bedford Gneiss Complex (Alavi, 1975), Mt. Prospect Complex (Cameron, 1943, 1951; Gates and Bradley, 1952), and the Hodges Mafic Complex (Gates and Christensen, 1965; Merguerian, 1979) are bodies along Cameron's Line that are largely composed of mafic and some ultra-mafic rocks. These rocks are interpreted to represent oceanic crust and the thinly laminated quartzites, coticules, and calcareous rocks that occur in the vicinity of their upper contacts are suggestive of associated sea floor deposits. The Croton Falls Complex and Peach Lake Complex, which are mafic and ultramafic complexes in New York (Prucha et al., 1968) are probably klippen of oceanic crustal rocks that, although not complete ophiolite sequences, are analogous to the ophiolite allochthons of western Newfoundland (Williams, 1971). The Cortlandt Complex is another such mafic-ultramafic body, near the Hudson River in New York, that might be considered to be allochthonous, but intrusive relations between it and Zone I autochthonous cover (Ratcliffe, 1968) indicate that it was intruded in its present setting and that it is not a klippe of oceanic crustal rocks. Serpentinite and associated ultramafic bodies in southern New England are almost entirely restricted to the terrane east of the western edge of Zone IIB. Exceptions to this occur mainly in the Precambrian of the basement anticlinoria. Paleozoic ultramafics are numerous and some of the larger examples of them are those on Staten Island in New York; west of Port Chester, New York; Chester, Massachusetts; Drury, Massachusetts; and East Dover, Vermont. All of these, and many others, presumably represent bits and pieces of oceanic crust or upper mantle that were tectonically emplaced by interaction, such as faulting, between oceanic crust and its overlying sedimentary and volcanic cover.

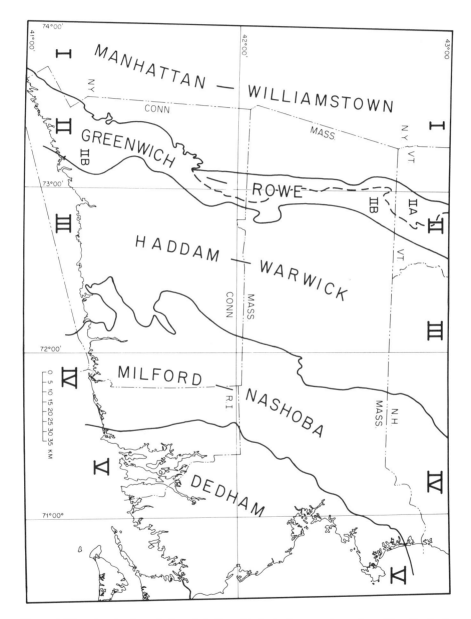

Figure 4. Map showing boundaries of stratigraphic-tectonic zones in southern New England.

Zones II and III

This boundary is drawn between the terrane of Avalonian and related basement in the cores of the Bronson Hill-type domes and the "North American" Grenville basement terrane. The boundary as drawn on the map (Fig. 4) is based on the easternmost occurrence of the Moretown Formation and its probable correlatives. This choice of boundary on the map is based on the fact that Cambrian and Lower Ordovician are well represented by cover rocks in Zones I and II whereas the oldest cover rocks in Zone III seem to be Middle Ordovician. Known exceptions to this occur in the vicinity of the Bristol and Waterbury domes (Figs. 1 and 2) where the Taine Mountain Formation, a probable Moretown equivalent, is in contact with Zone III basement. It is significant that no known gneissic basement occurs in Zone IIB and the inference is that Cambrian and Lower Ordovician strata in this realm were deposited on oceanic crust.

Zones III and IV

Zones III and IV have identical Late Precambrian gneissic rocks and gneissic rocks of uncertain age. The Middle Ordovician rocks are similar but have two distinctive differences: 1) Shaly strata of Zone III are dominantly sulfidic whereas those in Zone IV, although sulfidic in part, are dominantly an oxidized, magnetite-bearing facies. 2) There is evidence of intense pre-Silurian regional metamorphism in Zone IV, whereas in Zone III the pre-Silurian rocks were only weakly metamorphosed and locally invaded by shallow plutons (Highlandcroft). Silurian and Devonian strata lap across the zone boundary with no significant differences. Aside from these few differences then, what is the importance of the boundary and why has it been drawn? The answer lies in knowledge of the intensity of Alpine style tectonics, regional metamorphism, and plutonism, showing clearly that Acadian orogenesis reaches greatest intensity between the areas of exposure of Zone III basement to the west and Zone IV basement to the east.

In practice the boundary is most easily drawn in southern Connecticut where the two basement areas are separated by only a narrow syncline, Lundgren's Appendix (Lundgren and Thurrell, 1973). To the north the boundary skirts the west side of the Willimantic dome and Massabesic Gneiss, which are outliers of Zone IV basement (Figs. 1 and 2), and elsewhere arbitrarily follows axial surfaces of isoclines defined by Silurian-Devonian strata.

Zones IV and V

Zone V encompasses the westernmost occurrences of fossiliferous Cambrian strata, of sedimentary rocks of the Boston Basin, and of weakly metamorphosed volcanic and sedimentary rocks of probable Late Precambrian age. In principle the boundary separates unmetamorphosed Late Precambrian plutonic rocks in Zone V from gneissic Late Precambrian plutonic rocks and other metamorphic rocks in Zone IV. R.M. Goldsmith (pers. commun., 1978) thinks the boundary may, in principle, be a metamorphic gradient. In practice much of the boundary is drawn along faults. Beginning in the north, a fault that is a northern extension of the Bloody Bluff fault separates weakly metamorphosed Late Precambrian rocks and Newbury Volcanics from Paleozoic plutonic rocks of the Nashoba Terrane. The next segment is a fault between weakly metamorphosed and gneissic Late Precambrian rocks, and the next is between Boston Basin sedimentary rocks and Late Precambrian gneissic rocks. Southwest of the Boston Basin the importance of faulting is unclear and the boundary seems to be lost in the undifferentiated plutonic rocks of Rhode Island.

Southeast of the Narragansett Basin and the southeastern Massachusetts (Dedham) batholith another gneissic terrane has recently been recognized (R.M. Goldsmith, pers. commun., 1978) with a contact that is apparently a metamorphic gradient. This might have been called Zone VI in our scheme, but we prefer to leave it in Zone V until more is know about it.

TECTONIC SIGNIFICANCE OF ZONE BOUNDARIES

Generalized structural interpretations of southern New England are given in cross sections in Figure 5. The near surface portions are based on more detailed cross sections and are not schematic nor vertically exaggerated. The portions above and below this are necessarily speculative. A gravity profile for the northern section taken directly from the gravity map of Bromery (1967) is added for comparison, but was not used in construction of the sections. Structure and gravity sections both show two major boundaries that we interpret as the loci of two major episodes of convergent plate tectonics, the Taconian and the Acadian. These are the boundaries between Zones II and III and between Zones III and IV. Other zone boundaries discussed in the paper have subsidiary, or at least less well known significance.

Zones I and IIA

Zone I-IIA boundary is of paleogeographic significance because it marks the east edge of the Cambrian-Ordovician carbonate bank as it existed prior to the onset of Taconian deformation. Thus, this boundary was tectonically passive but important in the depositional history of the region. The boundary was eventually tectonically eroded in some places, becoming part of the allochthonous Taconian thrust sheets. In other places it was tectonically buried by overriding thrust sheets. This thrust fault activity commenced after normal faulting, erosion, and foundering of the carbonate bank in the Middle Ordovician and after the early stages of obduction of oceanic crust.

Zones I-IIA and IIB

This tectonic boundary began to develop early in the closing of the Cambrian-Lower Ordovician ocean when oceanic crust was obducted westward onto North American Grenville continental crust and its cover. This was the first phase of the long and complex history of development of Cameron's Line. The Cameron's Line fault zone seems to have behaved as the root zone of the allochthonous rocks associated with Taconian thrusting and gravity sliding in that rocks in the hanging wall, east of Cameron's Line, were thrust over the rocks of Zone IIA. At an unknown distance east of the Cameron's Line fault zone the first stages of the Zone II-Zone III boundary were developing, with oceanic crust and some of its cover being forced downward toward the east under sialic Avalonian and related basement of Zone III.

Zones II and III

With continued closing and eastward subduction, much volcanic activity developed as magma from melted subducted rocks worked its way upward to the surface through eastern or Avalonian basement. Thus the Middle Ordovician is marked by extensive volcaniclastic and volcanic rocks along with black shales which continued to be deposited across the zones during and after closure.

Late in the development of this boundary the ocean was completely closed and eastern basement was brought into thrust fault contact with, and thus sutured to, Zone I and II basement and cover rocks. Continued east-west compression caused the Zone III basement and its overlying tectonically passive cover rocks to override the western zones, to tectonically bury them, and to produce a high P-T environment. This resulted in high grade metamorphism and deformation that produced nappes of North American basement either by overfolding as in the Fordham Terrane or by a fold-thrust process as in the Berkshires in Massachusetts (Hall, 1968; Ratcliffe and Harwood, 1975). Further movement occurred along Cameron's Line, probably along with the development of the upper Taconian thrusts, due to the westward ramming of Zone III basement.

Zones III and IV

As considered above, the boundary between Zones III and IV is marked by only minor differences in pre-Silurian rock types. Nevertheless, the vicinity of the boundary is the position of the most intense deformation, plutonism, and metamorphism in New England. We believe these features are of the same magnitude as those of the European Alps, and have the appearance of being the product of strong convergence between two plates.

The generalized sequence of deformations that produced the structural features near the Acadian suture in Figure 5 is illustrated in cartoon form in Figure 6. The earliest recorded episode involving Lower Devonian strata was a series of west-directed regional nappes (Thompson et al., 1968). The strata had already been intruded by a series of quasi-concordant plutonic sheets at this stage, and there is ample evidence that structurally higher nappes were already at higher metamorphic grade than structurally lower nappes, and remained so in frontal regions throughout regional metamorphism. The *nappe stage* was followed by an intense stage of *backfolding* in which axial surfaces of nappes were overturned toward the east, in many cases into recumbent folds of large amplitude. The Colchester nappe of Dixon and Lundgren (1968) in southeastern Connecticut appears to be a feature of this stage. During backfolding structurally higher rocks with andalusite-bearing mineral assemblages were downfolded into much deeper zones to develop sillimanite-garnet-cordierite assemblages at the peak of regional metamorphism (Tracy et al., 1976). After or late in backfolding many high grade rocks seem to have been involved in cataclasis while still hot enough to undergo considerable recrystallization (Robinson et al., 1977). A number of plutonic masses appear to have been emplaced following the *nappe* and *backfold* stages but were severely deformed and metamorphosed during the following *dome stage*. The third major episode involved emplacement of a series of gneiss domes and related anticlines and synclines, driven in part by density contrast between light basement rocks and heavier cover. Much of the final metamorphic rock fabric, especially in Zone III, seems to have been determined in the *dome stage* even though many rocks reached a metamorphic peak earlier.

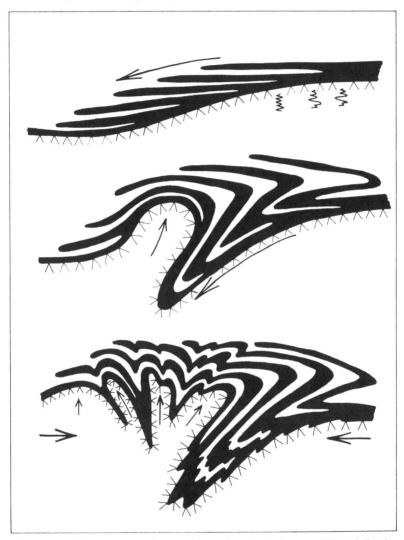

Figure 6. Cartoon of the sequence of major deformations in the Bronson Hill anticlinorium and Merrimack synclinorium, central Massachusetts. Top: Early Acadian nappes overfolded from east to west with heated rocks in the east overfolded onto cooler rocks in the west. Middle: Backfolding of early nappe axial surfaces with overfolding from west to east. Late in this stage there was extensive cataclasis due to high strain rates on a series of west-dipping surfaces. Bottom: Gneiss dome stage involving tight to isoclinal folding of earlier axial surfaces and the gravitational upward movement of low density basement gneisses into heavier mantling strata.

The details of how these structural, metamorphic, and plutonic episodes might relate to plate subduction and collision zones is explored elsewhere (Robinson and Hall, 1980). A point to be emphasized, however, is that despite the evidence of a Silurian-Devonian marine sedimentary trough, and of tectonics, plutonism (including gabbros) and metamorphism thought to characterize continental collision, no scrap of evidence has yet been found, in the form of mafic or ultramafic rocks of appropriate age, to suggest the existence of a Silurian-Devonian ocean along the boundary of Zones III and IV. Was the Acadian orogeny the result of compression along a break within a pre-existing single plate?

Zones IV and V

If one concentrates on the nature of pre-Silurian rocks, particularly on the similarity of Middle Ordovician strata across Zones I, II, III, and IV, and on the fact that Yonkers and Pound Ridge Gneiss in zone I are the same lithology and age as the Dry Hill and Sterling Gneisses in Zones III and IV, it is very tempting to look for a suture on the boundary between Zones IV and V. The fact that this region is heavily faulted may, of course, disguise the real evidence, and may juxtapose rocks in a sequence not directly analogous to their original positions. Nevertheless, the fact that neither Lower Cambrian nor Late Silurian-Early Devonian rocks have suffered significant deformation and metamorphism near this boundary suggests it has only minor significance to Acadian tectonics. At present we have the view that the boundary is essentially a faulted metamorphic gradient in Late Precambrian rocks that lies east of the zone of intense Acadian convergent tectonics.

CONCLUSIONS

We have subdivided southern New England into five stratigraphic tectonic zones to provide a better framework for understanding its tectonic history. Our two major conclusions are: 1) that the eastern Late Precambrian (Avalonian) basement and related rocks extend well west of the Connecticut River and that the Taconian orogeny was principally the result of suturing of this basement onto the Grenvillian basement of eastern North America, and 2) that the Acadian orogeny with its tectonics of collisional aspect took place in the position of a Siluro-Devonian sedimentary trough between two zones of very similar Avalonian and related basement, with no clear evidence of intervening oceanic crust.

ACKNOWLEDGEMENTS

The New York State Geological Survey, the State Geological and Natural History Survey of Connecticut, and the United States Geological Survey, Branch of Eastern Environmental Geology supported work by Hall and students in this region. Robinson and students have been supported by grants from the Geology and Geochemistry Programs of the National Science Foundation and from the United States Geological Survey, Branch of Eastern Environmental Geology. We thank these institutions for their financial support.
We have had the benefit of the ideas and advice offered by many colleagues over the years particularly during field trips with them in their areas of study. We wish to

specifically acknowledge M.P. Billings, H.R. Dixon, Richard Goldsmith, D.S. Harwood, N.L. Hatch, J.B. Lyons, P.H. Osberg, N.M. Ratcliffe, John Rodgers, R.S. Stanley, J.B. Thompson, Jr., and E-an Zen, all of whom provided us with much in the way of ideas as well as published and unpublished data. Lettering on the figures was done by Marie Litterer. The paper was improved as a result of review by R.D. Hatcher.

REFERENCES

Alavi, Mehdi, 1975, Geology of the Bedford Complex and surrounding rocks: Southeastern New York, Contribution No. 24, Geology Dept., University of Massachusetts, Amherst, 117 p.

Aleinikoff, J.N., 1978, Structure, petrology, and U-Th-Pb geochronology in the Milford quadrangle, New Hampshire: Ph.D. Thesis, Dartmouth College, 247 p.

Ashenden, D.D., 1973, Stratigraphy and structure, northern portion of the Pelham dome north-central Massachusetts: Contribution No. 16, Dept. Geology, University of Massachusetts, Amherst, 132 p.

Ashwal, L.D., Leo, G.W., Robinson, P., Zartman, R.E. and Hall, D.J., 1979, The Belchertown quartz monzodiorite pluton, west-central Massachusetts, a syntectonic Acadian intrusion: American Jour. Sci., v. 279, p. 936-969.

Billings, M.P., 1937, Regional metamorphism of the Littleton-Moosilauke area, New Hampshire: Geol. Soc. America Bull., v. 48, p. 463-566.

_____, 1956, The geology of New Hampshire, Part II bedrock geology: New Hampshire Planning and Development Commission, Concord, New Hampshire, 203 p.

Boucot, A.J., 1968, Silurian and Devonian of the Northern Appalachians: in Zen, E-an, White, W.S., Hadley, J.B., and Thompson, J.B., Jr., eds., Studies of Appalachian Geology: Northern and Maritime: New York, Interscience Publishers, p. 83-94.

Bromery, R.W., 1967, Simple Bouguer gravity map of Massachusetts: United States Geol. Survey, Geophysical Investigations, Map. GP-612.

Cameron, Barry and Naylor, R.S., 1976, General geology of southeastern New England: in Cameron, Barry, ed., A Guidebook for Field Trips to the Boston Area and Vicinity: New England Intercollegiate Geol. Conference Guidebook, p. 13-27.

Cameron, E.N., 1943, Origin of sulphides in the nickel deposits of Mt. Prospect, Conn.: Geol. Soc. America Bull., v. 54, p. 651-686.

_____, 1951, Preliminary report on the geology of the Mt. Prospect Complex: State Geol. and Natural History Survey Bull., No. 76, 44 p.

Clark, G.S. and Kulp, J.L., 1968, Isotopic age study of metamorphism and intrusion in western Connecticut and southeastern New York: American Jour. Science, v. 266, p. 865-894.

Crowley, W.P., 1968, The bedrock geology of the Long Hill and Bridgeport quadrangles: State Geol. and Natural History Survey of Connecticut Quadrangle Report, No. 24, 81 p.

Dietrich, J.H., 1968, Multiple folding in western Connecticut: A reinterpretation of structure in the New Haven-Naugatuck-Westport area: in Orville, P.M., ed., Guidebook for Field Trips in Connecticut: New England Intercollegiate Geological Conference Guidebook, p. D-2, 1 - D-2, 13.

Dixon, H.R. and Lundgren, L.W., 1968, Structure of eastern Connecticut: in Zen, E-an, White, W.S., Hadley, J.B., and Thompson, J.B., Jr., eds., Studies of Appalachian Geology: Northern and Maritime: New York, Interscience Publishers, p. 219-229.

Emerson, B.K., 1898, Geology of old Hampshire County, Massachusetts: United States Geol. Survey Monograph, v. 29, 790 p.

Eskola, P., 1949, The problem of mantled gneiss domes: Quarterly Jour. Geol. Soc. London, v. 104, p. 461-476.

Fairbairn, H.W., Moorbath, S., Ramo, A.D., Pinson, W.H. and Hurley, P.M., 1967, Rb-Sr age of granitic rocks in southeastern Massachusetts and the age of the Lower Cambrian at Hoppin Hill: Earth and Planetary Sci. Letters, v. 2, p. 321-328.

Faul, H., Stern, T.W., Thomas, H.H. and Elmore, P.L.D., 1963, Ages of intrusion and metamorphism in the Northern Appalachians: American Jour. Sci., v. 261, p. 1-19.

Fisher, D.W., Isachsen, Y.W. and Rickard, L.V., 1970, Geologic Map of New York: New York State Museum and Science Service, Map and Chart Series, No. 15.

Gates, Olcott, 1978, The Silurian-Lower Devonian marine volcanic rocks of the Eastport Quadrangle, Maine: in Ludman, A., ed., Guidebook for Field Trips in Southeastern Maine and Southwestern New Brunswick: New England Intercollegiate Geological Conference Guidebook, p. 1-16.

Gates, R.M. and Bradley, W.C., 1952, The geology of the New Preston Quadrangle with map: State Geol. and Natural History Survey, Miscellaneous Series, No. 5, 46 p.

Gates, R.M. and Christensen, N.I., 1965, The bedrock geology of the West Torrington Quadrangle with map: State Geol. and Natural History Survey, Quadrangle Rept., No. 17, 38 p.

Gates, R.M. and Martin, C.W., 1976, Pre-Devonian stratigraphy of the central Connecticut section of the western Connecticut highlands: in Page, L.R., ed., Contributions to the Stratigraphy of New England: Geol. Soc. America Memoir 148, p. 301-336.

Goldsmith, Richard, 1976, Pre-Silurian stratigraphy of the New London area, southeastern Connecticut: in Page, L.R., ed., Contributions to the Stratigraphy of New England: Geol. Soc. America Memoir 148, p. 211-275.

Grauert, Borwin and Hall, L.M., 1973, Age and origin of zircons from metamorphic rocks in the Manhattan Prong, White Plains area, southeastern New York: Annual Report of the Director, Department of Terrestrial Magnetism 1972-1973, Carnegie Institute Washington Year Book 72, p. 293-297.

Hall, B.A., Pollock, S.G. and Dolan, K.M., 1976, Lower Devonian Seboomook Formation and Matagamon Sandstone, Northern Maine: A flysch basin-margin delta complex: in Page, L.R., ed., Contributions to the Stratigraphy of New England: Geol. Soc. America Memoir 148, p. 57-63.

Hall, L.M., 1968, Times of origin and deformation of bedrock in the Manhattan Prong: in Zen, E-an, White, W.S., Hadley, J.B., and Thompson, J.B., Jr., eds., Studies of Appalachian Geology, Northern and Maritime: New York, Interscience Publishers, p. 117-127.

——————, 1976, Preliminary correlation of rocks in southwestern Connecticut: in Page, L.R., ed., Contributions to the Stratigraphy of New England: Geol. Soc. America Memoir 148, p. 337-339.

Hatch, N.L. and Stanley, R.S., 1973, Some suggested stratigraphic relations in part of southwestern New England: United States Geol. Survey Bull. 1830, 83 p.

——————, 1976, Geologic map of the Blandford Quadrangle, Hampden and Hampshire Counties, Massachusetts: United States Geol. Survey Geologic Quadrangle Map, G.Q. 132.

Hills, F.A. and Dasch, E.J., 1972, Rb/Sr study of the Stony Creek granite, southern Connecticut: a case for limited remobilization: Geol. Soc. America Bull., v. 83, p. 3457-3464.

Kaye, C.A. and Zartman, R.E., 1980, A late Proterozoic to Cambrian age for the stratified rocks of the Boston Basin, Massachusetts: in Wones, D.R., ed., Proceedings "The Caledonides in the U.S.A.", Memoir No. 2: Dept. of Geological Sciences, Virginia Polytechnic Institute and State University, Blacksburg, Virginia, p. 257-261.

Kroll, R.L., 1977, The bedrock geology of the Norwalk North and Norwalk South quadrangles: State Geol. and Natural History Survey of Connecticut Quadrangle Rept. No. 34, 55 p.

Long, L.E., 1969, Whole-rock Rb-Sr age of the Yonkers Gneiss, Manhattan Prong: Geol. Soc. America Bull., v. 80, p. 2087-2090.

Lundgren, L., Jr., 1966, Muscovite reactions and partial melting in southeastern Connecticut: Jour. Petrology, v. 7, p. 421-453.

Lundgren, L., Jr. and Thurrell, R.F., 1973, The bedrock geology of the Clinton quadrangle: Connecticut State Geological and Natural History Survey, Quadrangle Rept. No. 29, p. 1-21.

Lyons, J.B. and Livingston, D.E., 1977, Rb-Sr age of the New Hampshire plutonic series: Geol. Soc. America Bull., v. 88, p. 1808-1812.

McKerrow, W.S. and Ziegler, A.M., 1972, Silurian paleogeographic development of the Proto-Atlantic Ocean: 24th International Geol. Congress, Montreal, Section 6, p. 4-10.

Merguerian, C., 1979, Dismembered Ophiolite along Cameron's Line West Torrington, Connecticut: Geol. Society of America, Abstracts with Program, v. 11, no. 1, p. 45.

Moench, R.H. and Boudette, E.L., 1970, Stratigraphy of the northwest limb of the Merrimack synclinorium in the Kennebago Lake, Rangeley, and Phillips quadrangles, western Maine: in Boone, G.M., ed., New England Intercollegiate Geological Conference Guidebook for Field Trips in the Rangeley Lakes Dead River Basin Region, Western Maine, p. A-1, 1 – A-1, 25.

Moench, R.H. and Gates, O., 1976, Bimodal volcanic suites of Silurian and Early Devonian Age, Machias-Eastport area, Maine: Geol. Soc. America, Abstracts with Program, v. 8, p. 232.

Mose, D.G. and Hayes, J., 1975, Avalonian igneous activity in the Manhattan Prong, southeastern New York: Geol. Soc. America Bull., v. 86, p. 929-932.

Mose, D.G. and Helenek, H.L., 1976, Origin, age and mode of emplacement of Canada Hill Granite, Hudson Highlands, New York: Geol. Soc. America, Abst. with Programs, Northeastern Sec. and Southeastern Sec., v. 8, no. 2, p. 233.

Naylor, R.S., 1968, Origin and regional relationships of the core-rocks of the Oliverian domes: in Zen, E-an, White, W.S., Hadley, J.B., and Thompson, J.B., Jr., eds., Studies in Appalachian Geology, Northern and Maritime: New York, Interscience, p. 231-240.

_____, 1969, Age and origin of the Oliverian domes, central-western New Hampshire: Geol. Soc. America Bull., v. 80, p. 405-427.

_____, 1970, Radiometric dating of pennine-type nappes in the northern Appalachians: Eclogae Geologicae Helvetiae, v. 63/1, p. 230.

_____, 1975, Age provinces in the northern Appalachians: in Donath, F.A., et al., eds., Annual Review of Earth and Planetary Sciences, Palo Alto, California, p. 387-400.

Naylor, R.S., Boone, G.M., Boudette, E.L., Ashenden, D.D. and Robinson, P., 1973, Pre-Ordovician rocks in the Bronson Hill and Boundary Mountain anticlinoria, New England, U.S.A.: Trans. American Geophys. Union, v. 54, p. 495.

Norton, S.A., 1976, Hoosac Formation (Early Cambrian or older) on the east limb of the Berkshire massif, western Massachusetts: in Page, L.R., ed., Contributions to the Stratigraphy of New England: Geol. Soc. America Memoir 148, p. 357-371.

Osberg, P.H., 1978, Synthesis of the geology of the northeastern Appalachians, U.S.A.: in Caledonian-Appalachian Orogeny of the North Atlantic Region, Geol. Survey Canada Paper 78-13, p. 137-147.

Prucha, J.J., Scotford, D.M. and Sneider, R.M., 1968, Bedrock geology of parts of Putnam and Westchester Counties, New York, and Fairfield County, Connecticut: New York State Museum and Science Service, Map and Chart Series No. 11, 26 p.

Ratcliffe, N.M., 1968, Stratigraphic and structural relations along the western border of the Cortlandt intrusives: in Finks, R.M., ed., Guidebook for Field Excursions: 40th Annual Meeting of the New York State Geol. Assoc. 1968, p. 197-220.

Ratcliffe, N.M. and Harwood, D.S., 1975, Blastomylonites associated with recumbent folds and overthrusts at the western edge of the Berkshire massif, Connecticut and Massachusetts – a preliminary report: in Tectonic Studies of the Berkshire Massif, Western Massachusetts, Connecticut, and Vermont: United States Geol. Survey Prof. Paper 888A, p. 1-19.

Ratcliffe, N.M. and Zartman, R.E., 1976, Stratigraphy, isotopic ages, and deformational history of basement and cover rocks of the Berkshire massif, southwestern Massachusetts: in Page, L.R., ed., Contributions to the Stratigraphy of New England: Geol. Soc. America Memoir 148, p. 373-417.

Robinson, P., 1963, Gneiss domes of the Orange area, Massachusetts and New Hampshire: Ph.D. Thesis, Harvard University, 253 p. (separate volume of tables and appendices, portfolio of 16 plates)

_____, 1977, Interim geologic map of the Orange area, Massachusetts and New Hampshire, consisting of the Orange and Mt. Grace quadrangles and portions of the Northfield, Millers Falls, and Athol quadrangles: United States Geol. Survey Open File Report, 77-788.

Robinson, P. and Hall, L.M., 1980, Tectonic synthesis of southern New England: in Wones, D.R., ed., Proceedings "The Caledonides in the U.S.A.", Memoir No. 2: Dept. of Geological Sciences, Virginia Polytechnic Institute and State University, Blacksburg, Virginia, p. 73-82.

Robinson, P., Jackson, R.A., Piepul, R.G., Leftwich, J.T., Ashwal, L.D. and Jelatis, P.J., 1973, Progress bedrock geologic map, eastern part of Shutesbury quadrangle, central Massachusetts: Geology Dept., University of Massachusetts, Amherst.

Robinson, P., Tracy, R.J. and Ashwal, L.D., 1975, Relict sillimanite-orthoclase assemblage in kyanite-muscovite schist, Pelham dome, west-central Massachusetts, Abstract: Trans. American Geophys. Union, v. 56, p. 466.

Robinson, P., Tracy, R.J. and Pomeroy, J.S., 1977, High pressure sillimanite-garnet-biotite assemblage formed by recrystallization of mylonite in sillimanite-cordierite-biotite-orthoclase schist, central Massachusetts: Geol. Soc. America, Abstracts with Programs, v. 9, p. 1144-1145.

Rodgers, J., 1968, The eastern edge of the North American continent during the Cambrian and Early Ordovician: in Zen, E-an, White, W.S., Hadley, J.B., and Thompson, J.B., Jr., eds., Studies of Appalachian Geology: Northern and Maritime, New York, Interscience Publishers, p. 141-149.

Rodgers, John, Gates, R.M., Rosenfeld, J.L., 1959, Explanatory text for Preliminary Geological Map of Connecticut, 1956: Connecticut State Geol. and Nat. Hist. Survey Bull. No. 84, 64 p.

Rosenfeld, J.L. and Eaton, G.P., 1958, New England Intercollegiate Geological Conference: Wesleyan University, Middletown, Connecticut, Guidebook.

Shride, A.F., 1976, Stratigraphy and structural setting of the Newbury volcanic complex, northeastern Massachusetts: in Cameron, B., ed., Geology of Southeastern New England, A Guidebook for Field Trips to the Boston Area and Vicinity: New England Intercollegiate Geol. Conference Guidebook, p. 291-300.

Skehan, J.W. and Skehan, S.J., 1962, The Green Mountain anticlinorium in the vicinity of Wilmington and Woodford, Vermont: Vermont Geol. Survey Bull. No. 17, 159 p.

Skehan, J.W., Skehan, S.J., Murray, D.P., Palmer, A.R. and Smith, A.T., 1977, The geological significance of Middle Cambrian trilobite-bearing phyllites from Southern Narragansett Basin, Rhode Island: Geological Society of America, Abstracts with Programs, v. 9, no. 3, p. 319.

Stanley, R.S., 1964, The bedrock geology of the Collinsville quadrangle: State Geol. and Natural History Survey of Connecticut, Quadrangle Rept. No. 16, 99 p.

Steiger, R.H. and Jäger, E., 1977, Subcommission on geochronology: convention on the use of decay constants in geo- and cosmochronology: Earth and Planetary Sci. Letters, v. 36, p. 359-362.

Theokritoff, G., 1968, Cambrian biogeography and biostratigraphy in New England: in Zen, E-an, White, W.S., Hadley, J.B., Thompson, J.B., Jr., eds., Studies of Appalachian Geology: Northern and Maritime, New York, Interscience Publishers, p. 9-22.

Thompson J.B., Jr., Robinson, P., Clifford, T.N. and Trask, N.J., Jr., 1968, Nappes and gneiss domes in west-central New England: in Zen, E-an, White, W.S., Hadley, J.B., and Thompson, J.B., Jr., eds., Studies of Appalachian Geology: Northern and Maritime, New York, Interscience Publishers, p. 203-218.

Thompson, J.B., Jr. and Rosenfeld, J.L., 1979, Reinterpretation of nappes in the Bellows Falls-Brattleboro area, New Hampshire-Vermont: in Skehan, J.W., and Osberg, P.H., eds., The Caledonides in the U.S.A., Geological Excursions in the Northeast Appalachians: Dept. Geol. and Geophys., Weston Observatory, Boston College, Weston, Mass., p. 117-121.

Tilton, G.R. and Davis, G.L., 1959, Geochronology: in Abelson, P.H., ed., Researches in Geochemistry: New York, John Wiley and Sons Inc., p. 190-216.

Tracy, R.J., Robinson, P. and Thompson, A.B., 1976, Garnet composition and zoning in the determination of temperature and pressure of metamorphism, central Massachusetts: American Mineralogist, v. 61, p. 762-775.

Williams, H., 1971, Mafic-ultramafic complexes in western Newfoundland and the evidence for their transportation; a review and interim Report: Proceedings Geol. Assoc. Canada, v. 24, p. 9-25.

———————————, 1978, Tectonic lithofacies map of the Appalachian orogen: Memorial University of Newfoundland, Map No. 1.

Williams, H., Kennedy, M.J. and Neale, E.R.W., 1972, The Appalachian structural province: in Price, R.A. and Douglas, R.J.W., eds., Tectonic Styles in Canada: Geol. Assoc. Canada, Special Paper 11, p. 182-261.

Zartman, R.E. and Marvin, R.F., 1971, Radiometric age (Late Ordovician) of the Quincy, Cape Ann, and Peabody granites from eastern Massachusetts: Geol. Soc. America Bull., v. 82, p. 937-958.

Zartman, R.E. and Naylor, R.S., 1972, Structural implications of some U-Th-Pb zircon isotopic ages of igneous rocks in eastern Massachusetts: Geol. Soc. America, Abstracts with Programs, v. 4, p. 54.

Zartman, R.E., Hurley, P.M., Krueger, H.W. and Giletti, B.J., 1970, A Permian disturbance of K-Ar radiometric ages in New England: its occurrence and cause: Geol. Soc. America Bull., v. 81, p. 3359-3374.

Zen, E-an, 1967, Time and space relationships of the Taconic allochthon and autochthon: Geol. Soc. America Spec. Paper 97, 107 p.

Manuscript Received October 22, 1979
Revised Manuscript Received June 11, 1980

Major Structural Zones and Faults of the Northern Appalachians, edited by
P. St-Julien and J. Béland, Geological Association of Canada Special Paper 24, 1982

THE AVALONIAN AND GANDER ZONES
IN CENTRAL EASTERN NEW ENGLAND

John B. Lyons
Department of Earth Sciences, Dartmouth College, Hanover, NH 03755
Eugene L. Boudette and John N. Aleinikoff
U.S. Geological Survey, Denver, CO 80225

ABSTRACT

The Massabesic Gneiss, in southern New Hampshire, is a northeast-trending belt of Upper Precambrian (Hadrynian, 650 Ma) paragneiss and orthogneiss, with some Middle (?) Ordovician (480 Ma) orthogneiss. It has a structural setting similar to that of four additional Precambrian (?) blocks in coastal Maine and New Hampshire which appear to lie at the cores of anticlinoria, but in which stratigraphic units on either side of the anticlinoria (?) are distinctively different. Faults bound at least four of the five anticlinoria (?) on one or both sides.

We interpret the structure of the Precambrian (?) and its bordering formations as representing a series of folded thrust plates formed during the Taconian orogeny near the shelf-slope margin of the eastern Iapetus Ocean, and involving both Avalonian age basement and its Cambro-Ordovician cover. This zone of faults, first established during the Taconian links the Norumbega Fault of northern Maine to the Clinton-Newbury fault zone of Massachusetts, and was active throughout the Paleozoic. It lies entirely within the Gander Zone, rather than between the Gander and Avalon zone as is true of its northern extension (the Dover fault) in Newfoundland.

A soapstone belt interpreted as fragmented ophiolite occurs along the Connecticut Valley, and marks the approximate site of an Ordovician paleo-subduction zone. Another soapstone belt in central New Hampshire is of Acadian age, not of ophiolitic affinity, and seems to lie along a major zone of dislocation (the Concord tectonic zone) of uncertain significance.

The location and names of the major structural elements in New Hampshire are reviewed, and several proposals are made for revising several of Billings' (1956) structural subdivisions.

Paleozoic tectonic history in northern New England is summarized in the light of our current knowledge beginning with the opening of Iapetus in Early Cambrian time, and ending with Alleghenian plutonism and tectonism.

RÉSUMÉ

Le Gneiss de Massabesic occupe dans le sud du New Hampshire une zone de direction nord-est constituée de paragneiss et d'orthogneiss du Précambrien supérieur (Hadrynien; 650 Ma.) et en moindre proportion, d'orthogneiss de l'Ordovicien moyen (?; 480 Ma). Son contexte structural ressemble à ceux de quatre autres blocs de Précambrien(?) situés dans la zone cotière du Maine et au New Hampshire au coeur d'anticlinoria (?) mais dont les unités stratigraphiques de part et d'autre de ces structures sont nettement dissemblables. Au moins quatre des cinq anticlinoria (?) sont limités par des failles, d'un ou des deux cotés.

Nous proposons que cette structure de Précambrien (?) avec ses unités limitrophes représente une imbrication de chevauchements plissée au cours de l'orogénèse taconienne au voisinage de la jonction plateforme-pente continentale qui bordait à l'est l'Océan Iapetus; nous soumettons que le soubassement avalonien et sa couverture cambro-ordovicienne sont ici impliqués. Cette zone de failles taconiennes qui rattache la Faille de Norumbega du nord du Maine à la zone de failles de Clinton-Newbury au Massachussetts est demeurée active pendant tout le Paleozoique. Elle se confine entièrement à la zone de Gander plutôt que de séparer celli-ci de la zone d'Avalon comme il se produit plus au nord, à Terre-Neuve (faille de Dover).

Une zone de stéatite identifiée à un segment d'ophiolite longe la vallée de Connecticut et marque le site approximatif d'une paléosubduction ordovicienne. Une autre zone de stéatite, d'âge acadien et sise au centre de New Hampshire, n'a pas d'affinité ophiolitique et semble révéler une zone importante de dislocation (la zone tectonique de Concord) mais dont nous ignorons la signification.

La répartition et la nomenclature des principaux éléments structuraux du New Hampshire sont ici ré-examinés et plusieurs ré-interprétations des subdivisions structurales de Billings (1956) sont proposées.

Nous esquissons selon l'état actuel des connaissances l'évolution tectonique du nord de la Nouvelle Angleterre au Paléozoïque, depuis l'ouverture de l'Océan Iapetus, au Cambrien inférieur, jusqu'aux plutonisme et tectonisme alléghaniens.

INTRODUCTION

In this paper we examine linear and other tectonic features which have been identified in southern and central New Hampshire during the past 15 years, and attempt to explain them in terms of their regional structural pattern in New England. Our synthesis presents a contrast to that of Billings (1965) who interprets the geology of the state as being dominated by a major central synclinorium of the Lower Devonian (or possibly, also, somewhat older) Littleton Formation – the Merrimack Synclinorium – flanked on the west by the Bronson Hill Anticlinorium, and on the east by the Rockinham Anticlinorium (Figs. 1a and 1b).

Among the features of interest on Billings' map is the supposed Upper (?) Devonian Fitchburg Pluton about midway between the axes of the Rockingham Anticlinorium and the Merrimack Synclinorium. Besancon et al. (1977), Aleinikoff et al. (1979), Aleinikoff (1978), and Kelly et al. (1980) have now shown by Pb/U and Rb/Sr isotopic dating that the Fitchburg Pluton is a migmatite of Precambrian to Ordovician age. This result, coupled with Rb/Sr age dating in central New Hampshire (Lyons and Livingston, 1977) and with regional mapping, has allowed us to decipher a stratigraphy in part correlative with that in better-dated, less complicated strata of northwestern Maine. Thompson et al. (1968) have also proposed a reinterpretation of the geology of western New Hampshire involving at least three large nappes rooted east of the Bronson Hill Anticlinorium and transported westerly across it prior to the diapiric rise of the Oliverian domes along the Bronson Hill Anticlinorium. Extensive fault systems in eastern Massachusetts (Cuppels, 1961; Skehan, 1968; Castle *et al.*, 1976) and in coastal Maine (Hussey and Pankiwskyj, 1975; Wones and Thompson, 1979) are now known to correlate with similar structures in southeastern New Hampshire. Some of these faults are indicated on Billings' (1956) map, but their tectonic implications were not recognized, nor were all of the major faults identified.

Radiometric ages referred to in this paper or depicted on maps (Figs. 3 and 4) have been corrected to the new decay constants of Steiger and Jäger (1978).

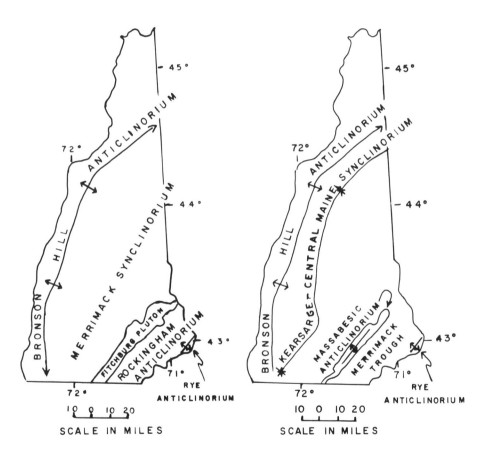

Figure 1a. Major tectonic features of New Hampshire, after Billings (1956).

Figure 1b. Major tectonic features of New Hampshire proposed in this paper.

Figure 2. Faults, folds, and major plutonic sequences in central eastern New England.

CONNECTICUT VALLEY SOAPSTONES

The tectonic significance currently attached to linear soapstone belts gives them an importance far beyond their limited areal distribution. Hitchcock (1877, 1878) noted two soapstone belts in New Hampshire, and those which were quarried a century ago are shown in Figures 2 and 3.

The western belt, approximately 30 km long, lies just west of the axis of the Bronson Hill Anticlinorium and the soapstones are enclosed in the Middle (?) Ordovician Ammonoosuc Volcanics. These soapstones are apparently younger than the Cambrian (?) Boil Mountain ophiolite which lies along strike to the north in western central Maine (Boudette, this volume), but are of the same age as a 40-km long belt of ultramafic pods to the south in Massachusetts (Robinson, 1967, Wolff, 1978). The Massachusetts soapstones also occur along or west of the Bronson Hill anticlinorium and are enclosed within Middle (?) Ordovician euximic schists of the Partridge Formation or within Middle (?) Ordovician Ammonoosuc Volcanics.

Because the Connecticut Valley ultramafics in both New Hampshire and Massachusetts occur only as conformable pods in Middle (?) Ordovician metamorphic rocks and are never found in nearby Silurian or Devonian metamorphics, a Middle (?) Ordovician age assignment for them seems reasonable. They may be of approximately the same age as the pre-Silurian Vermont ultramafic belt 50 to 75 km to the west, which lies east of the axis of the Green Mountain anticlinorium, but an exact age equivalence cannot be proven. Three of the Connecticut Valley soapstones in New Hampshire have been studied by Nielson (1974), who showed that they had the petrographic characteristics (blackwall enclosing a soapstone core), mineralogy (talc + chlorite + actinolite ± carbonate) and chemistry (Table I; analyses 1, 2, and 3) of typical metamorphosed ultramafics. Nearby pillowed metabasalts of the Ammonoosuc Volcanics, some with a capping of chert, have been shown by Aleinikoff (1977) to have chemical characteristics of both abyssal and island arc tholeiites on the basis of major-element bulk chemistry and Ti-Zr-Y-Sr minor-element chemistry. The association of oceanic and island arc tholeiites, cherts and euxinic schists with the ultramafics implies that the latter may be part of an ophiolite suite which has been dismembered, and which originally formed at or along a trench-island arc contact. The suggestion of dismembered ophiolites in Massachusetts was put forth by Wolff (1978), and was followed up by Robinson and Hall (1980). The latter authors visualized an easterly dipping subduction zone initiated in Early Ordovician time at a location just west of the present Bronson Hill anticlinorium. Ophiolite slabs were obducted westerly onto the oceanic and continental crust of proto-North America. By Middle Ordovician time a volcanic arc had formed at the present site of the Bronson Hill anticlinorium, and olistoliths of previously obducted ophiolites were being swallowed at the island arc-trench contact. Although the mechanism for emplacement of the ultramafic olistoliths is speculative, the general outline of the Robinson-Hall scheme appears to fit the geologic facts well; we agree with them that the present course of the Connecticut River is approximately coincident with a major suture of Ordovician age. This idea is further confirmed by the fact that Avalonian basement is exposed in the cores of some of the Oliverian domes of Massachusetts and Connecticut, but to the west the nearest basement exposures are Grenvillian (Robinson and Hall, 1980).

TABLE I

CHEMICAL ANALYSES AND CIPW NORMS OF SAMPLES FROM NEW HAMPSHIRE
SOAPSTONE QUARRIES

	Page	Orford-ville	Cotton-stone	Canter-bury	Mt. Misery	Frances-town
SiO_2	48.04	44.59	28.36	40.08	48.22	47.99
TiO_2	0.50	0.49	0.15	0.37	0.81	0.86
Al_2O_3	5.47	8.10	6.44	5.21	6.76	8.30
Cr_2O_3	0.08	0.40	0.31	0.47	0.21	0.17
FeO^*	11.79	10.19	7.39	14.64	10.58	8.90
MnO	0.20	0.16	0.12	0.20	0.25	0.16
NiO	0.05	0.11	0.18	0.04	0.05	0.13
MgO	21.02	21.65	33.97	29.37	20.79	21.35
CaO	7.83	9.36	0.69	2.90	4.48	4.98
Na_2O	0.26	0.12	0.09	0.37	0.17	0.22
K_2O	0.06	0.05	0.05	0.27	2.73	4.33
H_2O^+	3.16	4.43	1.53	5.17	2.83	1.82
H_2O^-	0.34	0.11	0.02	0.79	0.35	0.82
P_2O_5	0.06	—	—	0.07	0.10	0.13
CO_2	0.05	0.50	20.00	0.25	tr	0.25
Total	98.91	100.26	99.30	100.20	98.33	100.41

*Total Fe calculated as FeO.

Q	—	—	—	—	—	—
Or	0.35	0.30	—	1.60	16.13	25.59
Ab	2.20	1.02	—	3.13	1.44	1.86
An	13.58	21.42	3.61	11.76	9.62	8.87
Ne	—	—	0.41	—	—	—
Ks	—	—	0.17	—	—	—
C	—	—	6.24	—	—	—
Di	19.77	20.06	—	0.47	9.59	10.71
Hy	43.02	21.58	—	13.10	32.23	9.13
Ol	15.15	29.42	69.07	62.24	24.10	38.81
Il	0.95	0.93	0.28	0.70	1.54	1.63
Ap	0.14	—	—	0.16	0.23	0.30

Analyses from D.R. Nielson, 1974.

Osberg (1978), on the basis of regional geologic relations, has also proposed that a major suture zone (the boundary between his basement A and basement C domains) occurs in this region of New England, and that subduction was easterly. His boundary, however, is placed 5 to 40 km west of ours, closer to the eastern flank of the Green Mountain Anticlinorium.

Several authors (e.g. Naylor, 1968) had already proposed that the entire 400-km long belt of Oliverian domes represents a relict Middle Ordovician island arc, an idea which is consistent both with the tectono-stratigraphic setting of the domes, and with their petrochemistry. The island-arc lies along the southern extension of an Early to Middle Paleozoic tectonic hinge zone identified in central western Maine by Boudette and Boone (1976). In New Hampshire the hinge zone separates a thick pre-Silurian clastic and volcanic sequence on the west from a thinner (?) pre-Silurian clastic and volcanic sequence to the east. The latter group of rocks is poorly known because of a cover of Siluro-Devonian metasediments, and the presence of numerous sheets of Acadian-age plutons.

During the early Paleozoic the site of the present Bronson Hill Anticlinorium was close to the leading edge of a westerly-drifting microplate with a trench to the west in which, by Middle Ordovician time, sulphidic flysch (Partridge) and volcanics (Ammonoosuc) were accumulating. The Taconian closure of Iapetus along the Baie Verte-Brompton line (Williams and St-Julien, 1978) in the northern Appalachians was accompanied by both obduction and subduction of ophiolites. Along the Chain Lakes-Connecticut Valley belt which is 50 to 75 km east of the southerly extension of the Baie Verte-Brompton suture, the timing and method of emplacement of the ophiolites seems to have varied depending upon the times at which microplates made contact with one another. The Boil Mountain ophiolite (Boudette, this volume) was obducted westerly along a southeasterly-dipping thrust onto the 1600 Ma old Chain Lakes massif in Late (?) Cambrian time, during the first major tectonic pulse of the Taconian. The Bronson Hill ultramafic belt in New Hampshire and Massachusetts, by contrast, represents a series of olistoliths emplaced during the Middle (?) Ordovician, possibly during or close to the culmination of the Taconian orogeny, and involves crustal and oceanic plates of differing ages than those involved in the emplacement of the Boil Mountain ophiolite.

In our view, the Connecticut Valley soapstones occupy a structural position analogous to those of the Gander River ultramafic belt in Newfoundland (Williams, 1978), and the Vermont soapstones are comparable to the Humber Arm and Baie Verte ultramafics. Haworth *et al.* (1978) have shown geophysically that the upper mantle rocks of the Baie Verte and Gander River belts of Newfoundland are connected by a synclinorial fold, and that the Gander River belt then dips southeasterly, in a fashion similar to that which we postulate for the Connecticut River soapstones. The structural details in New England will, however, be much more complex because of the far greater degree of deformation.

CENTRAL NEW HAMPSHIRE SOAPSTONES

Hitchcock's (1978) central New Hampshire soapstone belt has been quarried in at least four localities (Figs. 3 and 4). These soapstones are not associated with volcanics nor a basement suture, and thus have a totally different tectonic setting from the Connecticut Valley soapstone belt. Three of the quarries have been studied in detail by Nielson (1974) and typical analyses of the rocks are shown in Table I, columns 4, 5, and 6. The soapstones occur as pods of varying size, generally conformable with the enclosing wallrock, which is of Siluro-Devonian (?) age. Inside a biotite blackwall the mineralogy is talc + chlorite + amphibole ± biotite or phlogopite ± plagioclase. The talc and chlorite are pseudomorphous after orthopyroxene or olivine, and in the Canterbury deposit (the northeasternmost on Figure 3) olivine is conspicious.

The mineralogy and chemistry of the Canterbury deposit (the northernmost) classify it as a metamorphosed ultramafic, but the Ni and Cr contents imply that it may have originally been a cumulate, rather than a true alpine ultramafic (Nielson,

1974). This suggestion of a cumulate origin is supported in the case of the Mt. Misery deposit (the third most southerly body in Fig. 3) where the chemistry is not that of a true ultramafic, and the soapstone is associated with a small metamorphosed monzodiorite. An unusual amount of phlogopite skews the chemistry of the Francestown quarry (the most southerly body) away from that of a typical ultramafic. This soapstone is underlain on its southeastern boundary by a pegmatite, which may have

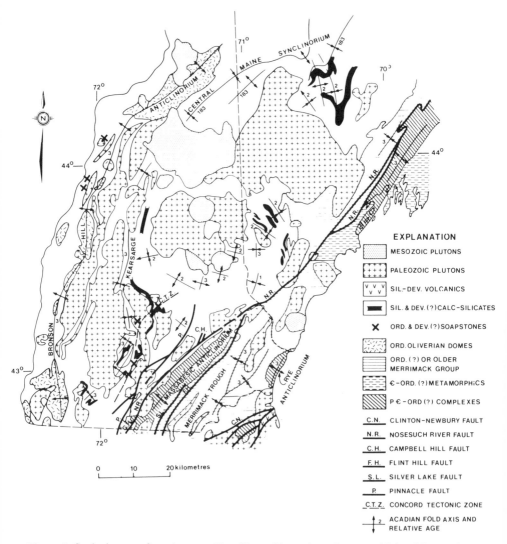

EXPLANATION

▦	MESOZOIC PLUTONS
++++	PALEOZOIC PLUTONS
v v v	SIL.-DEV. VOLCANICS
▬	SIL. & DEV. (?) CALC-SILICATES
✕	ORD. & DEV. (?) SOAPSTONES
▨	ORD. OLIVERIAN DOMES
▤	ORD. (?) OR OLDER MERRIMACK GROUP
▥	€-ORD. (?) METAMORPHICS
▧	P €-ORD (?) COMPLEXES

C.N.   CLINTON-NEWBURY FAULT
N.R.   NOSESUCH RIVER FAULT
C.H.   CAMPBELL HILL FAULT
F.H.   FLINT HILL FAULT
S.L.   SILVER LAKE FAULT
P.   PINNACLE FAULT
C.T.Z. CONCORD TECTONIC ZONE

↑2 / ⊥   ACADIAN FOLD AXIS AND RELATIVE AGE

Figure 3. Geologic map of southeastern New Hampshire and southwestern Maine. Map modified from Billings (1956), Hussey and Pankiwskyj (1975), and published and unpublished quadrangle reports.

been a source for the introduction of potassium. Nickel and chromiun contents are very high, as is the amount of talc. Because of the occurrence of tremolite, particularly near the margins of this soapstone, Greene (1970) concluded that it represented a metamorphosed dolomite, but the mineralogy and geochemistry strongly favor a protolith of mafic or ultramafic igneous composition. The remaining body which is very small (10 x 15 m) is poorly exposed, and apparently a xenolith within a 393 Ma (Lyons and Livingston, 1977) pluton.

In the Penobscot Bay region of Maine (Fig. 2) there are small ultramafic bodies quite similar to those in central New Hampshire, enclosed within Lower Paleozoic metasediments (Osberg and Guidotti, 1974). Many of them are along, or near faults.

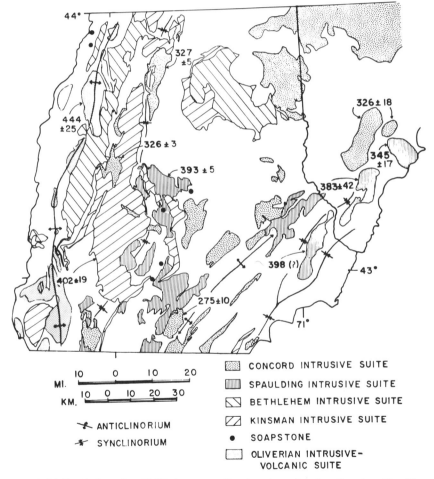

Figure 4. Rb/Sr whole rock or Pb/U zircon ages for plutonic rocks of southeastern New Hampshire and southwestern Maine. Data from Lyons and Livingston (1977), Gaudette *et al.* (1975), and Aleinikoff (1978, and unpublished).

Gaudette (in press) has recently dated zircons from a diorite in one of these ultramafic complexes at 410 ± 7 Ma, (Lower (?) Devonian), an age which would be reasonable (though speculative) for the central New Hampshire soapstones.

Because there are nearby volcanics which straddle the Siluro-Devonian time boundary, and because many (but not all) of the complexes lie along mapped fault traces, Osberg (1978) has proposed that the Penobscot Bay ultramafics lie along a major suture (the contact between his Basement "C" and Basement "D") representing the final closure of Iapetus and welding of Avalonia to North America, at the beginning of the Acadian orogeny. Williams (1979), by contrast, believes that Avalonia had been linked to North America prior to the beginning of the Silurian.

There is no evidence in this case of the central New Hampshire soapstones, that we are dealing with rocks of oceanic crustal affinity. It seems more likely that these are fragments of layered mafic plutons which crystallized at deep crustal levels, possibly during the early stages of the Acadian orogeny. Parts of the intrusives have been dragged upward along a tectonic slice or series of slices, which we here designate the Concord Tectonic Zone. We postulate that the zone dips easterly, in a listric fault zone similar to those now revealed by the Southern Appalachian COCORP line. A new COCORP line (Cook *et al.*, 1979) crossing the Concord Tectonic Zone is scheduled for the near future so we should soon be able to confirm or deny these speculations.

The wallrock of the small soapstone pod enclosed in an Acadian age pluton (Fig. 4) is unfortunately not exposed, but it is possible that this pod was rafted upward along with felsic Acadian-age magma rising toward the surface along a fault-controlled conduit.

THE BASEMENT IN NORTHEASTERN NEW ENGLAND

Figure 3 is a compilation of recent mapping in southeastern and central New Hampshire, and in southwestern Maine. The Maine geology is taken from Hussey and Pankiwskyj (1975) and Osberg (1974). The New Hampshire geology is modified from Billings (1956), from published and unpublished theses and research maps at Dartmouth College, and a map by Vernon (1971). Tectonic interpretations are those of the authors of this paper, and not necessarily those of the original map makers.

Of the basement complexes which have received close attention in recent years, one is the Massabesic Gneiss of southern New Hampshire – a migmatite which coincides approximately with what was mapped by Billings (1956) as the Upper Devonian (?) Fitchburg pluton. Besancon *et al.* (1977), Aleinikoff (1978), and Kelly *et al.* (1980) have shown by Pb/U and Rb/Sr isotopic dating and by geologic mapping that the protolith of the Massabesic paragneiss is approximately 650 Ma old. The paragneiss is cut by orthogneiss of similar age and also by 480 Ma old orthogneiss. The Massabesic paragneiss consists predominantly of felsic metavolcanics, but amphibolite, pelitic schist, calc silicate and quartzite are also present. The age of migmatization is uncertain but might be Taconian or younger, i.e. at one of the times during which anatectic granites were formed in New Hampshire. The Massabesic Gneiss apparently underlies a doubly-plunging anticlinorium, hereafter referred to as the Massabesic Anticlinorium.

The Nashoba metamorphic complex of northeastern Massachusetts (Fig. 3) lies between the Bloody Bluff and Clinton-Newbury fault zones and is cut by Silurian (?) granite. Olszewski (1978) has reported Pb^{207}/Pb^{206} ages of 742 Ma on euhedral zircons from its metavolcanic rocks, and 1550 Ma ages on anhedral, detrital zircons from the metasedimentary rocks. The Nashoba is shown on various maps as "Ordovician (?)", "pre-Silurian", or "Precambrian (?) to Ordovician (?)" but the isotopic ages now clearly favor the Precambrian age assignment. The Nashoba is on the northern edge of what is widely recognized as the Avalonian Platform of southeastern New England. It is interesting, therefore, that its age pattern is not greatly different than that in the Avalonian basement Greenhead Group of New Brunswick in which Olszewski and Gaudette (1980) have found 1640 Ma detrital zircon in paragneiss, 800 Ma orthogneiss, and a 470 Ma pegmatite (the latter very close to the 480 Ma orthogneiss of the Massabesic Gneiss).

Isotopic Pb/U and Rb/Sr dating of the Avalon platform in eastern Massachusetts yields ages ranging from 514 to 630 Ma (reviewed by Zartman and Naylor, in press) but the lower ages are Rb/Sr determinations which appear to have been lowered by later disturbances; therefore, an age span of approximately 600 to 800 Ma seems to encompass the more reliable ages on the now-recognized Avalonian basement. On this basis, the Massabesic Gneiss is clearly of Avalonian affinity, as are the Late Hadrynian (~570 Ma, Naylor *et al.*, 1971; Robinson and Hall, 1980) gneisses in the cores of some of the Oliverian domes of Massachusetts and Connecticut.

Another Avalonian age (Rb/Sr, 735 Ma) has been reported for a small sliver of Precambrian rocks of Ilseboro Island in Penobscot Bay (Stewart and Wones, 1974; Fig. 2). These rocks are correlated, on lithologic and structural grounds by Osberg (1974) with the Passagassawakeag Gneiss and with the Cushing Formation, both of which are shown by Osberg in the cores of anticlinoria, faulted along their western margins (Fig. 2). The Cushing is largely metavolcanic, and older than 528 Ma (Brookins and Hussey, 1978). One additional area of Precambrian (?) is the Rye anticlinorium in coastal New Hampshire (Fig. 3). The Rye Formation consists of a lower metasedimentary member and an upper metavolcanic unit, the latter not dissimilar to some of the metavolcanics of the Massabesic Gneiss. Billings (1956) assigned an Ordovician (?) age to the Rye, but there are no fossils, nor any isotopic ages.

Of the nine areas we have now cited of known or inferred Avalonian age (600 to 800 Ma) basement, two (southern New Brunswick and eastern Massachusetts) fit the classic Avalonian definition in that they have an unconformable cover of essentially unmetamorphosed Cambro-Ordovician rocks on top of a 600 to 800 Ma basement. Another (the Nashoba) lies at the faulted margin of the Massachusetts Avalonian platform, and the remainder lie on the southwest projection of Williams' (1978) Gander zone and have a cover of metamorphosed Cambro-Ordovician and younger rocks. What is important is that there seems to be no valid reason for separating the Avalonian and Gander basements on the basis of either lithology or age. The former seemingly represents the eastern cratonal platform of Iapetus in Cambrian and Early Ordovician time, and the latter its eastern shelf and slope.

A very significant fact, to which we shall return, is that in the areas of Figures 2 and 3, formations on either side of the Precambrian anticlinoria are never the same; there is no structural-stratigraphic symmetry.

THE MERRIMACK GROUP

A series of purple and black pelitic schists, light green calc-silicates and feldspathic quartzites of variable metamorphic grade occur in what we interpret as a synclinorial structure between the Rye and the Massabesic anticlinoria. These rocks were originally mapped by Hitchcock (1878) as the Merrimack Group and were subsequently subdivided in southwestern Maine (Katz, 1917) into the Kittery, Eliot, and Berwick Formations in what was thought to be the order of westerly-decreasing age. With the exception of the fact that the Elliot Formation has more metapelite than the others, distinctions among them are difficult to make, and as Billings (1956, p. 41) once observed, "Much of the difference between the Eliot and the Berwick formations is the result of a difference in the grade of metamorphism". The grade of metamorphism increases irregularly toward the Massabesic Gneiss.

The Merrimack Group traces northeasterly to the vicinity of Waterville, Maine, near the northeast edge of Figure 2, where it is continuous (?) across the Nonesuch River Fault with the Vassalboro Formation (Hussey and Pankiwskyj, 1975) a massive bluish greywacke with minor interbeds of grey phyllite. West of the Vassalboro are phyllites, wackes, and limestones of the Waterville and Mayflower Hill Formations, dated by graptolites as of late Llandovery to Ludlow age (late Early Silurian to Late Silurian). Osberg (1962), who worked out this succession, originally assigned the Vassalboro a Late Silurian (?) to Early Devonian (?) age, but now (1979, p. 42) believes that it underlies the Mayflower Hill and Waterville Formations and is of Llandoverian to latest Ordovician age. This age assignment is based, in part, on the presence of poorly preserved graptolites at one locality in the Vassalboro. The lower age boundary is set by the uncertain (Precambrian (?) to Ordovician (?)) age of the Rye Gneiss, if the Merrimack Group and the Vassalboro are continuous across the Nonesuch River fault.

In southern New Hampshire and northern Massachusetts the Merrimack Group is cut by isotopically dated granites (Zartman and Naylor, in press) several of which fall within the Silurian. Also, the fossiliferous and unmetamorphosed Newburyport Volcanics of northeastern Massachusetts (Fig. 3) which straddle the Siluro-Devonian time boundary (Shride, 1976), are in close juxtaposition with the regionally metamorphosed Merrimack Group rocks. These relations strongly suggest, but do not unequivocally prove, a pre-Devonian age of deposition and metamorphism of the Merrimack Group.

Some data point to the possibility of a Precambrian age for some of the rocks now mapped within the Merrimack Group. Metamorphic grade within it increases westerly, and near Manchester, New Hampshire, these rocks either rest unconformably upon the Massabesic Gneiss and are infolded into it, or grade into Massabesic paragneiss, and are therefore coeval with it. The latter possibility is buttressed by the fact that Pb^{207}/Pb^{206} ages on detrital zircons from the Berwick Formation of northern Massachusetts (1188 Ma, Aleinikoff, 1978) are substantially the same (1237 Ma, Aleinikoff, 1978) as those from quartzite of the Massabesic paragneiss of New Hampshire. More field work is necessary to resolve the question of the age of the Merrimack Group. Two of us (E.L.B. and J.N.A.) have a Precambrian bias; the third (J.B.L.) favors a Late (?) Ordovician age assignment.

Equivalents of the Merrimack Group in eastern Massachusetts are generally known as the Paxton and Oakdale Formations (Emerson, 1917) or as the Southbridge and upper part of the Hebron Formations (Barosh *et al.*, 1978). They speculate, as do most workers in the area, past and present, that the age of the Merrimack Group or its equivalents is Siluro-Devonian.

Whatever its age, the presumed westerly-younging homoclinal structure of the Merrimack Group (Katz, 1917) remains a problem, because the general geologic map and age pattern (Fig. 3) implies a synclinorial configuration. Billings (1956) labelled the area underlain by the Merrimack Group the Rockingham Anticlinorium, with the Rye Anticlinorium on its eastern side (Fig. 1a), but that designation is now inappropriate and should be abandoned. Therefore we propose reviving the somewhat less structurally specific term Merrimack Trough for this area, a term which was originally applied by Emerson (1917) to the region underlain by the Merrimack Group along strike in Massachusetts (Fig. 1b). This term is also appropriate because it applies to the original area in which Merrimack Group rocks were first defined (Hitchcock, 1878). It seems strange that the well-entrenched term Merrimack Synclinorium now applies to an area west of the Massabesic Gneiss in which no Merrimack Group rocks occur.

THE SOUTHERN NEW HAMPSHIRE FAULT ZONE: RELATIONS TO MAJOR SUTURES

Hussey and Pankiwskyj (1975) and Wones and Thompson (1979) have described the Nonesuch River-Norumbega Fault system of southern Maine (Fig. 2) which extends across the state from New Brunswick to New Hampshire. At the northeastern (Norumbega) end the fault zone is generally confined to the area underlain by the Vassalboro Formation, but as its southwestern (Nonesuch River) end it crosscuts several major formational boundaries, and also separates the Vassalboro Formation and the Merrimack Group (Fig. 3).

In eastern Massachusetts the Clinton-Newbury (Fig. 3) and numerous related faults are now well documented (Cuppels, 1961; Skehan, 1968; Castle *et al.*, 1976; Barosh *et al.*, 1978). The Clinton-Newbury traces southwesterly into the Lake Char fault new Worcester, Massachusetts, and thence into the Honey Hill Fault of Connecticut. These faults are generally regarded as marking the northwesterly boundary of the Avalon Platform in eastern Massachusetts.

However, Castle *et al.* (1976) and Barosh *et al.* (1978) have produced maps showing a series of faults splaying off to the north from the Clinton-Newbury Fault. These faults trace northerly into New Hampshire, and four such faults are shown on the map of Figure 3. All are well marked by linear depressions or lines of lakes, by bleached, retrograded and silicified rocks and by cataclasites or blastomylonites. The faults dip steeply or at high angles toward the west.

The four faults displayed on Figure 3 are, from west to east, the Pinnacle, Hall Mountain-Campbell Hill, Silver Lake, and Flint Hill Faults. The Flint Hill and Silver Lake Faults merge to the south in Massachusetts to become the Wekepeke Fault, and eventually join the Clinton-Newbury fault zone near Worcester, Massachusetts about 65 km south of the New Hampshire boundary. The Campbell Hill and Flint Hill

Faults merge indistinctly with the Nonesuch River Fault near the Maine-New Hampshire border, and thus form the connecting links for a continuous fault zone extending from New Brunswick to southwestern Connecticut – one of the master fault systems of the Appalachians.

The structural setting of the Hall Mountain-Campbell Hill Fault in New Hampshire is very similar to that along a portion of the Norumbega Fault in the Penobscot Bay region of Maine (Fig. 2); in both places the Precambrian abuts the fault along its southeast side. The Hall Mountain-Campbell Hill Fault cuts 275 Ma old granite at Milford, New Hampshire , but is cut by a quartz syenite dated by both K/Ar (J.T. Doherty, pers. commun.) and fission track (Aleinikoff, 1978) methods at 160 Ma. Thus this fault, and presumably most or all of the other faults related to the master fault system, have last moved during the Permian-Jurassic time span. We favor a Permian movement age, because Triassic-Jurassic faults in the Connecticut Valley region of New Hampshire and Massachusetts are characteristically mineralized normal faults, marked by fault gouge. Gouge is missing from the faults under discussion, implying a deeper, and older, fracturing.

Although Wones and Thompson (1979) believe there is at least 10 km of dextral offset along the Norumbega Fault in northeastern Maine, it is clear from the maps of Figures 3 and 4 that 330 Ma old Paleozoic granites in southwestern Maine are not appreciably offset by the Nonesuch River Fault, nor are we aware of any places in New Hampshire where lateral structural offsets may be unequivocally demonstrated along the four major mapped faults. However, Hussey and Newberg (1978) interpret the Nonesuch River Fault as a major right-lateral fault, with principal motion predating emplacement of the Paleozoic plutons. They also point out that on the Silver Lake Fault cleavage is rotated in a sense consistent with right-lateral displacement.

The abrupt increase in metamorphic grade west of the Flint Hill and Silver Lake Faults suggests that the Massabesic Gneiss may be a structural horst, raised several thousand metres with respect to the rocks on the east. Gravity studies over the Massabesic are hampered by low density contrasts between it and its wallrocks (Anderson, 1978; Aleinikoff, 1978), but the best-fit models show a relatively flat-bottomed Massabesic slab. In southernmost New Hampshire the slab is 2 km thick and dips westerly at 0° to 15° (Aleinikoff, 1978), but at its northeastern termination the slab has an irregularly horizontal bottom, and is less than 1 km thick (Anderson, 1978). Coupled with the mapped geology (formations on the western and eastern sides of the Precambrian (?) complexes are invariably different) an alternative to the horst or simple anticlinorial (Fig. 3) interpretations is that there is a series of folded thrust plates. In some places the Precambrian reaches the surface in folded thrusts, carrying on its back proximal (calcareous) or distal (pelitic) marine facies of Cambro-Ordovician or older age. Further evidence to support this general structural picture is provided by Osberg and Guidotti (1974) who have mapped three different stratigraphic sequences of Cambro-Ordovician (?) metamorphics in the western Penobscot Bay area (Fig. 2), and have shown two of them in a stacked and folded lower-angle overthrust package which has been thrust-faulted against the third sequence. Stewart and Wones (1974) have also pointed out that in the Penobscot Bay region there are at least six different structural-stratigraphic blocks, each of them probably representing different microplates accreted to the North American continent.

In coastal Maine (Doyle, 1967) and Massachusetts (Fig. 3) unmetamorphosed Siluro-Devonian volcanics rest with angular unconformity on Cambro-Ordovician or older metamorphics, or are in fault contact with them. There is a record of Ordovician igneous activity at 470 to 480 Ma both in southern New Brunswick (Olszewski and Gaudette, 1980) and in southeastern New Hampshire (Aleinikoff, 1978), as well as evidence of Ordovician magmatism and migmatization in coastal Maine (Wones, 1974). These data are consistent with the interpretation that there was considerable tectonic activity in this region during the time of Taconian deformation, much of it focussed along the transition zones from craton to shelf, and shelf to slope of the eastern Iapetus Ocean. Many of the fundamental breaks in the basement were initiated at this time. Subsequent tectonism during later Paleozoic time has broken, rotated, or reactived earlier Taconian faults, as well as opening new fractures. The southern New Hampshire Fault Zone (Fig. 3) apparently represents the break-up of the Massabesic thrust plate along reactivated faults. The Dover Fault Zone, which separates the Avalonian and Gander zones of Newfoundland has been projected by Williams (1979) to the north end of the Upper Devonian-Pennsylvania New Brunswick Basin. The Fredericton Fault emerges at the south end of that basin, on strike with the extension of the Dover Fault zone, and traces directly into the Norumbega-Nonesuch River Fault of Maine. Here, and in New Hampshire the fault zone lies entirely within the southward extension of Williams' (1978) Gander lithotectonic zone, but in central Massachusetts the Avalon and Gander zones are again in contact along the fault. The Dover Fault thus begins in Newfoundland at the craton-shelf boundary, migrates southward across a shelf-slope boundary, and then climbs back to the craton-shelf boundary in southern New England.

Recently, Kent and Opdyke (1978) have presented paleomagnetic evidence to show that coastal New England and the Canadian Maritime region were not at their present structural location until Late Carboniferous time, and that a northward transcurrent movement of at least 960 km during Late Devonian and Carboniferous time preceded the suturing of this continental block to North America. This would imply considerable left-lateral transport whereas the geology suggests, if anything, right-lateral motion. The maps of Figures 3 and 4 conclusively demonstrate that along what is believed to be the master suture zone of coastal northern New England the segments were locked into position prior to 330 Ma (Mississippian ?), and probably much earlier. There has been some Late Paleozoic faulting, but no provable transcurrent motion.

MAJOR FOLD PATTERNS

The stratigraphic and structural deductions reached thus far imply the need for some revision of the structural framework of New Hampshire. The Bronson Hill Anticlinorium on the west, and the Rye Anticlinorium on the east remain unchanged (Fig. 1b), but the Merrimack Synclinorium of Billings (1956) is now split into two unequal and non-coeval stratigraphic synclinoria by the Massabesic Anticlinorium. The smaller synclinorium of older metasediments on the east should be redefined as the Merrimack trough, for reasons we have discussed previously, and the name Rockingham Anticlinorium simply dropped.

The structure west of the Massabesic Anticlinorium is complex, and although many of the metasediments appear to be correlative with Siluro-Devonian rocks of northwestern Maine, Ordovician (?) or older rocks are also probably present. The youngest recognizable formation in the area shown in Figure 3 rests with either an angular unconformity or with a fault upon older Siluro-Devonian metasediments, and is a metaturbidite strongly resembling the Early Devonian (Siegenian) Seboomook Formation of northwestern Maine. The major and most continuous easterly outcrop belt of Seboomook-type rocks is shown as a synclinorial axis on Figure 3. This hinge is, as far as we now know, the major synclinorial axis between the Massabesic and Bronson Hill Anticlinoria. We propose the name Kearsarge-Central Maine Synclinorium for this structure because it passes beneath that conspicuous mountain in central New Hampshire, and traces northeasterly into central Maine. The Kearsarge-Central Maine synclinorium is the more precisely located hinge of what has been more loosely defined as the hinge of the Merrimack synclinorium, and has been placed 0 to 45 km to the east (Figs. 1a, and 1b).

A striking feature of the geology shown on Figure 3 is the divergence of major structural trends around the large Early (?) Devonian Winnipesaukee Pluton of eastern New Hampshire and the Sebago Pluton of western Maine (Fig. 3). At least three cycles of folding have affected the Siluro-Devonian (and older?) metasediments. The *first* of these is recumbent with axial directions which, at least locally, trend northwesterly. A *second* cycle of folding creates broad east-west folds, and the last fold cycle tightens up both sets of early folds about northeasterly axes. Some of the major folds, and the fold sequences, are shown on Figure 3; all are regarded as of Acadian age.

The earliest events of the deformational episode produced nappes which, at least in western New Hampshire, faced westerly. Rocks at the currently exposed erosion level became increasingly plastic because of tectonic thickening and rising P-T conditions. When they were invaded by felsic magmas originating from a deeper, higher-pressure environment, that yielded easily to the magmatic pressure, producing F2 folds. The Winnipesaukee and Sebago plutons may be regarded as giant mushrooms which have shouldered aside their wallrocks; the Winnipesaukee has a maximum thickness of less than 3.5 km (Nielson *et al.*, 1976) and the Sebago, with almost no gravity signature, must be wafer thin.

The *third* folding event (F3) presumably represents the later phase of the Acadina orogeny, and is older than Alleghenian-age (330 to 275 Ma) binary granites of Figure 4.

An unresolved question in this synthesis is why the Maine-New Hampshire border region seems to have been a focus for major Acadian-age plutonism, and why it is also the locus of the White Mountain batholith (Fig. 3). One answer would be that this area coincides with a sharp deflection from north to northeast in the regional fold pattern, and therefore may coincide with a zone of tensional fracturing. The flat shape of the syntectonic intrusives is due to the fact that the magmas exploited favoured stratigraphic horizons within the piled-up nappes. In central New Hampshire, for example, a rusty calc-silicate (the Francestown Formation) is the favored intrusive horizon for many of the Acadian-age plutons (Fig. 3).

MAGMA SERIES, METAMORPHISM, AND OROGENIC EPISODES

In western New Hampshire the calc-alkaline Highlandcroft Plutonic Series (Billings, 1956) occurs in a series of stocks along the Connecticut-River Valley from the west-central portion of the state northward. Geologically, the series is pre-upper Early Silurian, and is correlative with the Attean Quartz-Monzonite of central western Maine for which Boudette and Boone (1976) have reported a 470 Ma Pb/U isotopic age. A Taconian age for the series (or Early Caledonian) therefore seems well established. Naylor (1968) has shown by isotopic Pb/U and Rb/Sr dating that the core rocks of the Oliverian domes, which lie east of the Highlandcroft plutons, have ages of approximately 445 Ma (Late Ordovician) and are therefore probably somewhat younger than the Highlandcroft Series. In eastern New Hampshire, the age of some of the granite in the Massabesic Gneiss (480 Ma, Aleinikoff, 1978) is lower to middle Early Ordovician, suggesting that a Taconian orogeny may have commenced in that area prior to its initiation in the western part of the state. The suggestion of early Taconian orogeny in the east may also be deduced from the whole-rock Rb/Sr age of metamorphism (470 to 527 Ma, or late Cambrian-Early Ordovician; Brookins and Hussey, 1978) for the Casco Bay Group of coastal Maine (Fig. 2), although as Gilluly (1973) warns, magmatism and metamorphism are not necessarily coeval.

Recent Pb/U and Rb/Sr investigations in northeastern Massachusetts (Zartman and Naylor, in press) suggest that there plutonism may have been continuous throughout Taconian to Acadian time. The number of New England granites isotopically dated as Silurian is now impressively large and runs counter to our general notion that Taconian and Acadian orogenies were isolated deformational episodes of limited duration, provided that we assume some genetic association between orogeny and the generation of granitic magmas.

In central New Hampshire (Fig. 3) the bulk of the igneous rocks may be assigned geologically to the post-Early Devonian calc-alkaline (post-Siegenian) New Hampshire Plutonic Series and consists of a number of sheet-like plutons divisible into four intrusive suites. The Kinsman and Bethlehem suites range from tonalite to granite in composition, the Spaulding suite from gabbro to granite, and the Concord suite from granodiorite to granite. The first three suites are of early-tectonic to late-tectonic Acadian age; the last is post-Acadian. Rb/Sr or U-Th-Pb ages for the New Hampshire Plutonic Series (Lyons and Livingston, 1977; Gaudette et al., 1975; Aleinikoff, 1978, and written commun.) are displayed in Figure 4. The early to late tectonic series have an age range (402 to 383 Ma) consistent with what would be anticipated for Acadian-age plutonism. One pluton in southwestern Maine (345 ± 17 Ma) is younger than the norm.

Somewhat unexpectedly all of the highly peraluminous Concord suite plutons thus far dated by Rb/Sr or U-Th-Pb methods (Fig. 4) yield Hercynian ages of 327 to 275 Ma (Early Pennsylvanian to Early Permian). However, not all Concord-type granites may be of this age, and more dating is needed. For example, similar two-mica granites such as that at Barre in northeastern Vermont have 372 Ma isotopic ages (Naylor, 1971), although petrographically identical to the Concord-type rocks shown on Figure 3. Nevertheless, the volume and areal extent of the Hercynian-age granites is sufficiently impressive to imply tectonism during this time span. Fullagar and Butler (1979) have reported a group of 325 to 265 Ma old granites in the Piedmont of the Southern Appalachians, and granites in New Brunswick, Newfoundland and the Massif Central in France show the same age spread. What we are seeing, apparently, in the two-mica Concord-type granites is evidence for Alleghenian or Hercynian orogeny along much of the Appalachian-Hercynian chain prior to Mesozoic continental separation.

In New England the 275 Ma old Pb^{207}/Pb^{206} age of zircon from the granite at Milford, New Hampshire, is matched, only by the 275 Ma old Narragansett Pier Granite of southern Rhode Island. The Pennsylvanian sediments of the Narragansett Basin are regionally metamorphosed, generally in the lowest grades, but the sillimanite isograd is reached locally, particularly near the granite (Grew and Day, 1972). About midway between the Narragansett Pier Granite, and granite at Milford are metamorphosed Pennsylvanian phyllite, metaconglomerate and meta-anthracite at Worcester, Massachusetts (Grew, 1973), about 70 km south of Milford. Thus there is evidence that a stretch of eastern New England at least 120 km long has felt a late Hercynian (Alleghenian) orogeny, and evidence also that the orogency was accompanied by regional metamorphism and the intrusion of granites. Lead isotopic studies on the granite at Milford, New Hampshire (Aleinikoff, 1978) support the interpretation that the granite is an anatectic derivative of the 650 Ma metavolcanics of the Massabesic Gneiss.

As Gilluly (1973) has pointed out, the theory of plate tectonics is at some variance with the idea that orogenies are brief cataclysmic events. The record for the Cordillera of North America (Gilluly, 1973) demonstrates that magmatism has been almost continuous since the close of the Paleozoic, as one would anticipate on the basis of plate theory. For the New England Appalachians it has been customary to emphasize the Avalonian, Taconian, Acadian and Hercynian orogenies, yet magmatism is nearly continuous from 480 to 330 Ma. Admittedly, the volumes of magmas generated during any given interval of geologic time vary enormously in quantity. What is notable is that magmatism is strongly localized. Highlandcroft magmatism, for example, is concentrated in northwestern New Hampshire and in central western Maine. Oliverian plutonism and volcanism is essentially restricted to the Bronson Hill Anticlinorium and Silurian (430 Ma) magmatism to southeastern New Hampshire and northeastern Massachusetts. These examples illustrate what is probably true of the entire Appalachian orogen – that is, that orogeny may have been essentially continuous throughout much of Paleozoic time, but that deformation, plutonism, and metamorphism (?) were restricted in space and time, with the locus of activity shifting from place to place.

The identification of Late Paleozoic granites in northern New England also reopens another problem – the significance of the numerous K/Ar age determinations which are available for this region. Faul et al. (1963) had noted a strong clustering at 275 Ma, and had concluded that this was due to a degassing episode accompanying an Early Permian tectonic disturbance. Subsequently, Zartman et al. (1970) had argued, on the basis of many additional K/Ar age determinations, that the clustering was more apparent than real and that the age pattern was the result of sampling density and tectonic unloading; in short, the ages chiefly record a fossil isotherm for retention of argon. It now appears that both viewpoints may be correct, that there is both a Permian disturbance, as well as an unloading effect. The interpretation of individual K/Ar ages for any pluton in New England is obviously beset by formidable and probably irresolvable complexities, which means that most Paleozoic K/Ar ages must be interpreted with a considerable degree of caution.

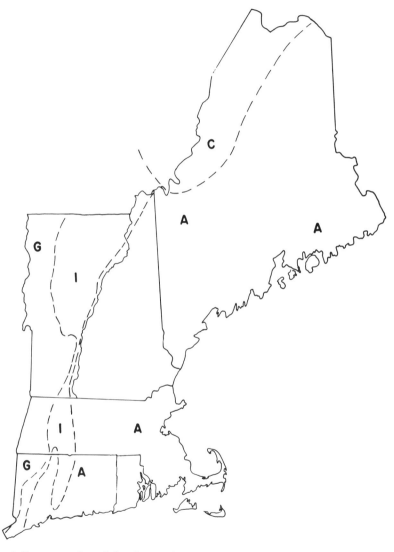

Figure 5. Reconstruction of plate boundaries in New England, modified extensively from Osberg (1978). C – Helikian (Chain Lakes); G – Grenvillian; A – Avalonian; I – Iapetus remnant (?).

DEVELOPMENT OF THE GANDER AND AVALON ZONES IN NEW ENGLAND

Our reconstruction of the history of this portion of the Appalachians is based upon the idea that at closure of Iapetus four fairly sizable microplates (and probably several smaller ones as well) were recognizable in New England (Fig. 5).

Osberg (1978) has identified two Avalonian age microplates in New England (basements C and D). Their contact coincides with the unmetamorphosed to weakly metamorphosed Siluro-Devonian volcanic belt of coastal Maine and Massachusetts, and with the slivers of mafic to ultramafic rocks in the Penobscot Bay area. Whether this is a major suture zone is uncertain, because the basement on either side is apparently the same (i.e. Avalonian), and we have not shown a similar subdivision of our basement A (Fig. 5).

The western boundary of our basement A lies along the western margin of the Bronson Hill Anticlinorium, where we have described a paleo-subduction zone. Osberg's equivalent boundary (his basement C) has been placed 5 to 40 km west of ours.

There is a zone (I on our diagram), not singled out by Osberg, where the basement is unexposed; it is concealed in large part below the Siluro-Devonian Connecticut Valley-Gaspé synclinorium. It may either be Grenvillian (G) as it is to the west, or a remnant of the Dunnage zone (the ancient floor of Iapetus), or, less likely, a sliver of Avalonian basement.

Basement B of Osberg and our basement C are equivalent, representing the 1600 Ma terrane of the Chain Lakes massif, and its presumed extension to the north.

Tectonic events which shaped Avalonian basement A now underlying the Gander and Avalon lithotectonic zones extended from 800 to 600 Ma. Continental rupture and the spreading of Iapetus commenced in Early Cambrian, but by Late Cambrian closure had begun, and at this time the Boil Mountain ophiolite was obduced into the Helikian Chain Lakes massif (Boudette, this volume). By the Middle Ordovician, closure of the basaltic ocean floor was well advanced. Along the western margin of the Avalonian basement an easterly-dipping subduction zone and an island arc (now the Bronson Hill Anticlinorium) were established. Plutonism and volcanism were active in this structural belt. Further west, along what is now the eastern flank of the Green Mountain Anticlinorium, obduction of the ocean floor onto Grenvillian basement caused some of the major deformation effects of the Taconian orogeny.

Coastal New England was also tectonically active at this time, particularly at the shelf-slope margin of the eastern Iapetus Ocean, where the crust and basement were disrupted by a series of westerly-directed thrust slices, and where plutonism was also locally important. A post-Cambrian (?) and pre-Middle Silurian low grade regional metamorphism in the Penobscot Bay (Stewart and Wones, 1974) and Casco Bay (Brookins and Hussey, 1978) areas can also be assigned to the Taconian orogenic cycle. The cratonal Avalonian terrane was apparently little affected during this deformational episode, but in eastern Massachusetts at least three anorogenic alkalic granites were emplaced (Zartman and Marvin, 1971).

During the Silurian, the collapsing eastern continental slope of the remnant Iapetus was the site of accumulation of a thick (> 5 km, Osberg, 1979) flysch sequence. Toward the southeast, at the margins of the Avalonian craton and shelf in northeastern Massachusetts, Silurian calc-alkalic magmatism was nearly continuous. With time, this magmatic activity migrated northward and became volcanic in character, so that a discontinuous group of shallow marine basaltic to rhyolitic Siluro-Devonian extrusives rest unconformably on the metamorphics of the eastern Gander belt, or are juxtaposed against the Avalon platform, as in northeastern Massachusetts (Fig. 3).

Continuing collapse of the last remains of Iapetus shifted the site of Early Devonian sedimentation westerly, where another thick (>4 km, Boucot, 1961) flysch sequence formed. Final destruction of the ocean, deformation, metamorphism and plutonism occurred toward the end of Early Devonisn (~ 400 to 380 Ma). The major effects are concentrated in central New Hampshire and Maine, dying out gradually toward the Green Mountains on the west, and more abruptly toward coastal New England, where the Siluro-Devonian is faulted and weakly metamorphosed, but not highly deformed.

Aside from the evidence of continuing magmatic activity throughout the Carboniferous (Fig. 4) there is little structural record of late Paleozoic events in northern New England. We do have evidence (Aleinikoff, 1978) suggesting that P-T conditions in the Massabesic Gneiss were sufficiently intense 275 Ma ago to produce anatectic granites, so Late Paleozoic tectonism is by no means trivial. We do know, also, that faulting continued at least into the Permian. Tracing the effects of Alleghenian deformation northward from southern New England remains one of the elusive tasks for the future.

ACKNOWLEDGEMENTS

Funds for the research on which this paper is based were provided to Lyons through National Science Foundation Grants DES72-01525 A01 and EAR77-20085. Aleinikoff was supported by the former grant, and also by the U.S. Geological Survey, which also sponsored all of the research by Boudette. We are indebted to Richard Naylor for constructive criticism of our earlier draft of this paper.

REFERENCES

Aleinikoff, J.N., 1977, Petrochemistry and tectonic origin of the Ammonoosuc Volcanics, New Hampshire – Vermont: Geol. Soc. America Bull., v. 88, p. 1546-1552.

───────────, 1978, Structure, petrology, and U-Th-Pb geochronology in the Milford quadrangle, New Hampshire: Ph.D. Thesis, Dartmouth College, 247 p.

Aleinikoff, J.N., Zartman, R.E. and Lyons, J.B., 1979, U-Th-Pb geochronology of the granite near Milford, south-central New Hampshire: new evidence for Avalonian basement and Taconic and Alleghenian disturbances in eastern New England: Contributions to Mineral. and Petrology, v. 71, p. 1-11.

Anderson, R.C., 1978, The northern termination of the Massabesic Gneiss, New Hampshire: M.A. Thesis, Dartmouth College, 111 p.

Barosh, P.J., Fahey, R.J. and Pease, M.H., 1978, The bedrock geology of the land area of the Boston 2° sheet; Massachusetts, Connecticut, Rhode Island and New Hampshire: United States Geol. Survey Open File Rept.

Besancon, J.R., Gaudette, H.R. and Naylor, R.S., 1977, Age of the Massabesic Gneiss, Southern New Hampshire: Geol. Soc. America, Abstracts with Programs, v. 9, no. 3, p. 242.

Billings, M.P., 1956, The geology of New Hampshire: Part II – Bedrock Geology: New Hampshire State Planning and Development Commission, 203 p.

Boucot, A.J., 1961, Stratigraphy of the Moose River synclinorium, Maine: United States Geol. Survey Bull. 1111-E, p. 153-188.

Boudette, E.L. and Boone, G.M., 1976, Pre-Silurian stratigraphic succession in central western Maine: in Page, L.R., ed., Contributions to the Stratigraphy of New England: Geol. Soc. America Memoir 148, p. 79-96.

Brookins, D.C. and Hussey, A.M., 1978, Rb-Sr ages for the Casco Bay Group and other rocks from the Portland-Orrs Islands area, Maine: Geol. Soc. America, Abstracts with Programs, v. 10, no. 2, p. 34.

Castle, R.O., Dixon, H.R., Grew, E.S., Griscom, A. and Zeitz, I., 1976, Structural dislocations in eastern Massachusetts: United States Geol. Survey Bull., 1410, 39 p.

Cook, F.A., Albaugh, D.S., Brown, L.D., Kaufman, S., Oliver, J.E. and Hatcher, R.D., 1979, Thin-skinned tectonics of the southern Appalachians; COCORP seismic reflection profiling of the Blue Ridge and Piedmont: Geology, v. 7, p. 563-567.

Cuppels, N.P., 1961, Post-Carboniferous deformation of metamorphic and igneous rocks near the Northern Boundary fault, Boston Basin, Massachusetts: in Geological Survey Research 1961; United States Geol. Survey Prof. Paper 424-D, p. D46-D48.

Doyle, R.G., ed., 1967, Preliminary Geologic Map of Maine: Maine Geol. Survey.

Emerson, B.K., 1917, Geology of Massachusetts and Rhode Island: United States Geol. Survey Bull. 597, 289 p.

Faul, H., Stern, T.W., Thomas, H.R. and Elmore, P.L.D., 1963, Age of intrusion and metamorphism in the northern Appalachians: American Jour. Sci., v. 201, p. 1-19.

Fullagar, P.D. and Butler, J.R., 1979, 325 to 265 m.y. old granitic plutons in the Piedmont of the Southern Appalachians: American Jour. Sci., v. 279, p. 161-185.

Gaudette, H.E., in press, Zircon isotopic age from the Union ultramafic complex, Maine: Canadian Jour. Earth Sci.

Gaudette, H.E., Fairbairn, H.W. and Kovach, A., 1975, Preliminary Rb-Sr whole-rock age determinations of granitic rocks in southwestern Maine: Geol. Soc. America, Abstracts with Programs, v. 7, no. 1, p. 62-63.

Gilluly, J., 1973, Steady plate motion and episodic orogeny and magmatism: Geol. Soc. America Bull., v. 84, p. 499-514.

Greene, R.C., 1970, The geology of the Peterborough quadrangle, New Hampshire: New Hampshire Dept. Resources and Economic Development, Bull. 4, 88 p.

Grew, E.S., 1973, Stratigraphy of the Pennsylvanian and pre-Pennsylvanian rocks of the Worcester area, Massachusetts: Amer. Jour. Science, v. 273, p. 113-129.

Grew, E.S. and Day, H.W., 1972, Staurolite, kyanite, and sillimanite from the Narragansett Basin of Rhode Island: United States Geol. Survey Prof. Paper 800-D, p. D151-D167.

Haworth, R.T., Lefort, J.P. and Miller, H.G., 1978, Geophysical evidence for an east-dipping subduction zone beneath Newfoundland: Geology, v. 6, p. 522-526.

Hitchcock, C.H., 1874-1878, Geology of New Hampshire: Vol. I, 1874; Vol. II, 1877; Vol. III, 1878, Atlas, Concord, New Hampshire.

Hussey, A.M. and Newburg, D.W., 1978, Major faulting in the Merrimack Synclinorium between Hollis, New Hampshire and Biddeford, Maine: Geol. Soc. America, Abstracts with Programs, v. 10, p. 48.

Hussey, A.M. and Pankiwskyj, K.A., 1975, Geologic map of southwestern Maine: Maine Geol. Survey Open File Map 1976-1.

Katz, F.J., 1917, Stratigraphy in southwestern Maine and southeastern New Hampshire: United States Geol. Survey, Prof. Paper 108, p. 165-177.

Kelly, W.J., Olszewski, W.J. and Gaudette, H.E., 1980, The Massabesic orthogneiss, southern New Hampshire: Geol. Soc. America, Abstracts with Programs, v. 12, p. 45.

Kent, D.V. and Opdyke, N.D., 1978, Paleomagnetism of the Catskill Red Beds: evidence for motion of the coastal New England-Canadian Maritime region relative to cratonic North America: Jour. Geophys. Research, v. 83, p. 4441-4450.

Lyons, J.B. and Livingston, D.E., 1977, Rb-Sr age of the New Hampshire plutonic series: Geol. Soc. America Bull., v. 88, p. 1808-1812.

Naylor, R.S., 1968, Origin and regional relationships of the core-rocks of the Oliverian domes: in Zen, E-an White, W.S., Hadley, J.B., and Thompson, J.B., Jr., eds., Studies in Appalachian Geology, Northern and Maritime: New York, Interscience Publishers, p. 231-240.

——————————, 1971, Acadian orogeny: An abrupt and brief event: Science, v. 172, p. 558-560.

Naylor, R.S., Boone, G.M., Boudette, E.L., Ashenden, D.D. and Robinson, P., 1971, Pre-Ordovician rocks in the Bronson Hill and Boundary Mountain anticlinoria, New England, U.S.A.: Trans. American Geophys. Union, v. 54, p. 495.

Nielson, D.R., 1974, Metamorphic diffusion in New Hampshire soapstone bodies and flecky gneisses: Ph.D. Thesis, Dartmouth College, 262 p.

Nielson, D.L., Clark, R.G., Lyons, J.B., Englund, E.J. and Borns, D.J., 1976, Gravity models and mode of emplacement of the New Hampshire Plutonic Series: in Lyons, P.C. and Brownlow, A.H., eds., Studies in New England Geology: Geol. Soc. America Memoir 146, p. 301-318.

Olszewski, W.S., 1978, U-Pb zircon ages for stratified metamorphic rocks of northeastern Massachusetts: Geol. Soc. America, Abstracts with Programs, v. 10, p. 79.

Olszewski, W.S. and Gaudette, H.E., 1980, Rb-Sr whole rock and U-Pb zircon ages from the Greenhead Group, New Brunswick: Geol. Soc. America, Abstracts with Programs, v. 12, p. 76.

Osberg, P.H., 1968, Stratigraphy, structural geology, and metamorphism of the Waterville-Vassalboro area, Maine: Maine Geol. Survey Bull. 20, 64 p.

——————————, ed., 1974, Geology of East-Central and North-Central Maine: New England Intercollegiate Geol. Conference Guidebook, 1974, University of Maine, 240 p.

——————————, 1978, Synthesis of the geology of the Northern Appalachians, U.S.A.: Geol. Survey Canada Paper 78-13, p. 137-147.

——————————, 1979, Geologic relationships in south-central Maine: in Skehan, J.W. and Osberg, P.H., The Caledonides in the U.S.A. Geological Excursions in the Northeast Appalachians: Weston Observatory, p. 37-62.

Osberg, P.H. and Guidotti, C.U., 1974, The geology of the Camden-Rockland area: in Osberg, P.H., ed., Geology of East-Central and North-Central Maine: New England Intercollegiate Geol. Conference Guidebook, p. 48-60.

Robinson, P., 1967, Gneiss domes and recumbent folds in the Orange area, west-central Massachusetts: in Robinson, P., ed., New England Intercollegiate Geological Conference, 59th Annual Meeting Guidebook, p. 17-47.

Robinson, P. and Hall, L.M., 1980, Tectonic synthesis of southern New England: in Wones, D.R., ed., Proceedings, the Caledonides in the U.S.A.: Department of Geological Science, Virginia Polytechnic Institute and State University Memoir No. 2, p. 73-82.

Shride, A.F., 1976, Stratigraphy and correlation of the Newbury volcanic complex, northeastern Massachusetts: in Page, L.R., ed., Contributions to the Stratigraphy of New England: Geol. Soc. America Memoir 148, p. 147-177.

Skehan, J.W., 1968, Fracture tectonics in southeastern New England as illustrated by the Wachusett-Marlborough Tunnel, east-central Massachusetts: in Zen, E-an, White, W.S., Hadley, J.B., and Thompson, J.B., Jr., eds., Studies in Appalachian Geology, Northern and Maritime: Interscience Publishers, New York, p. 281-290.

Stewart, D.B., 1974, Precambrian rocks of Seven Hundred Acre Island and Development of cleavage in the Ilseboro Formation: in Osberg, P.H., ed., Geology of East-Central and North-Central Maine: New England Intercollegiate Geological Conference, Guidebook, p. 86-98.

Stewart, D.B. and Wones, D.R., 1974, Bedrock geology of northern Penobscot Bay area: in Osberg, P.H., ed., Geology of east-central and north-central Maine: New England Intercollegiate Geological Conference Guidebook, p. 223-239.

Steiger, R.H. and Jäger, E., 1978, Subcommission on Geochronology: Convention on the use of decay constants in geochronology and cosmochronology: in Cohee, G.V., Glaessner, M.F. and Hedberg, H.P., The Geologic Time Scale: American Assoc. Petroleum Geol. Geology, no. 6, p. 67-72.

Thompson, J.B., Robinson, P., Clifford, T.N. and Trask, N.J., 1968, Nappes and gneiss domes in west-central New England: in Zen, E-an White, W.S., Hadley, J.B., and Thompson, J.B., Jr., eds., Studies in Appalachian Geology, Northern and Maritime: New York, Interscience Publishers, p. 203-218.

Vernon, W.E., 1971, Geology of the Concord quadrangle: in Lyons, J.B. and Stewart, G.W., Guidebook for Field Trips in Central New Hampshire and Contiguous Areas: New England Intercollegiate Geological Conference Guidebook, p. 118-125.

Williams, H., 1978, Tectonic lithofacies map of the Appalachian orogen: Memorial University of Newfoundland, Map No. 1.

Williams, H., 1979, Appalachian orogen in Canada; Canadian Jour. Earth Sci., v. 16, p. 792-807.

——————————, 1979, Appalachian orogen in Canada; Canadian Jour. Earth Sci., v. 16, p. 792-807.

Williams, H. and St-Julien, P., 1978, The Baie Verte-Brompton Line in Newfoundland and regional correlations in the Canadian Appalachians: Geol. Survey Canada Paper 76-1A, p. 225-229.

Wolff, R.A., 1978, Ultramafic lenses in the Middle Ordovician Partridge Formation, Bronson Hill Anticlinorium, central Massachusetts: Contribution No. 34, Dept. Geology and Geography, University of Massachusetts, 162 p.

Wones, D.R., 1974, Igneous petrology of some plutons in the northern part of the Penobscot Bay area: in Osberg, P.H., ed., Geology of East-Central and North-Central Maine: New England Intercollegiate Geological Conference Guidebook, p. 99-125.

Wones, D.R. and Thompson, W., 1979, The Norumbega Fault zone: a major regional structure in central eastern Maine: Geol. Soc. America, Abstracts with Programs, V. 11, no. 1, p. 60.

Zartman, R.E., Hurley, P.M., Krueger, H.W. and Giletti, B.J., 1970, A Permian disturbance of K-Ar radiometric ages in New England: Its occurrence and cause: Geol. Soc. America Bull., v. 81, p. 3359-3374.

Zartman, R.E. and Marvin, R.F., 1971, Radiometric age (Late Ordovician) of the Quincy, Cape Ann, and Peabody Granites from eastern Massachusetts: Geol. Soc. America Bull., v. 82, p. 937-957.

Zartman, R.E. and Naylor, R.S. (in press), Structural implications of some radiometric ages of igneous rocks in southeastern New England: MS to be submitted to Geol. Soc. America Bull.

Manuscript Received November 21, 1979
Revised Manuscript Received July 2, 1980

Major Structural Zones and Faults of the Northern Appalachians, edited by
P. St-Julien and J. Béland, Geological Association of Canada Special Paper 24, 1982

TACONIAN LINE IN WESTERN NEW ENGLAND AND ITS IMPLICATIONS TO PALEOZOIC TECTONIC HISTORY

———————

Norman L. Hatch, Jr.
U.S. Geological Survey, Reston, Va., 22092, U.S.A.

———————

ABSTRACT

A major stratigraphic break, here called the Taconian Line, between Middle Ordovician and older strata to the west and Middle Silurian and younger strata to the east trends roughly south across Vermont, Massachusetts, and Connecticut. The Middle Ordovician rocks immediately below this break are mostly metamorphosed black shales and volcanic rocks. Overlying them are thin lenses of a Middle Silurian unit of quartzite and marble. All these rocks are in turn overlain by grey schist, metagreywacke, and calcareous metagreywacke of Middle Silurian(?) to Early Devonian age.

Although rare local Taconian minor folds have been described west of the Taconian Line and increase in abundance to the west, no angular discordance along the line itself has been reported. Similarly, Taconian regional metamorphism, which is widespread and locally intense west of the Berkshire-Green Mountain Precambrian massifs, has only locally been recognized more than a few kilometres east of the Green Mountain-Berkshire axis.

These relations suggest the following model: an east-dipping subduction zone developed an Ordovician island arc on a continental fragment. Closing of the basin and resultant Taconian collision of the arc and North America emplaced the Vermont ultramafic belt by westward imbricate upthrusting and produced widespread westward thrusting, folding, and metamorphism west of the belt, with only mild deformation of the arc itself. After erosion, the still largely undeformed former arc became the site of disconformable deposition of an east-facing Middle Silurian shelf facies. After an increase in water depth, a thick sequence of grey greywacke-shale turbidites was deposited. The Acadian orogeny, which followed east-dipping subduction beneath westward advancing Avalonia, like the Taconian, deformed primarily west of its suture.

RÉSUMÉ

Une rupture stratigraphique, appelée dans ce texte la Ligne taconienne, et qui se situe entre l'Ordovicien moyen et assises plus anciennes à l'ouest, et le Silurien moyen et couches plus récentes à l'est, traverse en direction à peu près sud le Vermont, le Massachussetts et le Connecticut. L'Ordovicien moyen immédiatement sous la rupture est constitué surtout de shale noir et de roches volcaniques métamorphisés; immédiatement au-dessus de la rupture, le Silurien moyen est en minces lentilles de quartzite et marbre. Le tout est recouvert de schiste métamorphique gris, de métagrauwackes et de métagrauwackes calcareux, d'âge Silurien moyen (?) à Dévonien inférieur.

Quoiqu'on ait observé localement à l'ouest de la ligne quelques plis mineurs taconiens, plus fréquents vers l'ouest, aucune discordance angulaire n'a encore été vue à la ligne même. De même, le métamorphisme régional taconien, très manifeste et localement intense à l'ouest des massifs précambriens de Berkshire-Green Mountain, n'apparaît que rarement au delà de quelques kilomètres à l'est de l'axe Green Mountain-Berkshire.

Ces faits appellent le modèle suivant: une zone de subduction pentée vers l'est serait à l'origine d'un arc insulaire ordovicien formé à l'est sur un fragment continental. La fermeture du bassin et la collision taconienne entre l'arc et le continent nord-américain qui s'en est suivie, a donné lieu à la mise en place de la zone ultrabasique du Vermont par un chevauchement imbriqué vers l'ouest d'où sont résultés les multiples chevauchements pentés vers l'est, le plissement et le métamorphisme observés à l'ouest de la zone d'ultrabasites alors que l'arc, à l'est, ne montre, lui, qu'une faible déformation. Après érosion, l'arc toujours peu déformé, a été le lieu, au Silurien moyen, du dépôt discordant d'un facies de plateforme faisant face vers l'est; et après subsidence, s'y est accumulée une épaisse séquence turbiditique de grauwacke-shale gris. L'orogénèse acadienne, qui a succédé à la subduction vers l'est sous une Avalonia avançant vers l'ouest, s'est surtout manifestée, comme la taconienne, à l'ouest de la suture.

INTRODUCTION

A major hiatus has long been recognized in western New England between the Lower Paleozoic, probably Cambrian-Ordovician, rocks of the "East Vermont sequence" east of the Precambrian massifs and the Middle Paleozoic, probably Silurian-Devonian, metasedimentary rocks of the Connecticut Valley-Gaspé synclinorium. This hiatus and the trace of its surface on the present ground surface are referred to as the Taconian Line in this paper.

The Taconian Line in western New England is shown on Figure 1. It continues north and northeast across southeastern Quebec, re-enters the United States for a short distance in the northwest corner of Maine (Hussey et al., 1967), and can be traced northeast and east across the Gaspé Peninsula. This paper deals with this surface in Vermont, Massachusetts, and Connecticut where it is characterized by contrasts in stratigraphic, sedimentological, structural, and metamorphic styles. These data are used to deduce a geologic history of the area along the line, and a tectonic model based upon that history is suggested.

Figure 1. Generalized geologic map of western New England showing principal geologic units near and related to the Taconian Line. Symbols for geographic features are as follows: LM, Lake Memphremagog; WM, Worcester Mountains; NM, Northfield Mountains; RN, Randolph, Vermont; CD, Chester Dome; AD, Athens Dome; BR, Brattleboro, Vermont; CH, Charlemont, Massachusetts; NA, North Adams, Massachusetts; BK, Becket, Massachusetts; BL, Blandford, Massachusetts; NT, Newtown, Connecticut.

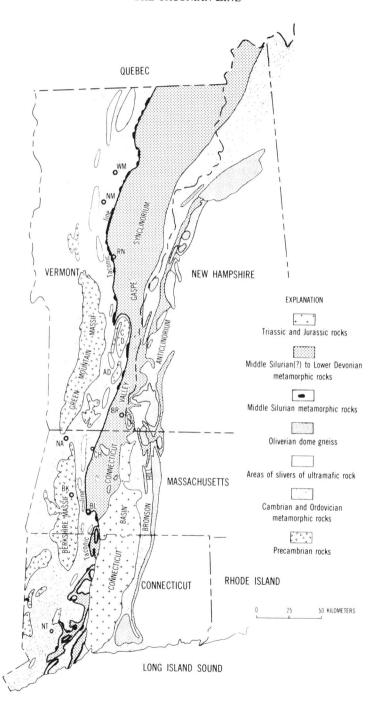

QUEBEC

WM

NM

SYNCLINORIUM

line

RN

Taconic

VERMONT

NEW HAMPSHIRE

GASPÉ

ANTICLINORIUM

EXPLANATION

Triassic and Jurassic rocks

Middle Silurian(?) to Lower Devonian
metamorphic rocks

Middle Silurian metamorphic rocks

Oliverian dome gneiss

Areas of slivers of ultramafic rock

Cambrian and Ordovician
metamorphic rocks

Precambrian rocks

MASSIF

GREEN MOUNTAIN

AD

C
D

VALLEY

BR

NA

CH

CONNECTICUT

BK

line

BL

Taconic

BASIN

BRONSON

MASSACHUSETTS

BERKSHIRE MASSIF

CONNECTICUT

RHODE ISLAND

CONNECTICUT

0   25   50 KILOMETERS

NT

LONG ISLAND SOUND

Across most of Vermont and Massachusetts, the Taconian Line follows a relatively straight course. In fact, between Lake Memphremagog on the Quebec border
and Blandford, Massachusetts, the only major deflections of this boundary are
around the Chester and Athens domes in southern Vermont (Fig. 1). South from
Blandford to Long Island Sound, the Taconian Line follows a tortuous path around a
series of domes, locally disappearing under the Mesozoic Connecticut Basin. The
tortuous path results from Acadian recumbent folding, refolding, and doming; considering the intensity of Acadian folding, the straighter sections of the Taconian Line
are more difficult to explain. That part of the Taconian Line immediately north of
the international border also is complexly deformed (Boucot and Drapeau, 1968;
St-Julien and Hubert, 1975).

STRATIGRAPHIC CONTRAST

The Taconian Line separates Middle Ordovician rocks of the Missisquoi Formation as used by Doll *et al.* (1961) in Vermont, correlative Hawley (Hatch, 1967) and
Cobble Mountain (Hatch and Stanley, 1973) Formations in Massachusetts, and an
assortment of correlated (see Hatch and Stanley, 1973, Plates 1 and 2 for details)
units in Connecticut below, from the Middle Silurian Shaw Mountain Formation of
Vermont, Russell Mountain Formation of Massachusetts, and correlated units in
western Connecticut (see Hatch and Stanley, 1973, Plates 1 and 2) above (Fig. 2).
Along much of the Taconian Line these Silurian units are missing and the largely
Lower Devonian(?) Northfield Formation in Vermont, Goshen Formation in Massachusetts, and Straits Schist (again see Hatch and Stanley, 1973, Plates 1 and 2 for
the interpretation followed here) in Connecticut directly overlie the Ordovician
strata.

Figure 3 portrays the rock units below the hiatus in western New England. Three
rock types characterize the Middle Ordovician section immediately below the Taconian Line: 1) schistose to gneissose felsic to mafic metavolcanic rocks; 2) grey to
black, graphitic, sulphidic, rusty-weathered schist and associated minor quartzite; and
3) light grey, nongraphitic, commonly feldspathic, nonrusty-weathered, granular
schist and lesser pelitic and semipelitic schist and gneiss.

Metavolcanic rocks in Vermont are sparse in this sequence north of Randolph
(Figs. 1 and 3). South from Randolph, the Barnard Volcanic Member of the Missisquoi Formation (as used by Doll *et al.*, 1961) abruptly appears immediately below the
Taconian Line and within a short distance achieves a thickness of 1000 to 2500 m
(Ern, 1963, p. 35-36; Chang *et al.*, 1965, p. 35). Although remarkably thinned (by
Acadian tectonism?) around the Chester and Athens domes, it continues south (with
a name change to metavolcanic member of the Hawley Formation at the Massachusetts State line) to within a few kilometres of Blandford, Massachusetts (Figs. 1
and 3). The maximum thickness of this volcanic unit of about 3000 m near Charlemont, Massachusetts, thins to only 100 m, 25 km to the south, suggesting an eruptive
volcanic centre somewhere near Charlemont. To the south in Connecticut, volcanic
units such as the Collinsville Formation (as used by Stanley, 1964), the Reynolds
Bridge Gneiss (Cassie, 1965), and the Hitchcock Lake Member of the Waterbury
Gneiss (Fritts, 1963) are locally present immediately below the Taconian Line and
presumably represent southern correlatives of the Barnard and Hawley (Fig. 2). The

AGE RANGE	SOUTHERN QUEBEC	VERMONT	NORTHERN MASSACHUSETTS	SOUTHERN MASSACHUSETTS	CONNECTICUT	NEW HAMPSHIRE
Early Devonian to Middle Silurian(?)	Glenbrooke Formation	Waits River Formation Northfield Formation	Waits River Formation Goshen Formation	Goshen Formation	Straits Schist	Littleton Formation
Salinic break						
Middle Silurian	Peasley Pond Formation	Shaw Mountain Formation		Russell Mountain Formation	Unnamed marble and calc. silicate rocks	Fitch Formation Clough Formation
Taconic break						
Middle Ordovician	Magog and Mictaw Groups Ascot and Weedon Formations	Cram Hill Formation *Barnard Gneiss Moretown Formation	Hawley Formation (schist) Hawley Formation (volcanic rocks) Moretown Formation	Cobble Mountain Formation Moretown Formation	*Schists of Collinsville and Rattlesnake Hill Formations *Volcanic rocks in Collinsville Formation, Hitchcock Lake Member of Waterbury Gneiss, and Reynolds Bridge Gneiss *Taine Mountain Formation	Partridge Formation Ammonoosuc volcanics Albee Formation

*Local usage

Figure 2. Correlation of major stratigraphic units above and below the Taconian Line in western New England and adjacent Québec and New Hampshire.

Ascot and Weedon Formations may be the correlatives of these units in southern Québec (St-Julien and Hubert, 1975, Fig. 1; Osberg, 1978, p. 143).

The second principal Middle Ordovician lithology immediately below the Taconian Line is grey to black, carbonaceous, sulphidic, and rusty-weathered phyllite and schist. These rocks are commonly intercalated with the metavolcanic rocks described above. They have been mapped under such names as Cram Hill Formation and carbonaceous schist member of the Missiquoi Formation (Doll *et al.*, 1961) in Vermont, black schist member of the Hawley Formation in Massachusetts, and parts of the Rattlesnake Hill Formation of Stanley (1964) and carbonaceous schist intervals of Stanley's Collinsville Formation in Connecticut (Fig. 2). The Magog and Mictaw Groups in southern Quebec are possible equivalents of these rocks (St-Julien and Hubert, 1975, Fig. 1).

The third rock type of presumed Middle Ordovician age below the Taconian Line is light-grey to brownish-grey, noncarbonaceous, nonsulphidic, generally well bedded and locally graded granulite and schist. Although these rocks most commonly are separated from the Taconian Line by sulphidic black schists and (or) metavolcanic rocks, they immediately underlie the Taconian Line along a 40-km interval in northern Vermont and across southernmost Massachusetts and much of Connecticut (Fig. 3). These rocks are commonly feldspathic and, particularly in Connecticut, some appear to grade laterally into some of the felsic metavolcanic gneiss described above. In southern Massachusetts, a complex north-south lateral facies change between light grey-brown granulite-schist and sulphidic black schist has been described in detail by Hatch and Stanley (1973, 1976) indicating the lateral juxtaposition of the depositional environments of the two rock types. Names that have been applied to these light grey non-carbonaceous rocks (Fig. 2) are Moretown Formation (or Moretown Member of the Missisquoi Formation) in Vermont, Moretown and Cobble Mountain Formations in Massachusetts, and such names as Taine Mountain and Satans Kingdom Formations of Stanley (1964), and Unit 1 of the Hartland Formation as used by Gates and Martin (1967) and Martin (1970) in Connecticut. (The interested reader is again referred to plates 1 and 2 of Hatch and Stanley, 1973 for interpreted correlations in Connecticut).

The Middle Ordovician age assignment of all of these rocks is very tenuously tied to the fossiliferous Magog Formation (Group, St-Julien and Hubert, 1975) at Magog, Quebec (Cooke, 1950; Berry, 1962): no fossils that stand up under the rigors of modern restudy have been reported from any of these rocks in this structural belt in New England. Thus, although the Middle Ordovician age for at least the upper part of this section of rocks in northernmost Vermont is based on correlation with the Magog fossiliferous rocks, the reliability of this correlation decreases southward.

Figure 3. Longitudinal section along the Taconian Line from the Québec border south to Long Island Sound. The three principal lithofacies of the Middle Ordovician below the Taconian Line are: c, carbonaceous schist; v, metavolcanic rocks, and g, noncarbonaceous granulite and schist. Small rectangles above the unconformity represent the approximate distribution of the Middle Silurian Shaw Mountain and Russell Mountain Formations and correlatives. Vertical exaggeration is 16X; thicknesses of Middle Ordovician units are taken as direct function of map outcrop width because beds are nearly vertical. Unlabeled area is pre-Middle Ordovician(?) strata.

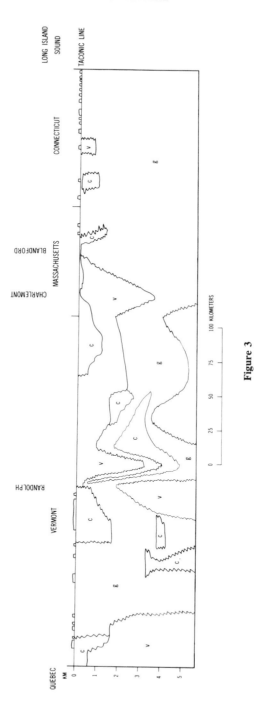

Figure 3

Although very few chemical analyses have been reported on the metavolcanic rocks of this sequence, the wide range of mineralogies from nearly black hornblende-plagioclase rocks to nearly white feldspar gneisses and schists and all gradations in between, strongly suggests a comparably wide range in chemical composition. This range in chemistry, combined with the close association of the metavolcanic rocks with metamorphosed euxinic black muds, seems to support the concept of these rocks as part of an island arc sequence. Furthermore, the quartzofeldspathic character of the rocks upon which they rest (Moretown and equivalents; gneisses of the domes) suggests that the arc was ensialic.

Above the Middle Ordovician rocks (above the Taconian break of the Taconian Line) throughout the length of the Taconian Line in western New England are thin lenses of a distinctive unit characterized by quartzite, calcareous quartzite, marble, and minor greenstone and quartz-sericite schist. These rocks have been named Shaw Mountain Formation in Vermont, Russell Mountain Formation in Massachusetts, and although generally unnamed in Connecticut, they have been correlated with the Russell Mountain by Hatch and Stanley (1973). Boucot and Thompson (1963) described fossils of Wenlockian age from the Shaw Mountain near Albany, Vermont. The unfossiliferous Russell Mountain and unnamed rocks in Connecticut have been assigned the same age solely on the basis of similar lithology and stratigraphic position.

The lithology of these units, particularly the quartzites, local conglomerates, marbles, and calcsilicate rocks, strongly suggests that they represent a Silurian shelf sequence, in rather sharp contrast to the island arc sequence below them.

East of, and stratigraphically above the Shaw Mountain – Russell Mountain rocks or, where they are absent, the Middle Ordovician rocks, is a thick sequence of grey, generally carbonaceous, moderately sulphidic to nonsulphidic, moderately rusty to nonrusty weathered metasedimentary rocks. At the base of this section throughout Vermont is the Northfield Formation which consists of relatively homogeneous, indistinctly bedded to nonbedded, dark-grey, crenulated phyllite. In northernmost Massachusetts, these rocks are much better bedded, are graded, and are called the Goshen Formation. Similar rocks in western Connecticut that have been interpreted by Hatch and Stanley (1973) to correlate with the Goshen are called Straits Schist. In contact to the east (either stratigraphically above or as lateral facies equivalents to the Northfield-Goshen-Straits) are grey, calcareous, shaly and sandy metasedimentary rocks (Waits River Formation-Conway Formation-Wepawaug Schist) and grey quartzitic and shaly metasedimentary rocks (Gile Mountain Formation). A thin (as much as 100+ m) metavolcanic marker unit, the Standing Pond Volcanics, is widespread in Vermont and northern Massachusetts at or near the Waits River-Gile Mountain contact. Because the structure of this section of grey rocks is locally known to be extremely complicated but is only locally understood, its thickness can only be roughly estimated to be a few kilometres. The age of this sequence of rocks has generally been thought to range between Middle Silurian (younger than the Shaw Mountain) and Early Devonian (pre-Acadian); no currently authenticated fossils have been found in these rocks in New England.

STRUCTURAL CONTRAST

In their excellent summary of the evidence for the Taconian orogeny in the northern Appalachians, Pavlides *et al.* (1968) indicated the area of the Taconian Line in Vermont and Massachusetts as being within a belt of disconformity; they indicated no data for the western Connecticut part. Although later studies have confirmed further the parallelism of both bedding and dominant foliation in the rocks of all three stratigraphic intervals (Middle Ordovician, Middle Silurian, and Middle Silurian(?) to Lower Devonian) near the Taconian Line throughout western New England, significant differences in structural history and style have been reported between the pre-Silurian and the Silurian and Devonian rocks in a few areas.

Rosenfeld (1968, p. 196) presented evidence from studies of rolled garnets in southern Vermont that schists of the Lower Cambrian Pinney Hollow Formation had developed a schistosity in a pre-Acadian event that he suggested was plausibly equated with the Taconian orogeny. Rosenfeld's samples came from the west side of the Chester and Athens domes only a few kilometres from the Taconian Line.

Osberg *et al.* (1971) and Hatch (1975) presented evidence for an episode of folding in pre-Silurian rocks a few kilometres west of the Taconian Line in northern Massachusetts that is not present in the Silurian-Devonian Goshen Formation in that area. Although only a very few such folds have been observed and none is large enough to affect the map pattern at 1:24,000 scale, Osberg and Hatch concluded that the diversity of orientations of the axes of the first set of Acadian folds in pre-Silurian rocks as compared with their relatively uniform orientation in Silurian-Devonian rocks results from deformation of the pre-Silurian beds by pre-Acadian folding. We have suggested a Taconian age for this pre-Acadian event inasmuch as the youngest strata involved are the Middle Ordovician(?) Moretown and Hawley Formations. Although interpreting the details of this deformation back through three Acadian episodes of folding is uncertain at best, the available evidence seems to suggest a Taconian event that imposed small-scale crumpling on the rocks immediately below the Taconian hiatus in much of western Massachusetts.

Stanley (1967, 1975) described similar structural relations from the Blandford-Woronoco area of southern Massachusetts and from the Newtown area of Connecticut. Although Silurian-Devonian rocks are not exposed in the Newtown area itself, they do crop out a few kilometres east thereof and Stanley has traced the fold generations with sufficient accuracy to correlate them between the Newtown and Blandford areas. In both areas, he has documented minor, and in the Newtown area, major, folds that predate the earliest folds present in the Silurian-Devonian Straits Schist and Goshen Formation.

Thus, although no angular unconformity has been reported in any of the many detailed studies of the rocks along the Taconian Line in western New England, a few of these studies have presented evidence for locally significant pre-Acadian folding in the pre-Silurian strata immediately west of the Taconian Line.

Zen (1968) summarized the effects of the Taconian orogeny in the Taconic area of westernmost New England and adjacent New York. Ratcliffe and Harwood (1975) documented a complex history of Taconian deformational events including multiple episodes of folding and faulting along the western margin of the Berkshire massif 20 to 30 km west of the Taconian Line in Massachusetts and northern Connecticut. Norton (1975) traced some of these Taconian structures eastward and integrated them into the tectonic history of the east side of the Berkshire massif. Norton (p. 26) described an eastward decreasing Taconian schistosity between North Adams and Becket, Massachusetts; farther east, this schistosity is not recognized. Norton also noted (1975, p. 26) that: "Eastward the frequency of occurrence and intensity of D_2 [first and strongest Acadian deformation along the Taconian Line] structures increase." Thus, we appear to have a situation, exemplified by western Massachusetts, where intense multiple Taconian deformations a few tens of kilometres west of the Taconian Line fade out markedly to the east to the extent that they are recognized only very locally and with difficulty in the pre-Silurian rocks immediately below (west of) the Taconian Line.

A second, quite different, kind of evidence has been presented that bears on the structural interpretation of the Taconian Line. A stratigraphy of relatively thin mappable units confirmed by abundant primary sedimentary tops has enabled Hatch and colleagues to document Acadian upright isoclinal folds having amplitudes of a few thousand metres and wavelengths of tens to a few hundreds of metres in the Silurian-Devonian Goshen Formation in Massachusetts (Hatch, 1968, 1975). Although minor folds of this generation are widespread and abundant in all the pre-Silurian units immediately below (west of) the Taconian Line, only one has been recognized to have an amplitude of more than a few metres or possibly tens of metres. Amplitudes of a few tens of centimetres are most common. Furthermore, the Taconian unconformity itself is not known to be folded by these folds in Massachusetts or Connecticut, and possible examples from Vermont are rare and ambiguous. As a result, a décollement has been inferred at the base of the Goshen Formation in Massachusetts. Although folds such as those mapped in the Goshen in Massachusetts have not been mapped in Connecticut, I have seen enough reversals in the direction of primary tops in the narrow synclines of Straits Schist in western Connecticut (Fig. 1) to be satisfied that it too must be isoclinally folded and that a similar décollement is probably present at the base of the Straits.

In Vermont, the stratigraphic equivalent of the Goshen, the Northfield Formation, locally contains the graded beds that enabled mapping of the large isoclinal folds in the Goshen. Because the Northfield includes both east- and west-facing beds with parallel attitudes, and because the Northfield and Goshen are lithically similar and thus presumably have similar competencies, I infer that Goshen-type isoclinal folding continues north into the Northfield of Vermont. By this same reasoning, I would thus conclude that the Taconian Line continues to represent a surface of décollement across much or possibly all of Vermont.

In summary, the structural evidence bearing on the unconformable nature of the Taconian Line is as follows: the well-developed Taconian structures in the Taconic Range and along the west margin of and within the Berkshire (and Green Mountain) massifs fade out eastward to the extent that they are only locally recognizable in the strata immediately west of the Taconian Line, and no angular discordance has been recognized anywhere along the Taconian Line. On the other hand, however, at the time of the first Acadian folding event, possibly at least in part as a result of Taconian events, the physical character of the rocks was such that the pre-Silurian strata failed to deform into the large upright isoclinal folds that formed so abundantly and extensively in their Silurian(?) to Lower Devonian cover.

In the preceding discussion, no distinction has been made between structures in the Middle Silurian rocks (Shaw Mountain and Russell Mountain Formations) and those in the overlying Silurian-Devonian strata. I am not aware, from my own field work in Massachusetts, from published reports, or from discussions with other workers, of any direct evidence of folding, cleavage, metamorphism, or other tectonic activity in the Middle Silurian formations not also recognized in the overlying Silurian and Devonian strata. However, both the Shaw Mountain and the Russell Mountain are very thin and discontinuous, and thus such structures could easily have been overlooked. Furthermore, I know of no evidence to indicate whether the discontinuous map pattern of these units results from local nondeposition, from later erosion, or from tectonic telescoping. If it results from erosion, a period of uplift, followed by erosion and relowering below sea level, clearly is indicated between the two sequences.

In his description of the Lower Devonian Seboomook Formation in eastern Maine, Pavlides (1973) reported conglomerate lentils containing cobbles with fossils of late Llandoverian C_3-C_5 age. He concluded that "The conglomerate probably represents local uplift during the Salinic disturbance" (Boucot, 1962). No structural evidence has been recognized along the Taconian Line which would either document or refute a comparable Salinic disturbance in western New England, although recently discovered conodonts of Pridoli age in the Fitch Formation near Littleton, New Hampshire (Anita G. Harris, written commun., 1979), suggest at least nearly continuous deposition across the Silurian-Devonian boundary in that area.

METAMORPHIC CONTRAST

In their synthesis of the Paleozoic regional metamorphism of New England, Thompson and Norton (1968, p. 325) treated the possibility of a Taconian metamorphic event with: "The possibility of pre-Silurian regional metamorphism in northern Vermont and southern Quebec is discussed by Albee. The isograds of Figure 24-1 may thus represent the combined effects of more than one metamorphic event, at least locally." Albee (1968, p. 331), in the paper referred to by Thompson and Norton, did indeed conclude that partial retrogression of kyanite-grade assemblages in the Worcester and Northfield Mountain anticlines in pre-Silurian rocks of northern Vermont probably resulted from superposition of Acadian regional metamorphism or an earlier regional metamorphic event having an age of 430 Ma or older. Lanphere and Albee (1974) confirmed this conclusion with a comprehensive $^{40}Ar/^{39}Ar$ age study. Both of these areas of retrogression are about 8 km west of the Taconian Line (Fig. 1).

In the discussion of structure above, I mentioned that Rosenfeld (1968) concluded from his study of rolled garnets that the schists of the Pinney Hollow Formation on the west side of the Athens dome had been metamorphosed to garnet grade prior to the earliest Acadian event. He stated (p. 196): "It seems plausible to equate the deformation indicated by the earliest rotation within such garnets with Zen's (1967, p. 44-69) Taconic diastrophism."

Norton (1975) described an early metamorphic event in Lower Paleozoic rocks in northwestern Massachusetts which he attributed to the Taconian orogeny. This metamorphism, which reached garnet grade in the Lower Paleozoic rocks immediately west of the north end of the Berkshire massif, decreased in intensity eastward and southeastward across the Hoosac nappe to the lower Paleozoic rocks immediately east of the north end of the Berkshire massif (Norton, 1975, p. 28-29). To the south, the grade of the later Acadian metamorphism increases (Norton, 1975, Fig. 24; Hatch, 1975, Fig. 44) and obscures the effects of the Taconian metamorphism.

In summary, evidence for Taconian metamorphism east of the western New England Precambrian massifs is available only from a few areas in Vermont and northern Massachusetts. Because this metamorphism was originally only very local or because of insufficiently detailed field studies, or because of subsequent retrogression in the Acadian, or because of the combined effects of all those factors, a general regional Taconian metamorphism higher than chlorite or possibly biotite zone cannot be substantiated more than a few kilometres east of the Precambrian massifs; the origin of the kyanite-grade metamorphism in northern Vermont noted above remains enigmatic.

DISCUSSION

The preceding sections of this paper have attempted to summarize the available data on the lithology and on the degree and style of structural and metamorphic deformation of the stratified rocks of Middle Ordovician, Middle Silurian, and Middle Silurian(?) to Early Devonian age in the immediate vicinity of the Taconian Line in western New England. These data should enable us to distinguish the effects of the Taconian orogeny, the Salinic orogeny (or "disturbance"), and the Acadian orogeny in this area. The full length of the Taconian Line between the Quebec border and Long Island Sound has been mapped at scales of either 1:62,500 or 1:24,000 over the past 30 years, and these studies have carefully described the stratigraphy, structure, and metamorphism of this belt of rocks. Nevertheless, opinions have differed as to the age of some of the rocks along the Taconian Line, and few of the published reports have assigned ages to the structural and metamorphic events. Furthermore, perhaps because most of the reports on these areas were written before the advent of plate tectonic theory, most contain insufficient discussion of sedimentary environments and their significance or of the significance of structural styles and their relations to metamorphic, plutonic, and volcanic events. The remainder of this paper will look at the Middle Ordovician to Early Devonian geologic history of this narrow belt of western New England where maximum data are available, and speculate upon the significance of these data in terms of tectonic models.

The similarity between Billings' (1937) stratigraphic sequence along the Bronson Hill anticlinorium in western New Hampshire and the Middle Ordovician and younger part of the stratigraphic section east of Green Mountain and Berkshire massifs in Vermont and Massachusetts has long been recognized (Fig. 2). The correspondence is particularly striking (1) between the Middle Ordovician(?) Moretown and Albee Formations and (2) between the Middle Ordovician Barnard Gneiss and Cram Hill Formation and the Ammonoosuc Volcanics and Partridge Formation. The former are both light-grey, non-carbonaceous quartz-plagioclase-mica granulites and schists characterized by paper thin partings of mica ("pinstripe"). The latter are dark-grey, carbonaceous, commonly sulphidic schists and intimately associated metavolcanic rocks. Aleinikoff (1977) concluded from a chemical study of a very local part of the New Hampshire belt of Ammonoosuc Volcanics that some are abyssal tholeiites and others are island-arc tholeiites; the chemistry of the felsic metavolcanic rocks indicated contamination by admixture of detritus and thus precluded petrogenetic classification. Field and thin section study of both the Vermont and the New Hampshire belts of metavolcanic rocks, however, suggests a wider range of compositions, with many apparently in the andesite-dacite range. For these reasons, and their close association with metamorphosed black shales, both belts are here interpreted as island-arc volcanic sequences.

The similarity between the eastern Vermont and the western New Hampshire pre-Silurian sequences noted above makes it very tempting not only to correlate the two but to infer their continuity under the large area of Middle Silurian(?) to Lower Devonian metasedimentary rocks of the Connecticut Valley-Gaspé Synclinorium between them (Fig. 1). The picture that thus evolves is of an area many hundreds of kilometres long and as much as 70 km wide in which a suite of island-arc volcanics and associated black muds was deposited on a basement of granitic gneisses and Moretown-Albee feldspathic greywackes through the eastern edge of which some of the gneisses of the Oliverian domes protruded (Robinson *et al.*, 1979).

If the Middle Ordovician volcanics were erupted upon a large mass of ensialic Albee-Moretown, two possible models come to mind. The first is the model of Bird and Dewey (1970) whereby with incipient closing of Iapetus in Middle Ordovician time, volcanism resulted from a subduction zone dipping westward beneath the east edge of North America. This model would place the Lower to Middle Ordovician continental margin, regardless of the amount of dip of the subduction zone, well east of the present Bronson Hill anticlinorium (even if allowances are made for later Acadian shortening). The present distribution of Cambrian and Ordovician shelf quartzite-carbonate rocks west of the Berkshire-Green Mountain massifs and of probably correlative deeper water (rise?) metasedimentary rocks immediately east of the massifs seems incompatible with a continental margin so located. A further problem with this model comes from the line of ultramafic rocks that extends from Quebec southward across all of western New England. These ultramafic rocks follow

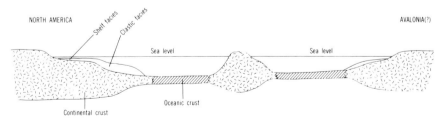

Figure 4. Configuration of continental masses in Late Cambrian time after late Precambrian rifting. Small continental fragment between North America and "Avalonia" is fragment which was broken off during rifting and which will become the site of the Ordovician island arc.

a narrow stratigraphic-tectonic interval east of the Precambrian massifs generally between the (Lower Cambrian or older?) Hoosac Formation and the Middle Ordovician(?) Moretown Formation (Fig. 1). If, as seems probable from their field relations, these ultramafic rocks are slices and fragments of oceanic crust mechanically emplaced cold into the host rocks, the Bird and Dewey model would require their derivation from a closing basin east of the present Bronson Hill anticlinorium. The absence of significant Taconian deformation and metamorphism in the Bronson Hill and the rocks immediately west of the Taconian Line, as well as the mechanical problems involved in transporting and emplacing a belt of ultramafic rocks many hundreds of kilometres long and a few kilometres wide across tens of kilometres of virtually undeformed terrain, make this model very difficult to accept.

A second model which seems to accommodate more of the field relations discussed in the first part of this paper is largely derived from that proposed by Osberg (1978). I would suggest that the island arc was built up during Ordovician time on a

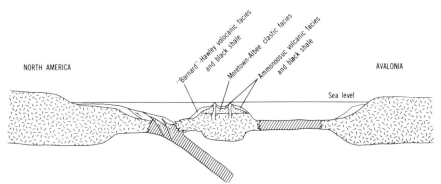

Figure 5. Configuration during Middle Ordovician time showing development of island arc on former continental fragment by eastward subduction and initial upward and westward thrusting of oceanic crust out of the closing basin up onto the North American continent.

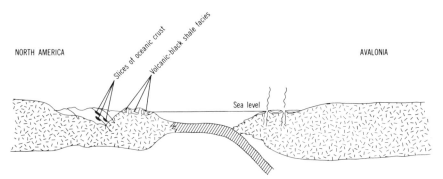

Figure 6. Configuration in Early Silurian time. Former island arc has been accreted onto North America during the Taconian orogeny. Upthrust slices of oceanic crust shown enclosed within metasedimentary host rocks. Incipient volcanism on Avalonia is future Newbury Volcanic Complex and coastal Maine volcanic belt.

fragment of continental crust broken off and isolated from the main continental masses during the period of late Precambrian rifting as schematically portrayed in Figure 4. With incipient closing of Iapetus, an eastward dipping subduction zone formed west of this continental fragment and east of North America. As closing continued, the Barnard-Hawley-Collinsille-Ammonoosuc volcanics formed into an island arc, and the Cram Hill-Hawley-Partridge black shales formed in the closing basin (Fig. 5). According to this model (see Osberg, 1978, p. 143), the culmination of the Taconian orogeny probably resulted from the collision of this island arc with continental North America. During the gradual closing of the basin, slivers of oceanic crust were imbricately upthrust into the Cambrian-Ordovician Hazens Notch-Pinney Hollow-Ottauquechee-Stowe-Rowe rocks and these imbricately sliced packages were in turn thrust up onto the continental margin. This model provides a local and logical source for the ultramafic rocks of the Quebec-western New England ultramafic belt. It also suggests that the Taconian suture was somewhere not far east of the present ultramafic outcrop belt, and thus offers an explanation for the observed distribution of Taconian structural deformation and metamorphism discussed in the first part of this paper. Figure 6 shows the postulated Early Silurian configuration immediately following the Taconian collision and suturing.

The Middle Silurian carbonate-quartzite Shaw Mountain-Russell Mountain rocks, which overlie the island arc rocks along the west margin of the Connecticut Valley-Gaspé synclinorium, have long been correlated with the Silurian Clough and Fitch Formations of western New Hampshire in the Bronson Hill anticlinorium. Accepting this correlation and following Osberg's (1978) model leads to the suggestion that after Taconian accretion, a period of erosion reduced the area of the former arc to a broad low continental shelf which presumably extended from the present Taconian Line eastward to at least the east side of the Bronson Hill anticlinorium as portrayed in Figure 7.

The lack of recognized angular discordance between the Middle Silurian shelf strata and the underlying Ordovician island-arc rocks further supports the model that the collision of the arc with the continent, as well as the subsequent uplift and erosion of the arc, all took place with only minor folding or gentle warping of the stratified rocks of the arc. Although these relations are difficult to explain, they seem more compatible with a model that places the suture west of the arc than one that would place the suture east of the arc and require that the arc transmit the forces that

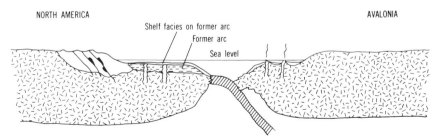

Figure 7. Configuration during Middle Silurian time showing deposition of shelf facies (Shaw Mountain-Russell Mountain-Clough-Fitch) on eroded former island arc.

generated the Taconian deformation to the west of the arc without itself being significantly deformed.

Because the age of the Northfield-Goshen-Straits grey metasedimentary rocks and the grey metasedimentary rocks to the east of them in the Connecticut Valley-Gaspé synclinorium has not yet been constrained by fossils, we cannot document the length, if any, of a Salinic break in western New England. Furthermore, with no reported evidence of angular discordance between the Middle Silurian rocks and the overlying grey rocks, we also have no documentation of Salinic deformation or metamorphism. In the absence of evidence for a time break, all that is required to achieve the situation portrayed in Figure 8 is an increase in water depth during the period of deposition of the grey (largely Lower Devonian?) sediments. The causal relations, if any, between these activities and the concurrent approach of Avalonia from the east (present direction) are not known.

The Acadian orogeny, which is somewhat beyond the scope of this paper, probably resulted from the collision of Avalonia with the east side of the former Middle Ordovician island arc. As was the case with the Taconian orogeny, most of the deformation and metamorphism in the Acadian orogeny appear to have taken place in

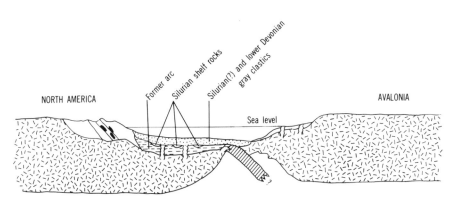

Figure 8. Configuration during Early Devonian time showing deposition of Northfield-Goshen-Straits on collapsed and partly eroded shelf.

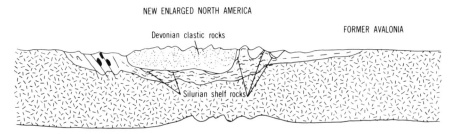

Figure 9. Accreted North America and Avalonia in Middle Devonian time following the Acadian orogeny.

the rocks west of the suture rather than east of it. Thus, the rocks of the Bronson Hill anticlinorium and the Connecticut Valley-Gaspé synclinorium were intensely folded and metamorphosed in at least three locally recognizable Acadian pulses, whereas the rocks of Avalonia appear to have been only mildly deformed and only slightly metamorphosed (Fig. 9).

Thus, by the model followed here, both the Taconian and the Acadian orogenies were preceded by eastward-dipping subduction that produced volcanism on the more easterly (present direction) of the two mutually approaching continental masses; both orogenies climaxed with collisions that caused intense folding, faulting, and high-grade metamorphism in the mass west of the suture and that only mildly deformed the mass east of the suture. Although the mechanics of this process are far from fully understood, the observed field data seem to support such a model.

ACKNOWLEDGEMENTS

The ideas presented here have evolved from many years of field work and discussion with co-workers and colleagues, particularly Rolfe S. Stanley, Philip H. Osberg, David S. Harwood, Nicholas M. Ratcliffe, Douglas W. Rankin, E-an Zen, Louis Pavlides, and Peter Robinson. The manuscript benefited significantly from thorough and thoughtful reviews by Louis Pavlides, Avery A. Drake, Jr., Robert B. Neuman and Wallace M. Cady.

REFERENCES

Albee, A.L., 1968, Metamorphic zones in northern Vermont: in Zen, E-an, White, W.S., Hadley, J.B., and Thompson, J.B., Jr., eds., Studies of Appalachian Geology – Northern and Maritime: New York, Interscience Publishers, p. 329-341.

Aleinikoff, J.N., 1977, Petrochemistry and tectonic origin of the Ammonoosuc Volcanics, New Hampshire-Vermont: Geol. Soc. America Bull., v. 88, p. 1546-1552.

Berry, W.B.N., 1962, On the Magog, Quebec, graptolites: American Jour. Sci., v. 260, p. 142-148.

Billings, M.P., 1937, Regional metamorphism of the Littleton-Moosilauke area, New Hampshire: Geol. Soc. America Bull., v. 48, p.463-566.

Bird, J.M. and Dewey, J.F., 1970, Lithosphere plate-continental margin tectonics and the evolution of the Appalachian orogen: Geol. Soc. America Bull., v. 81, p. 1031-1060.

Boucot, A.J., 1962, Appalachian Siluro-Devonian: in Coe, Kenneth, ed., Some Aspects of the Variscan Fold Belt: 9th Inter-University Geological Congress, Manchester University Press, p. 155-163.

Boucot, A.J. and Drapeau, Georges, 1968, Siluro-Devonian rocks of Lake Memphremagog and their correlatives in the Eastern Townships: Quebec Dept. Natural Resources, Spec. Paper 1, 44 p.

Boucot, A.J. and Thompson, J.B., Jr., 1963, Metamorphosed Silurian brachiopods from New Hampshire: Geol. Soc. America Bull., v. 74, p. 1313-1334.

Cassie, R.M., 1965, Evolution of a domal granite gneiss and its relation to the geology of the Thomaston quadrangle, Connecticut: University of Wisconsin, Madison, Wisconsin, Ph.D. Thesis, 109 p.

Chang, P.H., Ern, E.H. and Thompson, J.B., Jr., 1965, Bedrock geology of the Woodstock quadrangle, Vermont: Vermont Geol. Survey Bull. 29, 65 p.

Cooke, H.C., 1950, Geology of a southwestern part of the Eastern Townships of Quebec: Geol. Survey Canada Memoir 257, 142 p.

Doll, C.G., Cady, W.M., Thompson, J.B., Jr. and Billings, M.P., 1961, Centennial geologic map of Vermont: Vermont Geol. Survey, Montpelier, Vermont.

Ern, E.H., Jr., 1963, Bedrock geology of the Randolph quadrangle, Vermont: Vermont Geol. Survey Bull. 21, 96 p.

Fritts, C.E., 1963, Bedrock geology of the Southington quadrangle, Connecticut: United States Geol. Survey Geologic Quadrangle Map GQ-200, scale 1:24,000.

Gates, R.M. and Martin, C.W., 1967, The bedrock geology of the Waterbury quadrangle: Connecticut State Geol. and Natural History Survey, Quadrangle Rept. 22, 36 p.

Hatch, N.L., Jr., 1967, Redefinition of the Hawley and Goshen Schists in western Massachusetts: United States Geol. Survey Bull. 1254-D, 16 p.

_____, 1968, Isoclinal folding indicated by primary sedimentary structures in western Massachusetts: United States Geol. Survey Prof. Paper 600-D, p. D108-D114.

_____, 1975, Tectonic, metamorphic, and intrusive history of part of the east side of the Berkshire massif, Massachusetts: in Harwood, D.S., et al., Tectonic Studies of the Berkshire Massif, Western Massachusetts, Connecticut and Vermont: United States Geol. Survey Prof. Paper 888, p. 51-62.

Hatch, N.L., Jr. and Stanley, R.S., 1973, Some suggested stratigraphic relations in part of southwestern New England: United States Geol. Survey Bull. 1380, 83 p.

_____, 1976, Geologic map of the Blandford quadrangle, Hampden and Hampshire Counties, Massachusetts: United States Geol. Survey Geologic Quadrangle Map GQ-1312.

Hussey, A.M., Chapman, C.A., Doyle, R.G., Osberg, P.H., Pavlides, Louis, and Warner, Jeffrey, compilers, 1967, Preliminary geologic map of Maine: Maine Geol. Survey, Augusta, Maine, Scale 1:500,000.

Lanphere, M.A. and Albee, A.L., 1974, $^{40}Ar/^{39}Ar$ age measurements in the Worcester Mountains: Evidence of Ordovician and Devonian metamorphic events in northern Vermont: American Jour. Sci., v. 274, p. 545-555.

Martin, C.W., 1970, The bedrock geology of the Torrington quadrangle: Connecticut State Geol. and Natural History Survey Quadrangle Rept. 25, 53 p.

Norton, S.A., 1975, Chronology of Paleozoic tectonic and thermal metamorphic events in Ordovician, Cambrian, and Precambrian rocks at the north end of the Berkshire massif, Massachusetts: in Harwood, D.S., et al., Tectonic Studies of the Berkshire Massif, Western Massachusetts, Connecticut, and Vermont: United States Geol. Survey Prof. Paper 888, p. 21-31.

Osberg, P.H., 1978, Synthesis of the geology of the northeastern Appalachians: in IGCP Project 27, Caledonian-Appalachian orogen of the North Atlantic Region: Geol. Survey Canada Paper 78-13, p. 137-147.

Osberg, P.H., Hatch, N.L., Jr. and Norton, S.A., 1971, Geologic map of the Plainfield quadrangle, Franklin, Hampshire, and Berkshire Counties, Massachusetts: United States Geol. Survey Geologic Quadrangle Map GQ-877, Scale 1:24,000.

Pavlides, Louis, 1973, Geologic map of the Howe Brook quadrangle, Aroostook County, Maine: United States Geol. Survey Geologic Quadrangle Map GQ-1094, Scale 1:62,500.

Pavlides, Louis, Boucot, A.J., and Skidmore, W.B., 1968, Stratigraphic evidence for the Taconic orogeny in the Northern Appalachians: in Zen, E-an, White, W.S., Hadley, J.B., and Thompson, J.B., Jr., eds., Studies of Appalachian Geology – Northern and Maritime: New York, Interscience Publishers, p. 61-82.

Ratcliffe, N.M. and Harwood, D.S., 1975, Blastomylonites associated with recumbent folds and overthrusts at the western edge of the Berkshire massif, Connecticut and Massachusetts – A preliminary report: in Harwood, D.S., et al., Tectonic Studies of the Berkshire massif, Western Massachusetts, Connecticut, and Vermont: United States Geol. Survey Prof. Paper 888, p. 1-19.

Robinson, Peter, Thompson, J.B., Jr. and Rosenfeld, J.L., 1979, Nappes, gneiss domes, and regional metamorphism in western New Hampshire and central Massachusetts: in Skehan, J.S., and Osberg, P.H., eds., The Caledonides in the U.S.A.; Geological excursions in the northeast Appalachians − Contributions to the International Geological Correlation Program (IGCP) Project 27 − Caledonide Orogen: Boston College, Weston Observatory, Weston, Mass., p. 93-174.

Rosenfeld, J.L., 1968, Garnet rotations due to the major Paleozoic deformations in southeast Vermont: in Zen, E-an, White, W.S., Hadley, J.B., and Thompson, J.B., Jr., eds., Studies of Appalachian Geology − Northern and Maritime: New York, Interscience Publishers, p. 185-202.

St-Julien, Pierre and Hubert, Claude, 1975, Evolution of the Taconian orogen in the Quebec Appalachians: American Jour. Sci., v. 275-A, p. 337-362.

Stanley, R.S., 1964, The bedrock geology of the Collinsville quadrangle: Connecticut State Geol. and Natural History Survey Quadrangle Rept. 16, 99 p.

_____, 1967, Geometry and age relations of some minor folds and their relation to the Woronoco nappe, Blandford and Woronoco quadrangles, Massachusetts: in Guidebook for Field Trips in the Connecticut Valley of Massachusetts, New England Intercollegiate Geological Conference, p. 48-60.

_____, 1975, Time and space relationships of structures associated with the domes of southwestern Massachusetts and western Connecticut: in Harwood, D.S., et al., Tectonic Studies of the Berkshire Massif, Western Massachusetts, Connecticut, and Vermont: United States Geol. Survey Prof. Paper 888, p. 67-96.

Thompson, J.B., Jr. and Norton, S.A., 1968, Paleozoic regional metamorphism in New England and adjacent areas: in Zen, E-an, White, W.S., Hadley, J.B., and Thompson, J.B., Jr., eds., Studies of Appalachian Geology − Northern and Maritime: New York, Interscience Publishers, p. 319-327.

Zen, E-an, 1967, Time and Space Relationships of the Taconic Allochthon and Autochthon: Geol. Soc. America Bull., Spec. Paper 97, 107 p.

_____, 1968, Nature of the Ordovician orogeny in the Taconic area: in Zen, E-an, White, W.S., Hadley, J.B., and Thompson, J.B., Jr., eds., Studies of Appalachian Geology − Northern and Maritime, New York, Interscience Publishers, p. 129-139.

Manuscript Received April 27, 1979
Revised Manuscript Received January 30, 1980

Major Structural Zones and Faults of the Northern Appalachians, edited by
P. St-Julien and J. Béland, Geological Association of Canada Special Paper 24, 1982

GEOLOGY OF THE QUEBEC RE-ENTRANT: POSSIBLE CONSTRAINTS FROM EARLY RIFTS AND THE VERMONT-QUEBEC SERPENTINE BELT

Barry L. Doolan, Marjorie H. Gale, Peter N. Gale, and Robert S. Hoar
Department of Geology, University of Vermont, Burlington, Vermont 05405

ABSTRACT

Major orthogonal promontories and re-entrants along the early Paleozoic North American margin played significant roles in the pre- and post-collisional events in the Appalachian Orogen. Development of the Quebec re-entrant, centered around Montreal, controlled the location of early alkaline magmatism, Hadrynian to early Paleozoic clastic sedimentation and the northwest orientation of the shelf-slope sequence in northern Vermont. This rift-transform controlled re-entrant influenced the effects of later collisional tectonics in the Ordovician as evidenced by the present configuration of remnant ocean crust, volcanic arcs, and adjacent basement rocks. Despite earlier views that the Vermont serpentinites are ultramafic intrusions through sialic basement, field and chemical analyses support the view that the ultramafic and related rocks in northern Vermont are remnants of obducted oceanic crust with features similar to those reported in Quebec and Newfoundland ophiolite occurrences. Although a traceable tectonic stratigraphy along the serpentine belt is preserved across the Quebec-Vermont border, imbrication of remnant ocean crust with older clastic sequences and synchronous to post-emplacement ophiolitic and volcanogenic flysch increases sharply to the south. These differences are ascribed to compressional and subsequent left lateral transform tectonics during a diachronous collision of volcanic arcs against the irregular continental margin in the vicinity of the Québec re-entrant.

RÉSUMÉ

Dans l'orogène appalachien, au Paléozoïque Inférieur, des saillants et des rentrants majeurs, orthogonaux entre eux, le long de la marge continentale de l'Amérique du Nord, ont joué un rôle significatif dans les évènements antérieurs et postérieurs à la collision. Le développement du rentrant du Québec, dans la région de Montréal, est responsable de la localisation du magmatisme alcalin ancien, de la sédimentation clastique de l'Hadrynien-Paléozoïque inférieur et de l'orientation nord-ouest de la séquence plateforme-talus continental de la partie nord du Vermont. Ce rentrant résultant de la combinaison ride-faille transformante a influencé, plus tard, à l'Ordovicien, les effects de la collision tectonique. La distribution des lambeaux de la croûte océanique, des arcs volcaniques et des roches du socle adjacent supportent une telle interprétation. En dépit des opinions antérieurement exprimées voulant que les serpentinites du Vermont soient des roches intrusives ultramafiques injectées à travers un socle sialique, les évidences de terrain et les analyses chimiques supportent l'hypothèse que les roches ultramafiques et les roches connexes du nord du Vermont sont des lambeaux de croûte océanique obductée présentant les mêmes caractéristiques que celles des ophiolites du Québec et de Terre-Neuve. Bien qu'une stratigraphie tectonique puisse être suivie le long de la zone de serpentine de part et d'autre de la frontière Québec-Vermont, on remarque, vers le sud un plus grand nombre d'imbrications de lambeaux de croûte océanique avec des séquences de roches clastiques anciennes et aussi plus de flyschs ophiolitiques et volcanogéniques contemporains et postérieurs à l'emplacement des lambeaux de croûte océanique. Ce contraste, d'un côté à l'autre de la frontière internationale, est attribué à une compression suivie d'un décrochement sénestre le long d'une faille transformante résultant d'une collision diachrone entre des arcs volcaniques et la marge continentale irrégulière, au voisinage du rentrant du Québec.

INTRODUCTION

The northern Appalachians are now considered to be "two-sided" following the outline of Williams (1964) for Newfoundland, and the suggestion of Wilson (1966) that the Appalachian Orogen originated through opening and closing of a late Precambrian-Early Paleozoic Iapetus Ocean (Dewey, 1969; Bird and Dewey, 1970). Accordingly, the geologic framework of the Quebec-New England Appalachians is subdivided into the ancient eastern and western margins bordering the Iapetus Ocean (e.g., Williams, 1978).

The western margin of the Appalachian Orogen (i.e., the eastern margin of ancient North America) is well defined and built upon a "Grenville" age basement. The western margin consists of a shelf platform sequence overlain by Middle to Late Ordovician easterly derived flysch, continental slope deposits, and an Eo-Cambrian to Cambrian westerly derived clastic rise prism with associated rift volcanics formed during the initial opening of Iapetus (Bird and Dewey, 1970; St-Julien and Hubert, 1975; Williams et al., 1972, 1974; Fig. 1). The eastern margin is not as well defined but an Andean type margin has been suggested represented by a reworked Late Pre-cambrian to Ordovician (?) gneissic basement overlain by arc-related Ordovician volcanics (e.g., the Bronson Hill Anticlinorium in the New England Appalachians; Fig. 1; Williams and Doolan, 1978; Osberg, 1978; Hall and Robinson, this volume). Sutured between these contrasting basements are remnants of oceanic crust of ancient Iapetus (serpentine belt of Vermont and Eastern Townships).

The contrasting age of basement rocks of the presently juxtaposed ancient margins necessitates that plate motion during opening and/or closing of Iapetus was not perpendicular to the rifted margins and the "fit" of one margin against the other during collision was entirely random. Collision is thus expected to be most extreme where promontories of adjacent sialic basement collide and least where collision of adjacent re-entrants of continental margins occur (Williams and Doolan, 1978).

The purpose of this paper is to discuss the ancient rifted margin of North America marking the southwestern end of the ancient Quebec re-entrant (Thomas,

1977; axis of New England Salient of Cady, 1969). The presence of this ancient northwest trending plate margin strongly affected the Paleozoic collisional tectonics as evidenced by the present day reconstruction of the ultramafic belts of the Eastern Townships and Vermont which crosses this region. The origin of this ultramafic belt is discussed in light of the pre- and post-collision configuration of the Quebec re-entrant.

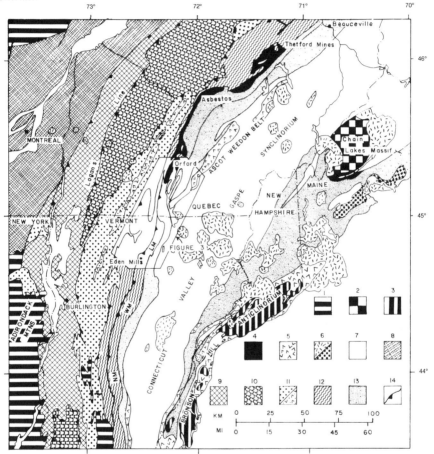

Figure 1. Geologic Map of the Quebec re-entrant showing major tectono-stratigraphic divisions and ophiolite localities: 1) "Grenville" basement; 2) 1.5 Ga old basement of Chain Lakes Massif; 3) Ordovician (?) to Late Precambrian basement of Bronson Hill Anticlinorium; 4) Ophiolite occurrences; 5) Island arc or "Andean" arc volcanism; 6) Felsic (dash) and mafic (squares) intrusive rocks undifferentiated in age; 7) Siluro-Devonian metasediments in Connecticut Valley-Gaspé Synclinorium; 8) St. Lawrence lowlands; 9) Cambro-Ordovician shelf and overlying flysch; 10) Taconic and Quebec Allochthons; 11) Early Paleozoic clastics (dots) and rift volcanics (v's); 12) Zone of margin/ophiolite/arc flysch; 13) Ordovician syn - to post orogenic flysch 14) Thrust fault; LM = Lowell Mtns.; WM = Worchester Mtns.; NM = Northfield Mtns.; (modified after Williams, 1978).

THE ANCIENT MARGIN OF EASTERN NORTH AMERICA – PRE-COLLISION CONFIGURATION OF THE QUEBEC RE-ENTRANT

The Quebec re-entrant originated in the early Paleozoic as a product of uplift and rifting of sialic basement. Although evidence for such early features in mountain belts are rarely preserved because of later collisional events (Burke and Dewey, 1973), such evidence for the Quebec reentrant has been provided by previous workers. Kumarapeli (1976, 1978) and Kumarapeli and Saull (1966), for example, state that the Mesozoic St. Lawrence rift system is largely a reactivated rift system which parallels the ancient ("eo-Appalachian") rift system of early Paleozoic Age. The age of an alkaline magmatic event of 565 Ma in the St. Lawrence rift system (Doig, 1970); Hadrynian (Hoffman, 1972) to Cambrian age arenites filling early graben structures (Kumarapeli, 1976); chemical data on early Paleozoic age Tibbit Hill volcanic rocks in southern Quebec suggesting extensional volcanism (Rankin, 1976) all support an early rift episode in the Quebec re-entrant.

Although Kumarapeli (1978) suggests that several hot spot uplifts were probably active in the Quebec re-entrant during rifting of the ancient margins (e.g., Montreal, Saguenay, Gulf of St. Lawrence), the evidence appears to be best documented in the Montreal region. The exact orientation of the tripartite rift in this region is conjectural; however, the Ottawa Graben (Kay, 1942) is generally agreed to represent the failed arm (aulacogen) of such a triple junction (Burke and Dewey, 1973; Kumarapeli, 1976, 1978; Rankin, 1976). The orientation of the two successfully rifted arms is less clear. Cady's (1969) northwest trending axis of the New England salient is the most likely orientation of the southern arm of the triple junction. The northwest trend of the Cambrian shelf edge in northern Vermont (Rodgers, 1968) also supports a similarly oriented rift margin in the early Paleozoic. This ancient carbonate bank edge was likely associated with the development of the Ottawa Graben (Rankin, 1976).

Direct evidence for the orientation of the northerly trending arm is lacking because of complete cover of the carbonate platform by the Quebec allocthons; however, the distribution of 565 Ma alkaline magmatic activity (Doig, 1970; Kumarapeli and Saull, 1966) and the present western limit of Appalachian fold belt supports the view of a rifted arm sub-parallel to the St. Lawrence River as suggested by Kumarapeli (1976, 1978).

A schematic configuration of the early rift stage of the Quebec re-entrant is shown in Figure 2 which incorporates the available information discussed above. While the plate geometry of Figure 2 is simplistic, the actual configuration was probably far more complex involving ridge jumps and changes in rotation poles. Unsuccessfully rifted blocks of Grenville basement sandwiching thick flows of rift volcanics would be expected near the triple junction as a result of such processes and may indeed explain strong geophysical anomalies in the Sutton Mountains (e.g. Kane et al., 1972). The available data does suggest, however, that the major "bend" in the Quebec re-entrant centres around the Montreal triple junction (Burke and Dewey, 1973) and that a stable block of Grenville basement extended south-eastward from Montreal from early Paleozoic time.

POST-COLLISION CONFIGURATION OF THE QUEBEC RE-ENTRANT

From the observations that adjacent margins have contrasting basements, it is apparent that the Wilson cycle for Iapetus was not a simple "open and shut case" (Dewey and Burke, 1974). The relative plate motions involved during the opening and closing of Iapetus will probably never be determined with any accuracy since the evidence for such determinations lies in the ocean basin itself which for the most part is completely destroyed. An appreciation for the complexity of plate interaction during ocean closure can be gained, however, by evaluation of the present arrangement of Paleozoic plate boundaries in major re-entrants such as Quebec where collision was most likely incomplete (Dewey, 1975; Dewey and Burke, 1974).

Paleozoic plate boundaries are approximated by considering the present locations of continental basement, arcs, and zones of imbricated ophiolites. These features are shown for the Quebec re-entrant in Figure 1. Two volcanic arc terrains are present: the Ascot-Weedon belt coring the Stoke Mountain anticlinorium (St-Julien and Hubert, 1975; Laurent, 1977; Lamarche, 1972; De Romer, 1979) and the Ammonoosuc volcanics and related rocks of the Bronson Hill anticlinorium. The basement of the former terrain is not known while the latter terrain is cored by Upper

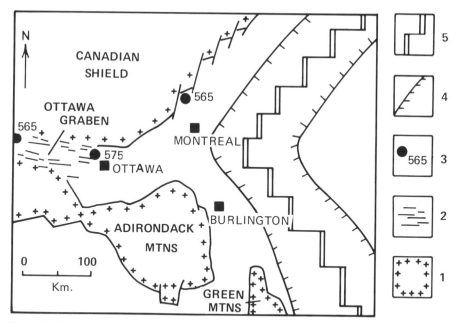

Figure 2. Late Precambrian-Early Cambrian reconstruction of the Montreal triple junction. 1) Approximate positions of Grenville Basement; 2) Probable Hadrynian (600-700 Ma) dike swarms; 3) Dated Early Cambrian alkaline carbonatite complexes and dike intrusion (number in Ma, Doig, 1970); 4) Approximate position of nascent carbonate bank edge (Rodgers, 1968); 5) Ridge transform system represents two successfully rifted arms of Montreal triple junction; Ottawa Graben (Kay, 1942) represents failed arm of Montreal triple junction (after Kumarapelli, 1978; Burke and Dewey, 1974).

Precambrian basement rocks and Ordovician gneisses considerably younger than Grenville age basement of the western margin (Naylor, 1968; Hall and Robinson, this volume). The Chain Lakes massif (1500 ± 20 m.y.; Naylor et al., 1973) is a third basement type which is separated from the western "Grenville" basement by a continuation of the Eastern Townships serpentine belt and from the eastern "Bronson Hill" basement by the Chain Lake ophiolite (Boudette and Boone, 1976; Boudette, this volume). The Chain Lakes massif is unconformably overlain on its northwestern side by Siluro-Devonian metasediments of the Connecticut Valley-Gaspé synclinorium. The northern extent of this basement is thus not known but a gravity low (Kane et al., 1972) between the serpentine belt northeast of Thetford Mines, Quebec, and the Chain Lakes ophiolite to the southeast suggest that the Chain Lakes "block" is at least 75 km wide and probably does not extend to the southwest as suggested by Osberg (1978). The occurrence of rocks similar to Chain Lakes basement rocks as a 10 km sliver in the St-Daniel formation in the vicinity of Beauceville, Quebec, supports this contention (St-Julien, 1980; Fig. 1).

A striking feature of the Quebec re-entrant is the occurrence of well preserved fragments of Iapetus (ophiolites) in the Eastern Townships of Quebec (St-Julien and Hubert, 1975; Laurent, 1973, 1975, 1978). These occurrences are similar in geologic setting and origin to ophiolitic remnants in the Baie Verte Peninsula, Newfoundland, and the entire belt is considered to be the westernmost root zone marking the closure of Iapetus Ocean against the ancient western margin (Baie Verte-Brompton Line, St-Julien et al., 1976; Williams and St-Julien, 1978; Williams and St-Julien, this volume).

The best preserved ophiolites in the Eastern Townships occur between Thetford Mines and Chagnon Mountain (Figs. 1, 3). Northeast of Thetford mines ultramafic bodies occur as highly serpentinized fragments with no overlying cumulate or high level intrusive or extrusive sequences collectively correlative with the "upper unit" of the Eastern Townships ophiolite belt (Laurent, 1978). Southwest of Chagnon Mountain and continuing into northern Vermont, the serpentine belt is again devoid of continuous sequences of cumulate and other upper level members characteristic of the Thetford-Chagnon belt. Indeed, these differences in setting of the ultramafic rocks from north to south have led previous workers in northern Vermont to suggest that the ultramafics originated as intrusive bodies into sialic basement rather than as oceanic remnants emplaced on a foundering continental margin (Cady et al., 1963; Chidester and Cady, 1972; Chidester et al., 1978).

In the following section, we summarize the tectonic setting and occurrences of ultramafic and related rocks along strike of an approximately 60 km segment of the ultramafic belt which straddles the International Border (Fig. 3).

TECTONO-STRATIGRAPHIC SUBDIVISION OF THE ULTRAMAFIC BELT IN THE VICINITY OF THE INTERNATIONAL BORDER

Introduction

The ultramafic bearing and related rock units in the vicinity of the International Border mark a zone of major discontinuity observable throughout the Appalachian

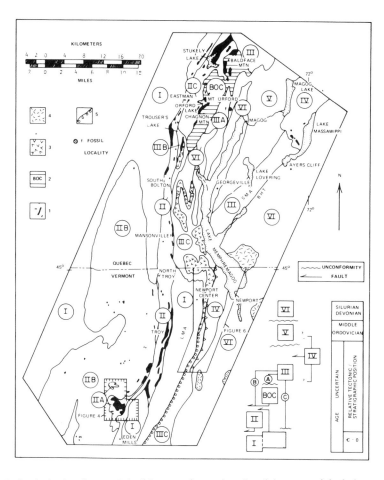

Figure 3. Geologic sketch map of the lithotectonic stratigraphy of the serpentinite belt across the International Border. Unit I: polydeformed and metamorphosed clastic rise sequence; Unit II: ultramafic-bearing schists, phyllites and metagreywackes undifferentiated; Unit IIA: Serpentinite, mafic schists and amphibolites; Unit IIB: Serpentinite-bearing black phyllite/schist structurally underlying Unit IIA; Unit IIC: metagreywacke-slate-phyllite imbricated with serpentinite; Unit III: Ordovician syn- to post-orogenic olistostromal black to grey slate unit undifferentiated; Unit IIIA, Unit IIIB, Unit IIIC explained in text; Unit IV: rocks of the Ascot-Weedon Island arc complex and correlatives; Unit V: Middle to Late Ordovician post-orogenic flysch (Magog Group); Unit VI: Silurian-Devonian metasedimentary rocks and intrusions. 1) Serpentinized ultramafic rocks in Units II, III and IV; 2) BOC = Baldface-Orford-Chagnon Ophiolitic complex; 3) Massive and pillowed metavolcanics imbricated (Unit IIIB) and interbedded (Unit IIIC) with Unit III metasedimentary rocks; 4) New Hampshire Plutonic Series intrusive rocks (Devonian); 5) Umbrella Hill Conglomerate. Abbreviations: B.H.T. = Bunker Hill Thrust; S.M.A. = Stoke Mountain anticlinorium; L.M.A. = Lowell Mountain anticlinorium. (After Cady et al., 1963; Doll et al., 1961; St-Julien, 1965; Cooke, 1950; Lamothe, 1979; DeRomer, 1979; and unpublished maps).

Orogen separating the clastic rise prism of ancient North America to the west from transported oceanic crust, island arcs and associated cover rocks to the east (Humber and Dunnage zones respectively of Williams, 1978; Baie-Verte-Brompton Line of St-Julien et al., 1976; Williams and St-Julien, this volume). According to these and other plate tectonic interpretations of Appalachian geology (e.g., Bird and Dewey, 1970; St-Julien and Hubert, 1975; Church, 1972; and Osberg, 1978) the ultramafic belt of the Eastern Townships of Quebec and Vermont are considered as major junctions of litho-tectonic units marking plate collisions in the Ordovician. Differences in the various models which are not summarized here deal with polarity of subduction, the position of the "root zone" of the ultramafic and related rocks, nature of the basement underlying the rocks on opposite sides of the ultramafic belt, and age relationships of the litho-tectonic units involved. With the exception of Osberg (1978), no attempt has been made to explain clear differences between tectonic and depositional history recorded in rocks with span the Quebec-Vermont border. In this section, we summarize some of our work along and east of the ultramafic belt in Vermont (Hollis-Gale, Gale and Hoar) and work presently in progress by Doolan and students at the University of Vermont in the Eastern Townships. We have drawn heavily from earlier compilations on both sides of the border (Cady et al., 1963; Cooke, 1950; St-Julien, 1965) and recent work by others some of which is ongoing (Rolfe Stanley, University of Vermont and students; Lamothe, 1979; DeRomer, 1976, 1979). Although this work is preliminary, results to date provide interesting constraints as to the timing of events and the correlation of lithotectonic units across the International Border. Units described below are depicted in a generalized geologic map of the Vermont-Eastern Townships Ophiolite Belt (Fig. 3).

Unit I: Polydeformed Metasediments of the
Ancient Continental Rise Prism

This unit underlies a major part of the Green Mountain-Sutton Mountain Anticlinorium in the vicinity of the International Border. These westerly derived metasediments are cored by the Pinnacle Formation, a rift-facies greywacke-conglomerate sequence and Tibbit Hill volcanics which on chemical and field criteria are related to extensional volcanism predating and possibly synchronous to formation of Iapetus (Pierratti, 1976; Rankin, 1976; Fig. 1). Within the confines of Figure 3, Unit I is comprised of Lower Cambrian (?) Underhill Formation in Vermont (Doll et al., 1961) and the Sutton-Bennet Schists in Québec (Clark, 1934; Osberg, 1965). The Underhill-Sutton-Bennett Schists are comprised of quartzo-feldspathic silvery-grey mica schist and phyllite, metagreywacke, greenstone and carbonaceous phyllite. Metamorphic grade lies entirely within the biotite zone. Lithically similar rocks included with Unit I in Figure 3 are the Jay Peak Formation (Cady et al., 1963) and parts of the Stowe Formation (Doll et al., 1961).

Earlier workers in Vermont consider the Stowe Formation to be considerably younger than Underhill Formation equivalents (Cady et al., 1963; Cady 1945; Doll et al., 1961) for two reasons: 1) the eastern flank of the Green Mountain anticlinorium is considered a stratigraphic sequence younging to the east; 2) the Stowe Formation is presumed to underlie with only local unconformity the Moretown Formation of presumed Middle Ordovician age (Cady et al., 1963; Cady, 1956). Since both of these tenets are now questioned for reasons discussed more fully below, we here correlate that part of the Stowe Formation coring the Lowell Mountain anticlinorium with lithically similar rocks of the Underhill-Jay Peak Formations lying west of the ultramafic-bearing formations on Figure 3.

All rocks of this unit as shown on Figure 3 are devoid of ultramafic pods, slivers or intrusions although all are believed to have been deformed simultaneously with ultramafic-bearing rocks structurally overlying them.

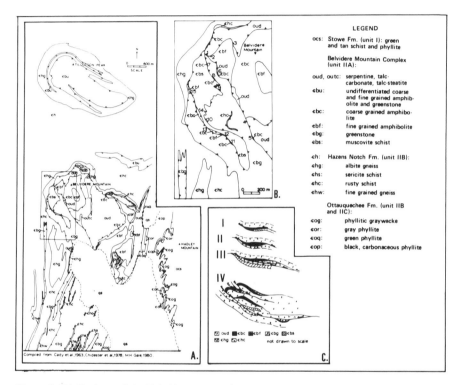

Figure 4. (A) Geology of the Belvidere Mountain area, Eden Mills, Vermont. Unit designations follow that of Cady *et al.* (1963); Chidester *et al.* (1978) and M.H. Gale (1980). Correlations of these units with units I and II of Figure 3 shown in legend.
(B) Geology of the west side of Belvidere Mountain showing details of fault contacts. Locations 1-14 discussed in text.
(C) Schematic model showing development (I to IV) of tectonic stratigraphy through series of imbricated thrust faults progressively underplated on serpentinite "sole" during westward transport of ophiolite sheet.

TABLE I

REPRESENTATIVE AND AVERAGE MAJOR AND TRACE ELEMENT ANALYSES
OF MAFIC ROCKS OF BELVIDERE MOUNTAIN, VERMONT,
AND THETFORD MINES, QUEBEC

	Foliated Greenstone		Amphibolite			Thetford
Wt %		Ave.	Fine-grain	Coarse-grain	Ave.	
ppm	MH298	1.	MH257	MH250	2.	A
SiO_2	46.62	46.83	44.90	49.28	46.35	42.52
TiO_2	1.03	1.25	0.91	0.78	1.00	1.12
Al_2O_3	14.17	13.87	15.80	13.20	17.09	16.33
FeO	4.47	5.74	5.45	5.22	9.46	2.65
Fe_2O_3	6.53	6.28	6.17	8.06	5.03	8.35
MnO	0.18	0.20	0.20	0.19	0.21	0.19
MgO	8.39	7.09	6.73	7.59	8.67	11.15
CaO	11.12	9.14	13.60	8.90	10.26	13.22
Na_2O	2.13	2.50	1.79	1.64	2.09	1.05
K_2O	0.03	0.19	0.34	0.20	0.25	0.19
P_2O_5	0.09	0.11	0.13	0.04	0.04	0.11
L.O.I.	3.78	5.79	2.99	3.60	3.98	2.52
Total	98.54	100.02	99.01	98.70	99.16	99.47
Nb	9	10	4	9	7	n.d.
Zr	65	89	62	57	54	n.d.
Y	29	34	28	31	30	n.d.
Sr	91	132	135	60	78	n.d.
Rb	—	9	6	3	5	n.d.
Pb	—	4	—	—	1	n.d.
Ga	20	19	19	18	20	n.d.
Zn	81	128	94	86	104	n.d.
Ni	74	93	128	73	94	208
Cu	46	78	261	68	121	n.d.
Ba	1	36	49	24	41	n.d.
V	300	345	290	335	322	n.d.
Cr	140	174	151	91	100	563
Ti	6175	6044	5455	4676	5130	

Notes. 1) is average of 34 analyses; 2) is average of 7 analyses of both fine and coarse-grained amphibolites; A is from garnet-bearing amphibolites reported by Laurent (1977); L.O.I. = loss on ignition except for A which includes H_2O^+ and H_2O^-. All analyses except A completed at Memorial University, Newfoundland; Gertrude Andrews and B. L. Doolan, analysts.

Unit II. Ultramafic-Bearing Rocks

A variety of ultramafic rocks including partially serpentinized harzburgites and dunites, fully serpentinized pods and slivers of varying sizes, talc schists and talc carbonates occur in distinct tectonostratigraphic positions in the Vermont-Québec ultramafic belt shown in Figure 3. Ultramafic-bearing rocks are subdivided into four subunits according to the nature of their host rocks and rock associations. The

TABLE II

MODAL ANALYSES[1] OF REPRESENTATIVE MAFIC ROCKS
AT BELVIDERE MOUNTAIN, VERMONT

	Foliated Greenstone	Fine-grained Amphibolite	Coarse-grained Garnet Amphibolite
	MH 298	MH 257	MH 250
Amphibole	39.9	44.2	57.4
Albite	15.5	3.3	5.6
Epid. Gp.	25.0	36.4	20.0
Chlorite	12.8	8.5	5.6
Calcite	0.1	—	—
Quartz	1.1	1.3	6.6
Biotite	0.1	1.1	1.6
Sericite	—	—	—
Garnet	—	—	2.4
Sphene	5.4	3.1	0.7
Opaque	0.1	2.1	0.1

[1]Determined on 1000 point counts on single thin sections.

division into these subunits is at this time crude and work is in progress to fully differentiate their areal extent, relative contact relationships and structural positions.

Unit IIA. Serpentinite-Foliated Aureole Complex. The basis for establishing a tectonic stratigraphy in the ultramafic-bearing rocks lies in the recognition of a basal aureole complex of dynamically metamorphosed greenstones, amphibolites and garnet amphibolites adjacent to serpentinite slabs in the Belvidere Mountain area (Figs. 3 and 4A). The dynamically metamorphosed rocks and associated imbricated serpentinite bodies are collectively referred to here as the Belvidere Mountain Complex (BMC). The BMC has an estimated maximum thickness of 500 m and consists of five units exhibiting fault contact relationships with each other. These units, from base to top are: a medium grained, tourmaline bearing, muscovite schist (εbs) which contains rounded and elongate fragments (1-20 cm in diameter) of greenstone and amphibolite; green to grey-green fine grained schistose greenstone (εbg); fine to medium grained, blue-grey to dark grey amphibolite and garnetiferous amphibolite (εbf); coarse grained amphibolite and garnetiferous amphibolite (εbc); serpentinite, and talc-carbonate, talc steatite, and quartz carbonate rock (oud, outc). Representative modes of these units and representative major and trace element analyses of the mafic rocks are listed in Tables 1 and 2 respectively. The chemical data supports the view that all mafic rocks have a common protolith which, according to the criteria of Pearce and Cann (1973), are oceanic tholeiites (Fig. 5).

Fault surfaces in the Belvidere Mountain area (Fig. 4B) are delineated on the basis of truncation of units along a common surface, truncation of structures, and physical features such as slickensides, fault slivers, and highly contorted talc-rich shear zones. A tight to isoclinal fold event (F_1) deforms the faults and together these structures constitute the earliest recognized deformation (D_1). Three subsequent fold events (F_2, F_3 and F_4) are superposed on the D_1 structures.

The fault contacts between subunits of the Belvidere Mountain Complex are discussed briefly below with reference to the locations shown on Figure 4B.

Asbestos-bearing serpentinite truncates the muscovite schist-greenstone contact (loc. 1), the fine-grained amphibolite-greenstone contact (loc. 1), and the fine-grained amphibolite-coarse-grained amphibolite contact (loc. 3). In addition round pods of extremely coarse grained chlorite-actinolite rock are found within the fine grained amphibolite at the serpentine-fine-grained amphibolite contact (loc. 2). The serpentinite-greenstone contact is marked by discontinuous occurrences of talc-rich shear zones (loc. 4) and fault slivers of coarse-grained amphibolite (loc. 5).

The contact between the fine-grained amphibolite and the greenstone is marked by discontinuous occurrences of schistose serpentinite, a 1 m thick fault sliver of coarse-grained amphibolite, actinolite-chlorite rock pods within the amphibolite, and a 2 to 15 mm "sheet" of talc carbonate rock (loc. 6).

The contact between the muscovite schist and greenstone is a folded (F_1) fault marked by numerous fault slivers of coarse-grained amphibolite (loc. 7, 8, 11, 13 and 14) and fine-grained amphibolite (loc. 12 and 15). In addition isolated occurrences of talc-carbonate rock and talc-rich shear zones are found (loc. 8, 9 and 10).

It is noteworthy that all fault contacts can be defined by truncation of units and by the presence of fault slivers of the upper units along successively lower fault surfaces.

The contact between BMC and the structurally underlying Hazens Notch Formation (Cady et al., 1963; Unit IIB, Fig. 4A) is also considered to be a fault contact since it truncates the serpentinite-muscovite schist contact and the folded (F_1) contact between the greenstone and muscovite schist units. In addition to truncation of units along this contact, isolated bodies of talc-carbonate, talc-steatite and quartz carbonate rock decorate the fault surface south of the confines of Figure 4 (M.H. Gale, 1980). Since the contact between BMC and the Hazens Notch Formation postdates both the fault construction of BMC and the early F_1 folding of these rocks, the Hazens Notch rocks are considered as part of a distinct tectono-stratigraphic subdivision (Unit IIB).

Detailed metamorphic petrography of BMC and the relationship between metamorphic events and geologic structures are reported elsewhere (M. H. Gale, 1980), however, it should be noted here that the variation in metamorphic grade (epidote-amphibolite and greenschist facies) is confined to the BMC and follows the fault contacts of the amphibolites and greenstones. The textural and mineralogical gradation from coarse-grained amphibolite to fine-grained amphibolite to greenstone reflects an early metamorphic history of cataclasis and retrograde metamorphism synchronous with D_1, but prior to both the incorporation of these rocks in the muscovite schist and the final emplacement of the entire BMC as a mélange with metasedimentary rocks of the underlying Unit IIB (Fig. 4C). Locally, glaucophane and omphacite are found in correlatives of the BMC on Tillotston Peak (Fig. 4A), suggesting that the early metamorphic event is characterized in part by high pressure and low temperature conditions (Laird and Albee, 1979).

A schematic reconstruction of the events which characterize the early deformation (D_1) and which is consistent with the field relationships is outlined in Figure 4C.

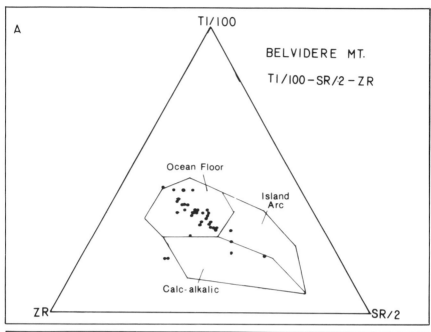

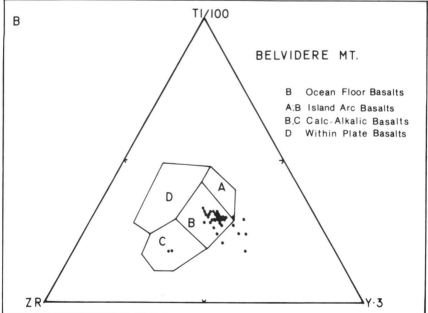

Figure 5. A) Ti/100-Sr/2-Zr and, B) Zr-Y·3 Ti/100 plots for mafic schists of the Belvidere Mountain area. Field boundaries after Pearce and Cann (1973).

Rocks possibly correlative with the BMC and reflecting similar highly trans-posed structures are found juxtaposed beneath the serpentinite-metapyroxenite-metagabbro-diabase-metavolcanic complex at Chagnon Mountain, Quebec (P. Winner, work in progress; Fig. 3). The overlying igneous complex similar to that exposed at Orford and Baldface Mountains to the north are referred to as the Baldface-Orford-Chagnon (BOC) ophiolite complex on Figure 3. The deformed mafic rocks juxtaposed beneath the serpentinite "sole" of this complex at Chagnon are highly transposed metagabbro bodies in structural contact with a metagreywacke-slate unit to the west (Unit IIC). Rocks of similar appearance to these transposed metagabbro bodies are found within the BMC correlatives near Warner Hill just south of North Troy, Vermont (Fig. 3; R.S. Stanley, Vermont Geological Society Field Trip, October, 1979), and west of a large serpentine body along the Sutton-Bennet Schist-Ottauquechee contact just northwest of Trouser's Lake, Quebec (Fig. 3; L. Martin, Work in Progress). Other occurrences of these dynamically metamorphosed mafic rocks may occur along numerous suspected faults in the region designated "undif-ferentiated Unit II" on Figure 3 which must await results from mapping still in progress.

Regionally, garnet amphibolites and greenstones are found juxtaposed beneath serpentinite soles of ophiolite occurrences to the north in Quebec (Thetford Mines and Asbestos; Laurent, 1978) and Newfoundland (Williams and Smyth, 1973; Jamieson, 1981).

Unit IIB: Serpentinite-Carbonaceous Schist/Phyllite-Quartzite-Albite Gneiss-Rusty Schist Mélange. Polydeformed metasedimentary rocks imbricated with numerous serpentinite and talc-steatite pods and slivers are structurally imbricated with Unit IIA in the vicinity of Belvidere Mountain (Fig. 4; Hazen's Notch Forma-tion and, in part, Ottauquechee Formation of Doll *et al.*, 1961). Although this unit is not differentiated from other ultramafic-bearing units on Figure 3 outside of the Belvidere Mountain area, correlative rocks in Québec include the Sweetsburg For-mation of Osberg (1968) and lower Mansonville Formation of Cooke (1950). This unit is in presumed fault contact with non-ultramafic bearing rocks of Underhill-Sutton-Bennet schists to the west (Unit I) and display complex imbricate contacts with other ultramafic rock types to the east. Where differentiated Unit IIB possesses many features of and occupies a similar structural position to mélange found beneath obducted ophiolites (Gansser, 1974). An ultramafic-bearing metagreywacke-phyllite/slate unit (Unit IIC) is in contact with Unit IIB along the eastern outcrop limit of the ultramafic belt shown in Figure 3. Where these units are not yet fully differentiated from other ultramafic bearing units they are shown as Unit II (undif-ferentiated) on Figure 3.

Unit IIC: Metagreywacke-Black, Grey, Green and Purple Slate Unit. Ultramafic bearing metagreywacke-slate-phyllite rocks outcrop along the eastern outcrop limits of ultramafic-bearing units shown on Figure 3. Although these rocks are only par-tially differentiated on Figure 3 from other Unit II rocks in Vermont and Québec, R.S. Stanley (University of Vermont, pers. communc., 1980) reports numerous metagreywacke occurrences between North Troy and Troy, Vermont (Fig. 3), and it may prove to be a consistent horizon to the Eden Mills area. Extensive exposures of similar but less deformed metagreywacke are found west of the BOC ophiolite complex in the Eastman-Orford area of Quebec (Fig. 3). Ultramafic pods are rare within this latter area; however, serpentinites are found on both the eastern contact with Unit IIA and BOC and western contact with Unit IIB and Unit II (undifferentiated; Fig. 3; G. Smith, University of Vermont, work in progress).

Well bedded metagreywacke west of Chagnon Mountain in Québec contains a large proportion of angular plagioclase and blue and milky quartz fragments. Plagioclase-quartz ratios are highly variable and K-feldspar is rare. Source rocks for these rocks are not known; however, their lithic similarity to metagreywacke of Underhill and Pinnacle Formation rocks in Vermont suggests that the original source area could ultimately be traced to the westerly derived clastic rise prism. The close approximation with ultramafic rocks and the structural position just below the Orford-Chagnon complex in Quebec suggests that these metagreywackes may be reworked rise prism clastics transported westward just prior to emplacement of ultramafic complexes. The lithic similarity of the Quebec greywackes to the analogous Blow-me-Down Formation in the Humber Arm Allochthon of western Newfoundland supports such an interpretation (W. S. F. Kidd, pres. commun., 1979).

Rocks delineated as Unit IIC in Québec (Fig. 3) correlate in part with the Brompton Formation of St-Julien (1965), the Miller Brook Formation of De Roiner (1963), and the upper part of the Caldwell Group of Cooke (1950). Other known occurrences in Vermont were mapped in part as Ottauquechee and in part as westernmost Stowe Formation by Cady et al. (1963).

Massive to well-bedded metagreywacke also occurs along the Bunker Hill antiform in the Stoke Mountain belt on the east side of Lake Memphremagog where it structurally underlies St-Daniel olistostromal black slate (Unit III, Fig. 3; De Romer, 1979). Thin beds of feldspathic metasandstone are also found within Unit III throughout the confines of Figure 3. Age relations and provenance of all these metagreywacke occurrences are at present poorly understood.

Other Ultramafic-Bearing Rock Units. Ultramafic rocks other than those described in Unit II are found in the region shown in Figure 3 in two additional tectonic positions: ultramafic slivers and "blocks (?)" in the St-Daniel Formation (Unit III) and ultramafic pods and slivers in the Stoke Mountain Belt (Unit IV). The latter occurrences are small in size and extent and were not investigated by us. The former occurrences are discussed more fully below.

Unit III: The St-Daniel Formation

Laminated and chaotically deformed black and grey slate with olistostromal horizons occur east and south of stratiform ophiolite occurrences throughout the Eastern Townships (St-Daniel Formation of St-Julien, 1970; St-Julien and Hubert, 1975; Laurent, 1975, 1978; Lamarche, 1972) and appear to be a consistent horizon along the entire length of the "Baie Verte-Brompton Line" in the Canadian Appalachians (St-Julien et al., 1976; Williams and St-Julien, this volume). Age and origin of this unit is uncertain; however, it is unconformably overlain by fossiliferous Middle Ordovician Magog Group (Unit V, Fig. 3) west of Magog, Quebec. The brecciated nature of these rocks is derived both from an unstable depositional environment disrupting intraformational argillites, shale and greywacke and from tectonic imbrication with dismembered ophiolites. Correlatives of the St-Daniel Formation in the vicinity of the International Border are shown in Figure 3, and is subdivided into three lithotectonic subunits discussed more fully below: Unit IIIA; chaotic grey and black slate; Unit IIIB; black and grey slate imbricated with ophiolitic and exotic igneous rocks; Unit IIIC; black and grey slate, metatuffs, pyroclastics and metavolcanics.

Unit IIIA. Chaotic Grey and Black Slate with Olistostromal Horizons. This subunit of the St-Daniel appears to be the most widespread throughout the Eastern Townships and lithically identical examples can be found from the type locality southeast of Thetford Mines (St-Julien, 1970) to the Orford Chagnon region: small to very large fragments of greywacke, argillite and shale, and impure quartzite are commonly found in olistostromal horizons. Locally, as in the Thetford Mines region, large blocks of serpentinite presumably derived from the ophiolite are also found. Within the region of Figure 3, however, ophiolite detritus are not recorded in this unit, and most of the clasts can be traced to intraformational slates and dark grey feldspathic quartzites occurring locally as beds ranging from several centimeters to approximately 0.5 m in width.

Unit IIIB: Serpentinite-Black Slate Tectonic Melange. Along the western contact of Unit II and Unit III in the Trouser's Lake-South Bolton area of Quebec (Fig. 3), black slate of Unit IIIA (as described above) is tectonically imbricated with serpentinite, metagabbro-metadiabase, greenstone and lithic rock types similar to those found in Units IIA and IIC (Lamothe, 1979; and unpublished maps). In addition, a wide variety of exotic igneous rocks including trondjhemite, monzonite and amygaloidal volcanics, trachytes, and agglomerates are found. The intrusive rocks themselves are locally highly brecciated by highly charged magma injection. Fragments within the brecciated quartz rich intrusives contain angular foliated metagabbro and hornblende syenite, a variety of plagioclase-rich volcanics and rounded coarse-grained trondjhemites and monzonites. No intrusive contacts with metasedimentary country rocks are observed; rather the brecciated bodies appear to be emplaced along and within sheared serpentinite bodies. Excellent examples of these intrusive rocks can be seen on the road cut east of Trouser's Lake and in numerous high standing knobs throughout this melange unit southeast of Trouser's Lake (Fig. 3). Some of the tectonically imbricated blocks within the mélange can be traced to similar rocks in the Orford igneous complex to the northeast (metagabbro, greenstone, serpentinite slivers); other blocks, especially some of the blocks within the intrusive breccias, however, are not known to occur in the vicinity and sources are thus difficult to ascertain.

Unit IIIC: Black and Grey Slate, Volcanogenic Metasedimentary Rocks and Metavolcanic Flows. Southeast of South Bolton, Quebec, clastic olistostromal black and grey slate are interbedded with massive to pillowed metabasalt (Fig. 3). This unit continues without a break across the International Border where it joins the black slate metavolcanic sequence at Bear Mountain and Coburn Hill (Cram Hill lithologies of Doll *et al.*, 1961). The metavolcanic rocks form spectacular exposures and prominent peaks rising above the low lying areas underlain by metasedimentary units. The origin and age of the volcanics (Bolton Lavas of Ambrose, 1945) has been a topic of much controversy by previous workers in the Eastern Townships (Ambrose, 1942, 1949, 1957; Cooke, 1950; Fortier, 1945). We interpret the volcanic rocks to be conformable and interbedded with the St-Daniel and Cram Hill rocks on the west side of Lake Memphremagog. De Romer (1979; pers. commun., 1979) reports occurrences of lithically and chemically similar pillowed metavolcanic rocks interbedded with St-Daniel "mélange" in the Fitch Bay region on the east side of Lake Memphremagog (Fig. 3).

The occurrence of interbedded metavolcanics with olistostromal black slate and phyllite appears to be a unique tectonostratigraphic subdivision for the St-Daniel

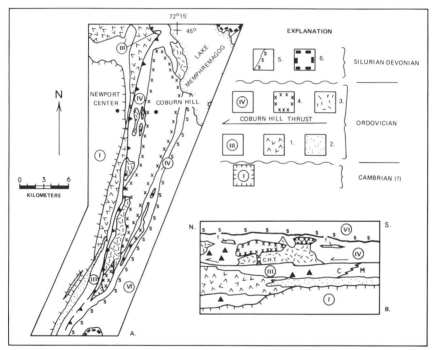

Figure 6. (A) Geologic sketch map of pre-Silurian rocks west of Lake Memphremagog, northern Vermont. I, III, IV represent metasedimentary units discussed in text and Figure 3. 1 = basic metavolcanic rocks (massive and pillowed); 2 = Umbrella Hill Conglomerate; 3 = Metagabbro bodies within unit IV; 4 = quartz diorite and related felsic intrusives rocks; 5 = Ordovician-Silurian(s) unconformity; 6 = Devonian felsic intrusives.
(B) Simplified N-S profile of geology in Fig. 6A (not to scale). Ornamentation same as that of Fig. 6A with the following additions: triangles = olistostromal horizons in the metasediment; f = felsites and silicic volcaniclastics; C and M refer to location of inferred facies transition from "Cram Hill" to "Moretown" in Unit III (Cady *et al.*, 1963). (Simplified from P.N. Gale, 1980; R. Hoar, in prep.)

unknown to us in other parts of the Eastern Townships. Further south in Vermont, this unit unconformably overlies the Stowe Formation of Unit I described earlier (Fig. 6). Locally, this contact is marked by the Umbrella Hill Conglomerate containing numerous white quartz clasts and angular phyllitic fragments within a phyllite matrix. Unlike Badger (1979), we interpret the source of quartz cobbles to be abundant white vein quartz found within the Stowe Formation implying an earlier history of metamorphism and uplift prior to deposition of Unit IIIC. The Umbrella Hill Conglomerate passes with gradational contact into black and grey slate of the Cram Hill Formation (Fig. 6). Further south, the conglomerate passes into more massive "pinstripe" feldspathic sandstones and argillite of the Moretown Formation (Cady et al., 1963; Doll et al., 1961).

The contact of Unit I-Unit IIIC is believed to be an unconformity where exposed along the Umbrella Hill Conglomerate. Further north, the contact is not exposed but volcanics and black and grey shale of Unit IIIC which stratigraphically underly the Umbrella Hill Conglomerate outcrop near the contact. Unit IIIC is thus interpreted as time transgressive from a basin opening to the north and west.

Unit IIIC is structurally overlain in northernmost Vermont by grey to rusty tan quartz rich phyllites, quartzites, chloritic quartzites, breccias and felsites intruded by both basic and felsic foliated igneous bodies (Unit IV discussed below). The Unit IIIC-Unit IV contact is characterized by lithic truncations of black phyllite in Unit III and is interpreted as a thrust fault (Coburn Hill thrust of P.N. Gale, 1980).

Unit IV: Ascot-Weedon Formation

A complex belt of metasedimentary, metavolcanic, hypabyssal, and granitic intrusive rocks outcrop in the northeast trending Stoke Mountain Anticlinorium east of Lake Memphremagog (Cooke, 1950; Fig. 3). Metasedimentary and metavolcanic rocks of Unit IIIC make up the bulk of the pre-Silurian section southwest of Lake Massawipi whereas north of Lake Massawipi mostly felsic and intermediate metavolcanic rocks, chloritic schists, cherts and metavolcaniclastic rocks of the Ascot Formation are found (De Romer, 1979; Fig. 3). These latter rocks correlate with the Weedon Formation east of Thetford Mines (Fig. 1; St-Julien and Hubert, 1975). These formations are interpreted as a Lower to Middle Ordovician island arc sequence based on the composition and rock associations of the extrusive rocks (St-Julien and Hubert, 1975; Laurent, 1975, 1978). The entire Stoke Mountain Belt is thrust eastward over Silurian age St. Francis Group (Unit VI, Fig. 3) along the Bunker Hill thrust.

We tentatively correlate the easternmost pre-Silurian section of northern Vermont with the Ascot-Weedon Formations (Figs. 3 and 6). These Vermont correlatives include tuffaceous felsite, quartzite and cherts, grey to rusty quartz-rich chloritic phyllites, breccias and grey slate intruded by foliated metadiabase-dioritetrondjhemite complex and a later foliated granite (Fig. 6). All intrusive rocks are considered pre-Silurian in age and are not found in the structurally underlying black and grey slate metavolcanic sequence of Unit IIIC. The contact between Unit IIIC and Unit IV in northern Vermont is a fault based on truncation of units along the contact and differences in igneous history (P.N. Gale, 1980; Hoar, in progress; Fig. 6). Trace element geochemistry of the foliated metadiabase-diorite bodies of Unit IV clearly differentiate these igneous rocks from adjacent metavolcanics of Unit IIIC (P. N. Gale, 1980).

Unit V: Magog Group (Beauceville Slates, St-Victor Flysch)

Graptolite-bearing black shales of the Magog Group unconformably overlie both the Orford-Chagnon complex and IIIA northwest of Magog, Quebec (St-Julien, 1965; St-Julien and Hubert, 1975; Laurent, 1978; Fig. 2). The Magog Group defines the axis of the St-Victor synclinorium and is interpreted to represent post-ophiolite flysch deposits (St-Julien and Hubert, 1975). Slates and siltstones overlying cherts and minor greywacke units near the base of the Magog Group at Castle Brook contain Middle Ordovician graptolites of the *Nemagraptus gracilis* and *Diplograptus Multidens* Zones (Riva, 1974; Berry, 1962; location F, Fig. 3) and represents the only pre-Silurian unit for which age is determined with certainty. Previous workers (Ambrose, 1942; Clark, 1934; Cooke, 1950) failed to separate the Magog slates from Unit III which led to the still widespread assumption that Unit III correlatives in Vermont (Cram Hill and Moretown Formations) are Middle Ordovician in age (e.g., Cady *et al.*, 1963, p. B-8; Laurent, 1978, p. 30). Since no correlatives of the Magog Group can be directly traced south across the International Border (Fig. 3), we conclude that none of the pre-Silurian stratigraphy of northern Vermont should be considered as younger than pre-Magog in age. The age differences between Unit IIIC and Unit V may not, however, be great.

Unit VI: Silurian-Devonian Metasediments and Intrusives

Silurian metasedimentary rocks (Unit VI) in Quebec lie in two narrow infolds parallel to the axis of Lake Memphremagog (Glenbrook Group) and a third belt (St. Francis Group) overridden by Ascot Weedon metasedimentary rocks along the Bunker Hill thrust (Clark, 1934; Cooke, 1950; St-Julien, 1965; Fig. 3). The Glenbrook Group consists of the Peasley Pond Conglomerate at the base overlain by the Glenbrook Slate and the Sargent Bay limestone (Cooke, 1950). The lower two units correlate by lithic similarity and position to the Shaw Mountain Conglomerate, the Northfield Slate (Doll *et al.*, 1961) whereas the overlying Ayers Cliff Member of the Waits River Formation in Vermont is directly traceable to calcareous slates and limestones of the St. Francis Group. In Québec, the Glenbrook Group overlies Units IIIA, IIIC, and V with angular unconformity; however, the unconformity in Vermont appears only on a regional scale and locally is a disconformity (Figs. 1, 3 and 6).

The Silurian metasedimentary rock shown in Figure 1 are cut by Devonian diorite and granodiorite intrusives. These intrusives appear to be epizonal with well-defined contact aureoles and xenoliths of Silurian metasedimentary rocks and are assumed to have accompanied the Acadian Orogeny.

COMPARISON OF TECTONIC STRATIGRAPHY: SOUTHERN QUEBEC-NORTHERN VERMONT

A correlation of the tectonic stratigraphy in southern Quebec and northern Vermont is presented in Figure 7. The age relationships between Units I, II, III, and IV described in the previous section are uncertain because of suspected structural contacts. For this reason, the names assigned to the units are not given the stratigraphic importance of previous workers. In general, however, the lithic descriptions follow

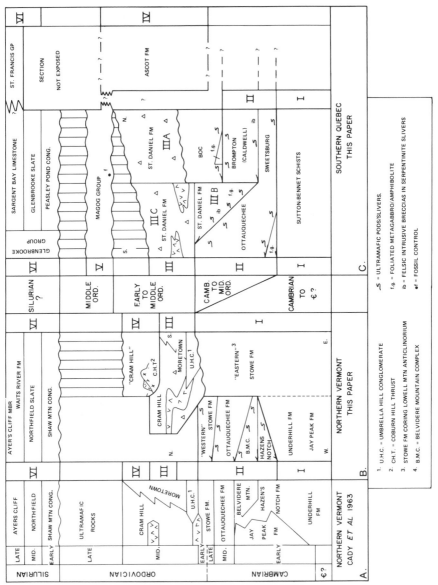

Figure 7. Correlation of tectonic stratigraphy between northern Vermont and southern Quebec. Column A: adapted from Cady *et al.*, (1963); Column B: Schematic tectonic stratigraphy of northern Vermont; Column C: Schematic tectonic stratigraphy of Eastern Townships, Quebec. Correlations of unit designations of this paper are approximate but not exact to formational names of earlier workers. I - VI designates units shown on Figure 3 and described in text; x symbol in Unit IV (Vermont) designates unnamed felsic intrusives rocks; random dash symbol designates gabbroic intrusive rocks in Unit IV of Vermont (See Fig. 6); triangles in Unit III designates olistostromal horizons in black slate; v symbol – metavolcanic rocks in unit III.

the usage of St-Julien (1965), St-Julien and Hubert (1975), and Laurent (1978) for Québec, and Cady *et al*. (1963) and Doll *et al*. (1961) for Vermont.

The most striking difference in the tectonic stratigraphy from north to south is the loss of stratiform ophiolite occurrences (Baldface-Orford Chagnon complex of Fig. 3) and the occurrences in Vermont at the same structural position of quartz-rich sericite chlorite-albite schists (Unit I) containing no ultramafics. The unit overlying these contrasting lithologies (Unit III) also appears to undergo significant change from north to south.

Unit III in Vermont differs from Quebec correlatives by the presence of a basal quartz pebble conglomerate unit (Umbrella Hill Conglomerate) and the greater abundance of massive and pillowed metavolcanics, and minor metatuffaceous units (Cram Hill Formation) which further south undergoes a facies transition to quartz arenites and phyllites of the Moretown Formation.

In addition to these lithologic variations, Unit III changes in tectonic setting from north to south: in the Chagnon-Orford area, this unit is largely olistostromal containing many sedimentary fragments of intraformational origin and consistently lies east of disrupted ophiolite fragments. To the south of Place Mountain and in the Trouser's Lake region, this unit incorporates tectonic slivers of metagabbro metavolcanic rocks and serpentinite cut by a variety of intrusive breccias containing mafic and felsic metavolcanic rocks (some of which were assuredly derived from the Baldface-Orford-Chagnon complex) and sodic and potassic intrusives of unknown origin. The juxtaposition of Unit III and Unit II in Quebec appears to represent a structurally distinct tectonic environment from the setting observed in the Orford-Chagnon region to the north and the Vermont section to the south (Fig. 3). Elaboration of these differences is not warranted here since work in this region is presently in progress. However, some speculations are offered in the Discussion.

ORIGIN OF THE ULTRAMAFIC AND RELATED ROCKS

Two contrasting models for the origin of ultramafic rocks across the Quebec-Vermont border have been proposed. Ultramafic and associated rocks in the Eastern Townships are explained as obducted slabs of ocean crust emplaced on a foundering Ordovician continental margin (Laurent, 1973, 1975, 1978; St-Julien *et al*., 1976; Williams and St-Julien, 1978). This model is based both on structural and petrologic studies at Thetford Mines (Laurent, 1973; St-Julien, 1972), Asbestos (Laurent, 1978; Lamarche, 1972; Laurent *et al*. 1979) and Chagnon-Orford, Quebec (De Romer, 1963; Laurent, 1977) and regional correlations with Newfoundland (St-Julien *et al*., 1976; Williams and St-Julien, 1978; Williams and St-Julien, this volume). Vermont correlatives of the ultramafic belt are considered, on the other hand, to be intrusive ultramafic bodies into sialic basement (Chidester and Cady, 1972; Chidester *et al*., 1978).

The following evidence suggests to us that all the ultramafic rocks across the International Border have a common origin and represent ophiolite fragments in an imbricated continental margin ocean crust east dipping obduction surface:

1) Aureole rocks consisting of greenstones, fine and coarse-grained amphibolites and garnet amphibolites are found along truncated structural contacts imbricated below serpentinized harzburgites and dunites at the Eden Mills ultramafic body in

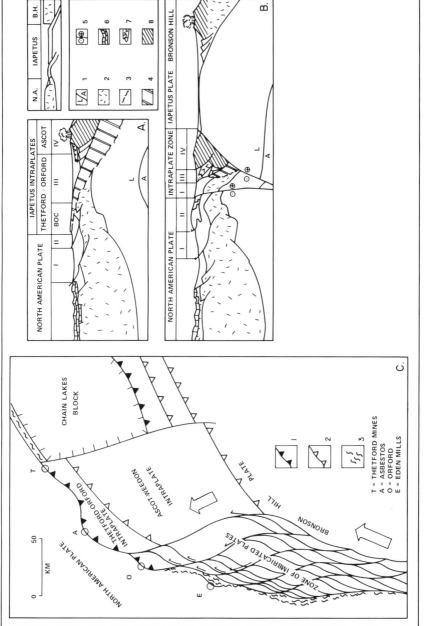

Figure 8

northern Vermont (Belvidere Mountain Complex of Figs. 3 and 4). Major and trace element analyses of these mafic aureole rocks are consistent with the view that these rocks originated as gabbro and/or volcanic members of oceanic crust overridden by oceanic mantle during obduction. The occurrence of blueschist assemblages in the Belvidere Mountain Complex at Tilloston Peak is not inconsistent with this model (Laird and Albee, 1979; Fig. 4).

2) Polydeformed metasedimentary rocks structurally underlying the aureole/ultramafic rock association (Unit IIB where differentiated on Figure 3) contain numerous fragments of serpentinized harzburgites and dunites. These rocks occupy similar structural position and possess many of the features of ophiolite mélange found beneath obducted ophiolites (Gansser, 1974).

3. Blanketing and possibly synchronous with both the emplacement and subsequent deformation of obducted ophiolite is a consistent cover of black and grey slate with olistostromal horizons (i.e. Units IIIA, IIIB, and IIIC, Fig. 3).

These features and correlations support a common origin for the tectonic evolution of the Eastern Townships-northern Vermont ultramafic belt. However, the following major differences are observed in the lithotectonic units described earlier between Quebec and Vermont: 1) the lack of a coherent "upper" cumulate-hypabyssal-extrusive cap on the lower harzburgite in Vermont; 2) the juxtaposition of polydeformed Unit I rocks against serpentinized and imbricated ophiolite fragments on the eastern side of the ultramafic belt in Vermont; 3) the change in character in Unit III rocks which suggest a pre- to syn-ophiolite emplacement character north of Chagnon Mountain to a syn- to post-ophiolite emplacement age south of Chagnon Mountain; and , 4) the juxtaposition of Unit III rocks imbricated with ophiolitic fragments against Unit II rocks north of the International Border and south of the Chagnon Mountain (Fig. 3).

We propose that these features which mark the major differences between the characteristics of the Eastern Townships ophiolite belt and the Vermont section and

---

Figure 8. Plate tectonic model for the Eastern Township-Vermont ultramafic belt depicting diachronous island arc-continental margin collision.
(A) Schematic pre-Silurian model of plate interaction during early stage of arc-ophiolite-margin imbrication (e.g., the Thetford-Asbestos region of Fig. 8C and Fig. 1).
(B) Schematic model of arc-ophiolite margin imbrication (e.g., in region between Orford and Eden Mills of Figs. 8C and 3).
Units shown in Figure 8A and 8B (I-VI) are defined in text and depicted in Figures 3, 6, 7. Ornamentation on Figures 8A and 8B as follows: 1) lithosphere (L) – asthenosphere (A) boundary; 2) "basement" of North American Plate and Bronson Hill Plate; 3) Ophiolite metamorphic aureole; 4) Oceanic crust and mantle; 5) Transform fault showing motion towards ($\cdot$) and away from (+) reader; 6) Shelf sequence of ancient North America; 7) Arc volcanics; 8) Island arc and "Andean" arc volcaniclastic sediments, volcanics and intrusives.
(C) Ornamentation on Figure 8C
1. – The western limit of ophiolite obduction as coherent slab;
2. – Arc-trench system of both island arc and "Andean-type";
3. – Ophiolite remnant.
Large open arrows depict plausible plate motion of Iapetus and Bronson Hill plate with respect to a fixed North American plate, consistent with a diachronous south to north arc-margin collision.

have led to contrasting theories as to origin of the ultramafic rocks, is a result of diachronous collision of a migrating trench-arc complex (Ascot-Weedon belt, Unit IV, Fig. 3) against an irregular North American continental margin. The present position of the Ascot-Weedon belt with respect to the rapid change in the litho-tectonic stratigraphy across the International Border suggests to us a causal relation-ship. The pre-arc collision relationship north of the border are preserved according to this model because of the presence of the Québec re-entrants in which Ordovician collision was incomplete.

These ideas are shown schematically in Figure 8 and briefly discussed below as an interpretation of a pre-Silurian plate mosaic in the Québec re-entrant. We stress that this mosaic has been strongly influenced by the pre-ophiolite emplacement (e.g., early rifting) stage of the ancient North American continent.

DISCUSSION

The geology of Québec re-entrant suggests that the present distribution of lithotectonic units in the vicinity of the International Border resulted from diachron-ous impingement of irregular adjacent margins during continent-arc collisional events. The first stage of closure was marked by obduction of oceanic crust over the continental margin of ancient North America resulting in westward migrating ophiol-ite bearing thrusts and nappes and underlying aureole-mélange units (the Thetford-Orford Intraplate of Fig. 8A). Subsequent collision of the ophiolite imbricated margin with the Ascot-Weedon arc (Ascot-Weedon Intraplate of Fig. 8A) resulted in change of plate interaction from shallow east-dipping subduction (Fig. 8A) to steep left-lateral trench-transform motion (Fig. 8B). This second stage collisional event pro-duced "sideways-driven transform splinters" (Dewey and Burke, 1974, p. 59) of underlying metasediments which are interpreted as formerly constituting part of the ancient rise prism of eastern North America (Unit I, Fig. 3). Coeval with this event is the development of "sideways-feeding flysch fans" (Dewey and Burke, 1974, p. 59) and further dismemberment of the ophiolite sheets to cryptic serpentinite sutures well represented by the Vermont ultramafic belt. Greywackes, metatuffs, and olis-tostromal black slate coeval with extrusion from the encroaching island arc complex (Unit IIIC, Fig. 3) are products of the flysch fans which migrated northward onto deformed, obducted oceanic crust and subsequently imbricated with more highly dismembered ophiolite fragments (Unit IIIB). The Lowell, Worcester, and North-field ranges of Vermont (Fig. 1) are interpreted as products of a progressively north-ward lengthening imbricated zone of reactivated continental margin transform slivers and deformed overlying flysch. The apparent termination of well developed ophiolite sequences south of Chagnon Mountain, Quebec, is, thus a result of increased arc-continent collisional activity to the south reflecting both the geometry and relative plate motions of the adjacent margins, and associated island arcs.

The termination of the Thetford-Orford Intraplate and the Ascot-Weedon In-traplate to the north (Figs. 1 and 8) is interpreted as truncation along transcurrent faults formed during closure of Iapetus against the Chain Lakes Massif. The origin of the Chain Lakes Massif is problematical; it could represent a local "promontory" of the ancient western margin, or, as implied in Figure 8C, a continental block from unknown sources juxtaposed between contrasting basements during opening or clo-sure of Iapetus.

All of the diachronous collisional events described here may have occurred along the entire western margin in New England. However, subsequent collision effects of the Bronson Hill "Andean-type" margin may mask any attempts to reconstruct these earlier events to the south where continent-continent collisional effects are much more pronounced (Robinson and Hall, this volume).

These speculations warrant further work and discussion on both sides of the International Border and both sides of the ancient ocean. Undoubtedly, this will result in refining (and in some cases redefining) the tectonic units briefly described here. In light of the numerous similarities in the tectonic history of the Québec-Vermont ultramafic belt and the reasonable interpretation that the Québec re-entrant is a feature inherited from the earliest stages of the Wilson Cycle, the hypotheses set forth here provide a framework for further testing in the field.

ACKNOWLEDGEMENTS

The authors are grateful to R. S. Stanley, Dana Roy, Deen Bryan, and Philip Winner at the University of Vermont for discussions and access to unpublished data on their ongoing work in the ultramafic belt of northern Vermont and southern Quebec, and to Daniel Lamothe, Cambridge University, for discussions in the field. The senior author extends thanks to Pierre St-Julien who first introduced him to the geology of the Eastern Townships some years ago and for ongoing communications. Special thanks also acknowledged to Harold Williams, Dave Strong, and John Malpas of Memorial University, Newfoundland, who have provided stimulating discussions of ophiolite genesis and regional syntheses of Appalachian geology during a sabbatical leave of absence at Memorial in 1977-1978 (B.L.D.), Gertrude Andrews and David Press of the Memorial University support staff are gratefully acknowledged for their expertise and excellent supervision in obtaining the analytical data presented in Table I. This paper has greatly benefited from critical reviews from R.S. Stanley, Pierre St-Julien, D.W. Rankin and an anonymous reviewer. Field support in 1978 (for M.H.G.) was provided by U.S. Geological Survey under Project No. 9510-02108 to R. S. Stanley. This study has been supported by NSF Grant EAR 77-14577 (B.L.D.).

REFERENCES

Ambrose, J.W., 1942, Preliminary map, Mansonville, Québec: Geol. Survey Canada Paper 42-1.
————————, 1949, Are the "Bolton" lavas of post-Devonian age? (abst.): Royal Soc. Canada Proceedings, 3rd Ser., v. 43, p. 239.
————————, 1957, The age of the Bolton lavas, Memphremagog district, Quebec: Naturaliste Canadien, v. 84, p. 161-170.
Badger, R.L., 1979, Origin of the Umbrella Hill Conglomerate, north-central, Vermont: American Jour. Sci., v. 279, p. 692-702.
Berry, W.B.N., 1962, On the Magog, Quebec, graptolites: American Jour. Sci., v. 260, no. 2, p. 142-148.
Bird, J.M. and Dewey, J.F., 1970, Lithosphere plate-continental margin tectonics and the evolution of the Appalachian Orogen: Geol. Soc. America Bull., v. 81, no. 4, p. 1031-1059.
Boudette, E.L. and Boone, G.M., 1976, Pre-Silurian stratigraphic succession in Central Western Maine: Geol. Soc. America Memoir 148, p. 79-96.

Burke, K. and Dewey, J., 1973, Plume-generated triple junctions; Key indicators in applying plate tectonics to old rocks: Jour. Geol., v. 81, p. 406-433.

Cady, W.M., 1956, Bedrock geology of the Montpelier Quadrangle, Vermont: United States Geol. Survey, Geologic Quadrangle Map, GQ-79.

——————————, 1969, Regional tectonic synthesis of Northwestern New England and adjacent Quebec: Geol. Soc. America Memoir 120, 181 p.

——————————, 1945, Stratigraphy and structure of west-central Vermont: Geol. Soc. America Bull., v. 56, p. 515-587.

Cady, W.M., Albee, A.L. and Chidester, A.H., 1963, Bedrock geology and asbestos deposits of the upper Missisquoi Valley and vicinity, Vermont: United States Geol. Survey Bull. 1122-B, p. B1-B78.

Chidester, A.H. and Cady, W.M., 1972, Origin and emplacement of alpine-type ultramafic rocks: Nature, Physical Sciences, v. 240, no. 98, p. 27-31.

Chidester A.H., Albee, A.L. and Cady, W.M., 1978, Petrology, structure and genesis of the asbestos-bearing ultramafic rocks of the Belvidere Mountain areas in Vermont: United States Geol. Survey Prof. Paper 1016, 95 p.

Church, W.R., 1972, Ophiolite: its definition, origin as oceanic crust and mode of emplacement in orogenic belts with special reference to the Appalachians: in Irving, E., ed., The Ancient Oceanic Lithosphere: Dept. Energy Mines and Resources, Ottawa, Earth Physics Branch Publ., v. 42, p. 71-85.

Clark, T.H., 1934, Structure and Stratigraphy of Southern Quebec: Geol. Soc. America Bull., v. 45, p. 1-20.

Cooke, H.C., 1950, Geology of a southwestern part of the Eastern Townships of Quebec: Geol. Survey Canada Memoir 257, 142 p.

De Romer, H.S., 1963, Differentiation in Chagnon-Orford-Intrusive-Komplex, Sudostteil der Provinz Quebec, Kanada: Geologischen Rundschau, Bd 52, 2, p. 825-835.

——————————, 1976, Preliminary report on the Fitch Bay area, Orford County: Ministère des Richesses Naturelles, Québec, Document Public 343, 23 p.

——————————, 1979, Cambro-Ordovician Stratigraphy and tectonics south of Sherbrooke area: Geol. Assoc. Canada and Mineral. Assoc. Canada, Annual Meeting, Quebec, Guidebook for Excursion A-12, 17 p.

Dewey, J.F., 1969, Evolution of the Appalachian-Caledonian orogen: Nature, v. 222, no. 5189, p. 124-129.

——————————, 1975, Finite plate implications for the evolution of rock masses at plate margins: American Jour. Sci., v. 275-A, p. 260-284.

Dewey, J.F. and Burke, K., 1974, Hot spots and continental breakup: Implications for collisional orogeny: Geology, v. 2, no. 2, p. 57-60.

Doig, R., 1970, An alkaline rock province linking Europe and North America: Canadian Jour. Earth Sci., v. 7, p. 22-28.

Doll, C.G., Cady, W.M., Thompson, J.B., Jr. and Billings, M.P., 1961, Centennial geologic map of Vermont, scale 1:250,000: Vermont Geol. Survey.

Fortier, Y.O., 1945, Preliminary May, Orford, Eastern Townships, Quebec: Geol. Survey Canada Paper 45-8, 5 p.

Gale, M.H., 1980, Geology of the Belvidere Mountain Complex, Eden and Lowell, Vermont: M.S. Thesis, University of Vermont, 169 p.

Gale P.N., 1980, Geology of the Newport Center area, north central Vermont: M.S. Thesis, University of Vermont, 126 p.

Gansser, A., 1974, The ophiolite melange, a worldwide problem on Tethyan examples: Eclogae geologicae Helvetiae, v. 67/3, p. 479-507.

Hofmann, H.J., 1972, Stratigraphy of the Montreal area: 24th International Geol. Congress, Montreal, Guidebook for Excursion B-03.

Jamieson, R.A., 1981, The formation of metamorphic aureoles beneath ophiolite suites – evidence from the St. Anthony Complex, Northwestern Newfoundland: Geology, v. 8, p. 150-154.

Kane, M.F., Simmons, G., Diment, W.H., Fitzpatrick, M.M., Joyner, W.B. and Bromery, R.W., 1972, Bouger gravity and generalized geologic map of New England and adjoining areas: United States Geol. Survey, Geophysical Investigations Map GP-839.

Kay, M., 1942, Ottawa-Bonnechere graben and Lake Ontario homocline: Geol. Soc. America Bull., v. 53, p. 585-646.

Kumarapeli, P.S., 1976, The St. Lawrence rift system, related metallogeny, and plate tectonic models of Appalachian evolution: in Strong, D.F., ed., Metallogeny and Plate Tectonics: Geol. Assoc. Canada Spec. Paper 14, p. 301-320.

_____, 1978, The St. Lawrence paleo-rift system: a comparative study: in Ramberg I.B. and Newman, E.R., eds.: Tectonics and Geophysics of Continental Rifts, D. Reidel Publishing Company, Dordrecht, Holland, p. 367-384.

Kumarapeli, P.S. and Sauli, V.A., 1966, The St. Lawrence Valley system: A North American equivalent to the East African rift valley system: Canadian Jour. Earth Sci., v. 3, p. 639-658.

Laird, Jo and Albee, A.L., 1975, Polymetamorphism and the first occurrence of glaucophane and omphacite in northern Vermont: Geol. Soc. America Abstracts with Programs, v. 7, p. 1159.

Laliberté, R., Spertini, F. and Hébert, R., 1979, The Jeffrey Asbestos Mine and the ophiolitic complex at Asbestos Quebec: Geol. Assoc. Canada and Mineral. Assoc. Canada, Annual Meeting, Québec, Guidebook for Excursion B-3, 18 p.

Lamarche, R.Y., 1972, Structural geology of the Sherbrooke area: 24th International Geol. Congress, Montreal, Guidebook No. 24, Part B-05, 17 p.

_____, 1972, Ophiolites of southern Quebec: in The Ancient Oceanic Lithosphere: Dept. Energy, Mines and Resources, Ottawa, Earth Physics Branch Publication 42, Part 3, p. 59-65.

_____, 1973, Géologie du complexe ophiolitique d'Asbestos, Cantons de l'Est: Ministère des Richesses Naturelles, Québec, Document Public GM-28558, 9 p.

Lamothe, D., 1979, Région de Bolton-Centre: Ministère des Richesses Naturelles, Québec, Document Public 687, 14 p.

Laurent, R., 1973, The Thetford-Mines Ophiolite, Paleozoic "flake" of oceanic lithosphere in the Northern Appalachians of Quebec: Northeastern Section, 8th Annual meeting, Geol. Soc. America, Abstracts with Program, v. 5, no. 2, p. 188.

_____, 1975, Occurrences and origin of the ophiolites of southern Quebec, northern Appalachians: Canadian Jour. Earth Sci., v. 12, no. 3, p. 443-455.

_____, 1977, Ophiolites from the Northern Appalachians of Quebec: in Coleman R.G., and Irwin, W.P., eds., North American Ophiolites: State of Oregon Dept. Geology and Mineral Industries, Bull. 95, Portland, p. 25-40.

Laurent, R., Hébert, R. and Hébert, Y., 1979, Tectonic setting and petrological features of the Quebec Appalachian Ophiolites: in Malpas, J., and Talkington, R.W., eds., Ophiolites of the Canadian Appalachians and Soviet Urals: Memorial University of Newfoundland, Geology Dept. Rept. 8, p. 53-77.

Naylor, R.S., 1968, Origin and regional relationships of the corerocks of the Oliverian domes: in Zen, E-an, White, W.S., Hadley, J.B., and Thompson, J.B., Jr., eds., Studies in Appalachian Geology, Northern and Maritime: New York and London, Wiley-Interscience, p. 231-240.

Naylor, R.S., Boone, G.M., Boudette, E.L., Ashenden, D.D. and Robinson, Peter, 1973, Pre-Ordovician rocks in the Bronson Hill and Boundary Mountains anticlinorium, New England, U.S.A. (abst.): EOS, American Geophys. Union Trans., v. 54, p. 495.

114 DOOLAN ET AL.

Osberg, P.H., 1965, Structural Geology of the Knowlton-Richmond area, Quebec: Geol. Soc. America Bull., v. 76, p. 223-250.

——————, 1978, Synthesis of the Geology of the Northeastern Appalachians, U.S.A.: in E.T. Tozer and Paul E. Schenk, eds., Caledonian-Appalachian Orogen of the North Atlantic Region: Geol. Survey Canada Paper 78-13, p. 137-147.

Pearce, J.A. and Cann, J.R., 1973, Tectonic setting of basic volcanic rocks determined using trace element analysis: Earth and Planetary Sci. Letters, v. 19, p. 290-300.

Pieratti, D.D., 1976, The origin and tectonic significance of the Tibbit Hill metavolcanics, northwestern Vermont: M.S. Thesis, University of Vermont, 136 p.

Rankin, D.W., 1976, Appalachian salients and recesses; late Precambrien continental breakup and the opening of the Iapetus Ocean: Jour. Geophys. Research, v. 81, no. 32, p. 5605-5619.

Riva, J., 1974, A revision of some Ordovician graptolites of eastern North America: Palaeontology, v. 17, Part 1, p. 1-40.

Rodgers, J., 1968, The eastern edge of the North American continent during Cambrian and Early Ordovician: in Zen, E-an, White, W.S., Hadley, J.B. and Thompson, J.B., Jr., eds., Studies in Appalachian Geology, Northern and Maritime: New York, Wiley-Interscience, p. 141-149.

St-Julien, P, 1965, Geologic map of the Orford-Sherbrooke area: Quebec Dept. Natural Resources Map No. 1619, scale 1:50,000.

——————, 1970, Geology of the Disraeli region, eastern half, Frontenac, Wolfe and Megantic counties: Quebec Dept. Natural Resources, Preliminary Rept. No. 587, 23 p.

——————, 1972, Appalachian structure and stratigraphy, Quebec: 24th International Geological Congress, Montreal, Guidebook for Excursion A56-C56, 35 p.

St-Julien, P. and Hubert, C., 1975, Evolution of the Taconian orogen in the Quebec Appalachians: American Jour. Sci., v. 275-A, p. 337-362.

——————, 1980, Structural setting of the Thetford Mines Ophiolite Complex: Geol. Assoc. Canada and Mineral. Assoc. Canada, Annual Meeting, Quebec, Guidebook for Excursion B-10, 27 p.

St-Julien, P., Hubert C. and Williams, H., 1976, The Baie-Verte Brompton line and its possible tectonic significance in the Northern Appalachians: Geol. Soc. America, Abstracts with Program, Northeastern and Southeastern Sections, Arlington, Virginia, P. 259-260.

Strong, D.F. and Williams, H., 1972, Early Paleozoic flood basalts of northwestern Newfoundland: Their petrology and tectonic significance: Geol. Assoc. Canada Proceedings, v. 24, no. 2, p. 43-52.

Thomas, W.A., 1977, Evolution of the Appalachian-Ouachita salients and recesses from reentrants and promontories in the continental margin: American Jour. Science, v. 277, p. 1233-1278.

Williams, H., 1964, The Appalachians in Northeastern Newfoundland; A two sided symmetrical system: American Jour. Science, v. 262, no. 10, p. 1137-1158.

——————, 1978, Tectonic Lithofacies Map of the Appalachian Orogen, Memorial University of Newfoundland, Map. No. 1.

Williams, H. and Doolan, B.L., 1978, Vestiges and Margins of Iapetus in the Appalachian Orogen (abst.): Geol. Soc. America and Geol. Assoc. Canada Joint Meeting, Toronto, Abstracts with Program, v. 10, no. 7, p. 517.

Williams, H., Kennedy, M.J. and Neale, E.R.W., 1972, The Appalachian Structural province: in R.A. Price and R.J.W. Douglas, eds., Variations in Tectonic Styles in Canada: Geol. Assoc. Canada Spec. Paper 11, p. 181-261.

——————, 1974, The Northeastward termination of the Appalachian Orogen: in The Ocean Basins and Margins: v. 2, The North Atlantic: New York, Plenum Press, p. 79-123.

Williams, H. and St-Julien, P., 1978, The Baie-Verte Brompton Line in Newfoundland and regional correlations in the Canadian Appalachians: Geol. Survey Canada Paper 78-1A, p. 225-229.

Williams, H. and Smyth, W.R., 1973, Metamorphic aureoles beneath ophiolite suites and alpine peridotites: tectonic implications with west Newfoundland examples: American Jour. Sci., v. 273, no. 7, p. 594-621.

Wilson, J.T., 1966, Did the Atlantic close and then re-open?: Nature, v. 211, no. 5050, p. 676-681.

Manuscript Received May 10, 1980
Revised Manuscript Received October 3, 1980

Major Structural Zones and Faults of the Northern Appalachians, edited by
P. St-Julien and J. Béland, Geological Association of Canada Special Paper 24, 1982

TACONIAN AND ACADIAN STRUCTURAL TRENDS IN CENTRAL AND NORTHERN NEW BRUNSWICK

L.R. Fyffe
Mineral Resources Branch, New Brunswick Department of Natural Resources,
P.O. Box 6000, Fredericton, N.B. E3B 5H1

ABSTRACT

Structural trends in central and northern New Brunswick record the effects of the Taconian and Acadian Orogenies. Polyphase Taconian deformation is intense in the northern Miramichi Highlands and decreases toward the southwest whereas Acadian deformation is uniform along the length of the Highlands. Rocks of the Matapedia Basin to the northwest, and the Fredericton Trough to the southeast, have undergone a generally homogeneous northwest-southeast compression during the Acadian Orogeny. Cleavage trends deviate from the normal along the faulted margins of the Highlands.

Southward verging Taconian folding is compatible with closing of an Ordovician Ocean in northern New Brunswick. Upright Acadian folds and high-angle reverse faults resulted from further continental convergence.

RÉSUMÉ

Les directions structurales dans le centre et le nord du Nouveau-Brunswick sont les effets des orogénèses taconienne et acadienne. Les déformations polyphasées dues à l'orogénèse taconienne sont intenses dans le nord des Hautes Terres du Miramichi mais diminuent vers le sud-ouest. Par contre, la déformation acadienne est uniforme dans toutes les Hautes Terres. Les roches du bassin de Matapédia, au nord-ouest, et celles de la fosse de Frédéricton, au sud-est ont subi, durant l'orogénèse acadienne, une compression généralement homògène de direction nord-ouest-sud-est. L'orientation du clivage diffère toutefois le long des bordures faillées des Hautes Terres.

Le déversement vers le sud des plis taconiens est compatible avec l'hypothèse de la fermeture, dans le nord du Nouveau-Brunswick, d'un océan ordovicien. Les plis droits acadiens et les failles inverses à fort pendage résulteraient d'une convergence plus poussée des continents.

IGCP Project 27   *Canadian*
Caledonide   *Contribution*
Orogen   *No. 25*

INTRODUCTION

This paper covering the structural geology of the northern two-thirds of New Brunswick is based on data obtained over the past eight years while working on projects sponsored by the Department of Regional Economic Expansion, Ottawa and the Department of Natural Resources of New Brunswick. The description is necessarily brief since most of the data was collected during the mapping of large areas over limited periods of time. Future detailed mapping on specific areas would be required in order to attempt to resolve several problems. However, the advantage the writer has in having seen at least something of the entire area may compensate for the lack of a more thorough documentation in the field. Nevertheless, the summary presented is a considerable simplification of the region's complex history.

One problem of prime interest is whether Precambrian basement is exposed within central New Brunswick (Rast et al., 1976). The concordancy between the deformed older granitic rocks and their host rocks and their subsequent remobilization under high grade metamorphic conditions makes it difficult to confirm or deny the existence of such basement. Rb-Sr age data on these rocks have so far yielded Ordovician ages but dating of zircons currently in progress by the Geological Survey of Canada may resolve the problem.

We also encountered difficulty in attempting to correlate structural events over large distances in the polydeformed terrane. Style has been used as a main guide and from this it is evident that structures of different styles but similar orientations have developed in separate areas at various times. These problems of correlation, however, should not greatly affect the overall picture presented here.

Central and northern New Brunswick can be divided into four major paleogeomorphic zones which are, from northwest to southeast: the Matapedia Basin, the Elmtree inlier, the Miramichi Highlands, and the Fredericton Trough (Fig. 1). The geologic history of each belt as determined from the stratigraphy is summarized below and is followed by a description of their structure. Finally, an attempt is made to explain the geologic events in terms of a plate tectonic model.

GEOLOGIC HISTORY

Cambro-Ordovician

In the Late Precambrian, rifting disrupted the pre-Hadrynian craton of New Brunswick, a remnant of which is exposed in the Caledonia Highlands in the south of the province. During the Cambrian and Early Ordovician, quartzose turbidites were deposited off the northern margin of the craton in the area that would eventually form the Miramichi Highlands (Fig. 2). In Early Ordovician, the sedimentary regime changed abruptly with deposition of a thin conformable unit of calcareous siltstone and minor felsic tuff in the Hayesville area of central New Brunswick (Poole, 1963) that changes to an unconformable unit of quartz-pebble conglomerate a short distance to the southwest (Potter, 1969).

In the Hayesville area, brachiopods and conodonts from the calcareous siltstones are of Late Arenig-Llanvirn age (Neuman, 1968; Nowlan, 1979). Near Tetagouche Falls in the Bathurst area, brachiopods of the same age are found in a slightly calcareous slate conformably overlying quartzose greywacke. In the Water-ville area, 50 km to the southwest of the Hayesville locality, conodonts from lime-stone interbedded with volcanoclastic greywacke are of Late Llanvirn-Early Llan-deilo age (Nowlan, 1979). The above fossil occurrences apparently occur in the shore area of volcanic islands (Neuman, 1968) and indicate that volcanic activity did not begin before Late Arenig-Llanvirn time. No Cambrian volcanics are known in the Miramichi Highlands; thus the plate tectonic model presented by McBride (1976) needs modification.

By the Caradocian a medial belt of north-northeasterly trending volcanic islands had developed in the Miramichi area. The volcanic sequence begins with red and black manganiferous siltstone and chert that conformably overlies the Lower Or-dovician units. In central New Brunswick thin flows of mostly basaltic composition are interlayered with the manganiferous rocks and are overlain by greywacke.

The volcanic pile thickens to the northeast where, in the Bathurst area, quartz-

Figure 1. Paleogeomorphic zones of New Brunswick.

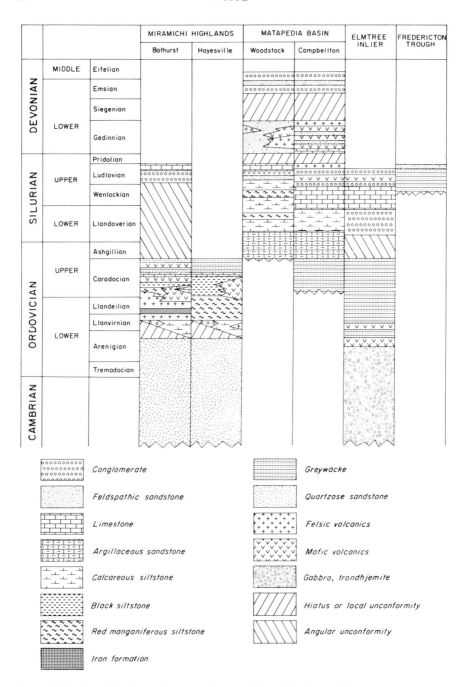

Figure 2. Stratigraphic columns for central and northern New Brunswick.

feldspar porphyry and rhyolite interfinger with, and are overlain by, a thick succession of pillow basalt and associated gabbro. Stratabound sulphide deposits overlain by iron formation occur in the acid rocks. The mafic volcanic rocks are intercalated with manganiferous slate, limestone, and greywacke.

Black slate interbedded with basalt near the mouth of the Tetagouche River contains graptolites of Caradocian age (Skinner, 1974). To the west near Camel Back Mountain, limestone interbedded with basalt contains conodonts of Middle Caradocian age (Kennedy et al., 1979). These localities support the stratigraphic and structural evidence that indicate the basaltic rocks are younger than the felsic volcanics as suggested by Helmsteadt (1971). Some confusion has arisen in this regard because Skinner (1974) shows an anticline north of Bathurst that would cause the basalts to underlie quartzose turbidites. However, new highway exposures containing abundant pillows confirms that the sequence youngs to the north away from the turbidites. Skinner's single south-facing locality is north of the new exposures; also reexamination of this locality reveals poorly formed pillows with a probable north-facing direction.

Chemically, the volcanic sequence forms a bimodal suite of calc-alkaline dacite and rhyolite, and spilitic tholeiite exhibiting a pronounced iron-enrichment trend. Alkali basalt is less common and andesite is rare. The abundance of acid volcanic rocks and associated pre-Taconian granitic intrusions suggest that a thick sialic basement underlies much of the volcanic terrane. This interpretation is in direct conflict with that of Whitehead and Goodfellow (1978) who suggest that the tholeiitic basalts represent oceanic crust whereas the alkali basalts represent within-plate oceanic islands. Unfortunately they follow McBride (1976) in assigning an older age to the basalts than their contained fossils. Strong (1977, p. 77) points out the hazards of indiscriminate use of chemical data to classify volcanic rocks as to geologic setting without due regard to the field evidence. Based on geologic relationships, Hynes (1976) considers Ordovician alkali basalt in Maine to have formed in an island arc environment.

The Matapedia Basin separates the Miramichi Highlands from Ordovician volcanics of the Bronson Hill and Stoke Mountain Anticlinoria. Whereas the Miramichi area with its high proportion of felsic volcanics is likely underlain by continental crust, the Matapedia Basin and volcanics to the northwest may have formed partially on oceanic crust (Bird and Dewey, 1970; St-Julien and Hubert, 1975). The island arc volcanics of the Central Mobile Belt of Newfoundland, interpreted to have formed on oceanic crust, differ from those of the Miramichi area by consisting largely of basalt and basaltic andesite (Strong, 1977). However, the Roberts Arm Group, occurring south of the Lukes Arm Fault and dated as Llandeilo-Caradoc (Bostock et al., 1979), is a bimodel assemblage that chemically resembles the Miramichi volcanics.

During the Caradocian, greywacke turbidites were being deposited in the Matapedia Basin to the northwest of the volcanic islands (St. Peter, 1978). The area to the southeast now covered by Silurian rocks of the Fredericton Trough was also probably a site of Caradocian greywacke deposition judging from the abundance of exposed greywacke on the southeast margin of the Miramichi Highlands. Both basins are areas of subgreenschist grade metamorphism and little igneous activity; they were apparently formed where the sialic crust had been greatly distended during rifting.

Rocks found in the Elmtree area northwest of Bathurst indicate that oceanic crust lies to the north and perhaps to the northwest of the Miramichi Highlands. The Elmtree inlier contains deformed gabbro and interbanded amphibolite injected by trondhjemitic dykes. These rocks are overlain by pillow basalt of subgreenschist grade which are in turn overlain by quartzose greywacke of probable Caradocian age. Gravity measurements suggest that the inlier is allochthonous.

The presence of oceanic crust to the north of the Ordovician volcanic islands suggests that they are products of southward subduction (Pajari *et al.*, 1977) although scarcity of andesite contrasts with recent arcs. A southeasterly dipping subduction zone has also been postulated in Newfoundland (Church and Stevens, 1971).

Siluro-Devonian

Uplift, subsequent to the Taconian Orogeny, formed a landmass along the former site of the Ordovician volcanic islands. The Elmtree inlier with its oceanic crust forms a northern extension of the Taconian landmass separated from the Miramichi Highlands by a graben of Silurian sedimentary rocks that is bound on the south by the Rocky Brook-Millstream Fault.

The Fredericton Trough lying in fault contact with, and to the southeast of the Miramichi Highlands, contains Wenlockian to Ludlovian greywacke turbidites. No shelf deposits are found along the Trough's northwestern margin but they do occur on its southeastern margin. (Pickerill, 1976).

The Fredericton Trough separates the Miramichi Highlands from the Precambrian Caledonia Highlands. Although some uplift occurred to the southeast of the Fredericton Trough during the Taconian Orogeny, the degree of Taconian deformation appears to be significantly less than to the northwest (Ruitenberg *et al.*, 1977).

The Matapedia Basin was the site of continuous deposition during the Taconian Orogeny. Within the deeper portion of the basin, Caradocian greywacke is conformably overlain by Ashgillian calcareous turbidites and a Llandoverian to Ludlovian sequence of interbedded calcareous siltstone and shale, red and green siltstone, and minor volcanic rocks (St. Peter, 1978).

During the Llandovery, the Matapedia sea transgressed the deformed Elmtree inlier, depositing a basal conglomerate. The conglomerate is overlain by Wenlockian shallow water nodular limestone (Noble, 1976) which, in turn, is overlain by a regressive sequence of Silurian red sandstone, conglomerate, and basaltic, andesitic and rhyolitic volcanics (Greiner, 1970). The volcanics are a bimodal calc-alkaline suite possessing a gap from 57 to 67 per cent SiO_2.

Gedinnian basalt, andesite and minor rhyolite interbedded with calcareous sandstone and limestone conformably overlie the Silurian rocks. The Devonian volcanics form a unimodal calc-alkaline to mildly alkaline suite. Upper Gedinnian continental sedimentary rocks unconformably overlie the Devonian volcanics near Campbellton (Dineley and Williams, 1968).

Transgression and initiation of extensive volcanism in the south (Fig. 2) occurred later than to the north. In the Woodstock area, conglomerate unconformably overlies deformed Ordovician rocks indicating that the Matapedia sea transgressed over the margin of the Miramichi Highlands during the Late Silurian (Lutes, 1979). This margin became the locus of volcanic activity in the Early Devonian (Gedinnian).

These southwesterly trending Gedinnian volcanics are interbedded with shallow water quartzose sandstone and siltstone. Acid ignimbrites are abundant in the southeast whereas basaltic rocks become more abundant to the northwest. The interbedded sedimentary rocks contain plant debris derived from the adjacent landmass. Northwestward the volcanics thin out and Gedinnian feldspathic sandstone, siltstone, and minor conglomerate overlie Ludlovian nodular limestone of the Matapedia Basin.

In the Brighton Mountain area of central New Brunswick, the southwesterly trending volcanic belt swings abruptly to the southeast forming a graben between the main belt of the Miramichi Highlands to the northeast and a separate block of Cambro-Ordovician quartzite to the southwest. A deltaic assemblage of Emsian conglomerate and sandstone containing abundant volcanic detritus and plant debris unconformably(?) overlies these volcanics. The southeasterly trending volcanic belt extends for approximately 30 km and then swings back to the southwest. Deformation within this narrow southwestern graben is intense in contrast to the gentle to moderate dips found in the rest of the belt.

Numerous subvolcanic stocks of granite and gabbro intrude both the Silurian and Devonian volcanic terranes. Several large northerly trending gabbro bodies are localized along the faulted northwestern boundary of the Miramichi Highlands and Elmtree inlier. Late tectonic to post-tectonic Devonian plutons are exposed along the length of the Highlands.

STRUCTURE

Miramichi Highlands

The Miramichi Highlands can be divided into three main structural zones separated by faults (Fig. 3): a *northern zone* (Zone 1) to the north of the Catamaran Fault, a *central zone* (Zone 2) to the south of the Catamaran Fault but to the northeast of the southeasterly trending graben of Devonian volcanics and a *southern zone* (Zone 3) to the southwest of the graben.

The *northern zone* exhibits the most complex deformational history. The first phase of folding in this zone was accompanied by subgreenschist facies metamorphism in the Bathurst area. Metamorphism increases to biotite grade to the southwest (Helmsteadt, 1973a) and reaches amphibolite grade just north of the Catamaran Fault (Fyffe and Cormier, 1979). Macroscopic first folds with a southern vergence have been mapped along contacts between sedimentary and volcanic rocks in the northwest (Helmsteadt, 1971).

A composite fabric is produced by coplanar crenulation of first schistosity. Tight to isoclinal folds accompanying the crenulation cleavage are the most common folds observed. The orientation of the composite fabric varies systematically from steep to subhorizontal across the northern zone. In the northeast, the steep composite fabric is folded about north-northeasterly to northeasterly trending, steeply plunging, open macroscopic folds with associated steeply dipping axial planar cleavage. North-northwesterly trending macroscopic folds with steep plunges occur to the northwest and may postdate the more northeasterly trending folds (Davis, 1972). In the southwest, the subhorizontal composite fabric is folded by north-northwesterly trending macroscopic folds with gentle plunges.

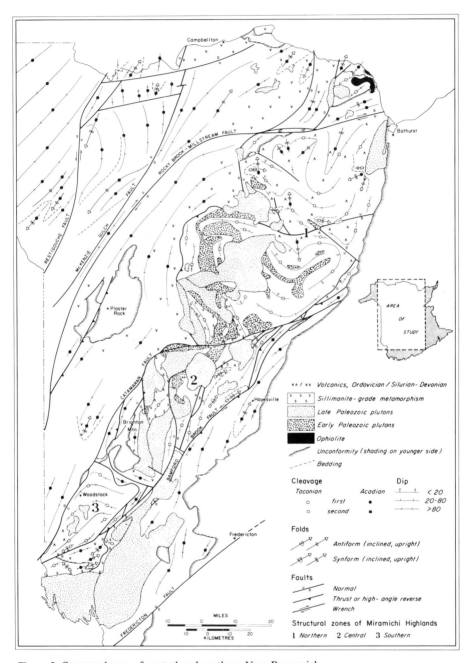

Figure 3. Structural map of central and northern New Brunswick.

Devonian granitic plutons in the northern zone are generally oval-shaped with narrow cordierite-andalusite contact aureoles that statically overprint the shallow-dipping composite fabric. However, fibrolite occurs in the southern portion of the zone along the eastern margin of an elongate northwesterly trending pluton. Fibrolite also occurs to the west of the pluton but is related to high grade regional metamorphic events in that area.

The boundary between the Miramichi Highlands and Matapedia Basin is, in most places, faulted. Uplift along the faulted boundary is more extensive in the southern part of the northern zone, where sillimanite-grade paragneiss of the Highlands is juxtaposed against subgreenschist Devonian volcanic rocks on the margin of the Matapedia Basin. High level granite and gabbro which intrude the fault zone have been cataclastized (Fyffe and Cormier, 1979).

Structure in the *central zone* is relatively simple. A composite fabric strikes rather consistently to the northeast and dips steeply to the northwest. In the vicinity of the fault separating the Miramichi Highlands from the Fredericton Trough, (Bamford Brook Fault), the cleavage swings to the south-southwest. Metamorphism is subgreenschist grade in the southeastern portion of the central zone. To the northwest the low grade regional metamorphism increases to sillimanite-grade. The fibrolite fabric, which grew during the development of the composite cleavage, is deformed by later kink folds. Aureole rocks surrounding the northerly trending Devonian granite plutons contain robust sillimanite that overprints the regional fibrolite. Some flattening of the composite cleavage postdates the development of andalusite in these aureoles. Narrow cordierite-andalusite aureoles surround small circular late tectonic granite stocks in the area.

Faulting and uplift along the boundary between the central zone and Matapedia Basin has juxtaposed deep level Devonian granite plutons and high grade Cambro-Ordovician rocks against low grade Devonian volcanics. Mylonitic zones in the pre-Taconic granite along this boundary dip gently to the east, and indicate thrusting of the Miramichi rocks to the west over rocks of the Matapedia Basin.

In the *southern zone,* a composite cleavage strikes northeasterly in the west and swings to the southeast in the east defining a steeply plunging, northerly closing fold. This style of folding is similar to the north-northwesterly trending folds in the northern block. Metamorphism is generally of lower greenschist grade. A narrow selvage of sillimanite-grade rocks possessing a continuous northeasterly cleavage trend forms the syntectonic aureole of an upfaulted northeasterly trending granite pluton lying on the southeastern margin of the zone.

The timing of the folding within the Miramichi Highlands is a matter of considerable debate (Helmsteadt, 1970, 1971, 1973a, b; Davis, 1972, 1973; Rast *et al.*, 1976; O'Brien, 1977). However, it is certain that the composite fabric, at least in the northern zone, is of Taconian age (Helmsteadt, 1971). Later fold structures are probably of Acadian age. It is tempting to relate the southerly verging, recumbent folds in the north to closing of the Ordovician ocean whereby the northeastern part of the Miramichi arc is sutured against the continental rise prism of the Canadian Shield. The lack of recumbent structures in the central and southern zones is in accord with a suture zone to the north.

Fredericton Trough

Within the Fredericton Trough, the Acadian slaty cleavage trends generally north-northeasterly and is steep. The associated folds are close to tight with shallow plunges. In the vicinity of the Bamford Brook Fault, the slaty cleavage trends northerly and is overprinted by a later northeasterly trending crenulation cleavage with moderate dips to the southeast.

The above trends are consistent with an initial northwest-southeast oriented stress field that produced an early phase of upright Acadian folds. This was followed by thrusting of the Silurian rocks over the Taconian basement of the Highlands to the northwest as indicated by the overturned second folds. This caused the northeasterly trending granite pluton within the faulted boundary to be preferentially uplifted along its northwestern margin. Rotation of the slaty cleavage into a more northerly direction may be related to subsequent left-lateral slip along the Bamford Brook Fault.

Matapedia Basin

Within the Matapedia Basin, the Acadian slaty cleavage trends about north-northeasterly along most of the belt, but swings northeasterly north of the Rocky Brook-Millstream Fault. The cleavage appears to mold itself against the margin of the Taconian basement; for instance, along the western margin of the Elmtree inlier, the cleavage strikes northerly, whereas along its southern margin it strikes easterly. Associated folds are generally tight, upright, and shallow plunging. Within the volcanic terrane, structures are more open.

A horst within the Matapedia Basin, bounded on the east by the McKenzie Gulch Fault and on the west by the Restigouche Fault, contains a northerly to northwesterly trending cleavage. This anomalous cleavage trend suggests left-lateral movement on the Restigouche Fault (St. Peter, 1978).

This regional picture of Acadian deformation is complicated near the boundary between the central and southern zones of the Miramichi Highlands (Zones 2 and 3). The intervening southeasterly trending graben of Devonian volcanic rocks forms an embayment into the Taconian basement of the Highlands and separates the southern zone from the central zone. Rocks within the graben possess a slaty cleavage that strikes parallel to the graben's margins and forms an anomalous Acadian trend, apparently resulting from compression between the two separate Taconian blocks. Silurian and Devonian sedimentary rocks of the Matapedia Basin have been thrust eastward across the entrance of the embayment where a series of at least three westerly dipping thrust sheets, bounded by Silurian limestone, has been recognized (St. Peter, pers. commun., 1979).

A few kilometres to the southeast, the cleaved rocks in the graben change trend and strike southwesterly. In this area, an exposed fault along the northwestern margin of the graben shows that the Cambro-Ordovician quartzite of the southern zone has been thrust southward over the graben.

Two major wrench faults, expressed by pronounced topographic lineaments, are present in central and Northern New Brunswick: the Catamaran and the Rocky Brook-Millstream Faults. The Catamaran Fault strikes westerly across the core of the Miramichi Highlands and swings to the southwest along the contact between the

Highlands and the Matapedia Basin. Right-lateral slip on the fault has been estimated at 7 km (Anderson, 1972). The main period of movement displaces Devonian granite and the Bamford Brook Fault. Carboniferous rocks are downfaulted along the fault zone. The Rocky Brook-Millstream Fault strikes westerly across northern New Brunswick where it defines the northern limit of the Miramichi Highlands. It then strikes southwesterly where, for most of its length, it separates Devonian volcanic and sedimentary rocks to the southeast from Ashgillian limestone to the northwest. The fault records 12 km of right-lateral slip, as indicated by offset of Devonian redbeds.

Some thrusting has occurred in the Carboniferous; Mississippian rocks of the Plaster Rock Basin have been thrust northwestward over Devonian sandstone (St. Peter, 1979).

DISCUSSION

Paleogeographic consideration sets certain constraints on plate tectonic models for the New Brunswick Appalachians. An ocean existed in northern New Brunswick during the early Ordovician but the region had become land area by the Llandovery. By the early Gedinnian, the Matapedia Basin had developed into a restricted seaway with a characteristic brachiopod fauna (Johnson and Dasch, 1972) and was separated from marine waters of the Fredericton Trough by the Miramichi Highlands. Thus no oceanic crust was being generated in the Matapedia Basin when extensive Devonian volcanism was occurring along the northwestern margin of the Highlands.

It has been suggested that the Fredericton Trough is the site of a former Silurian ocean (McKerrow and Ziegler, 1971) but no pelagic or trench deposits are found in the area. Similarly, there are no oceanic sediments of Silurian or Devonian age in the Meguma Basin of Nova Scotia to the south of the Caledonian Highlands (Williams, 1979).

The Matapedia Basin represents a remnant seaway persisting after closure of the Ordovician ocean during the Taconian Orogeny; whereas, the Fredericton Trough represents a graben developed within what was once a continuous Taconian land-mass formed by the Miramichi and Caledonian Highlands.

Since no trench environment existed in the New Brunswick Appalachians during the Siluro-Devonian, plate models that attempt to relate volcanism of this age to subduction must be in error. This explains the lack of volumetric increase and composition change that would be expected in the volcanics across New Brunswick as the hypothetical trenches are approached.

Acadian tectonism may be related to continued continental convergence following closing of the Ordovician ocean during the Taconian Orogeny. Acadian folds are upright in contrast to the recumbent Taconian structures of northern New Brunswick. Shallow-dipping second Acadian cleavage is localized along fault boundaries. Most fault indicate high-angle reverse movement toward the northwest, but local interaction between fault blocks complicate the pattern.

Volcanism along the northwestern margin of the Miramichi Highlands, and high grade metamorphism and plutonism within the Highlands can then be related to crustal thickening accompanying continental collision (Dewey and Burke, 1973). The locus of volcanic extrusion and plutonism is largely fault-controlled. During convergence the Highlands were thrust to the northwest over the adjacent downfaulted volcanic margin. Fault adjustments continued on a lesser scale into the Carboniferous.

ACKNOWLEDGEMENTS

The writer wishes to thank G.E. Pajari (University of New Brunswick) and C. St. Peter (New Brunswick Mineral Resources Branch) for discussions on the geology of the area. R. Phillips drafted the figures.

REFERENCES

Anderson, F.D., 1972, The Catamaran fault, north-central New Brunswick: Canadian Jour. Earth Sci., v. 9, p. 1278-1286.

Bird, J.M. and Dewey, J.F., 1970, Lithosphere plate-continental margin tectonics and the evolution of the Appalachian orogeny: Geol. Soc. America Bull., v. 81, p. 1031-1060.

Bostock, H.H., Currie, K.L., and Wanless, R.K., 1979, The age of the Roberts Arm Group, north-central Newfoundland: Canadian Jour. Earth Sci., v. 16, p. 599-606.

Church, W.R. and Stevens, R.K., 1971, Early Paleozoic ophiolite complexes of the Newfoundland Appalachians as mantle-oceanic crust sequences: Journal of Geophysical Research, v. 76, p. 1460-1466.

Davis, G.H., 1972, Deformational history of the Caribou stratabound sulphide deposit, Bathurst, New Brunswick, Canada: Econ. Geol., v. 67, p. 634-655.

_____, 1973, Deformational history of the Caribou stratabound sulphide deposit, Bathurst, New Brunswick, Canada – a reply: Econ. Geol., v. 68, p. 572-577.

Dewey, J.F. and Burke, K.C.A., 1973, Tibetan, Variscan, and Precambrian basement reactivation, products of continental collision: Jour. Geol., v. 81, p. 683-692.

Dineley, D.L. and Williams, B.P.J., 1968, The Devonian continental rocks of the lower Restigouche River, Quebec: Canadian Jour. Earth Sci., v. 5, p. 945-953.

Fyffe, L.R. and Cormier, R.F., 1979, The significance of radiometric ages from the Gulquac Lake area of New Brunswick: Canadian Jour. Earth Sci., v. 16, p. 2046-2052.

Greiner, H.R., 1970, Geology of the Charlo area, 21-0/16, Restigouche County, New Brunswick: Mineral Resources Branch, Dept. Natural Resources, New Brunswick, Map Series 70-2.

Helmstaedt, H., 1970, Geology of Head of Middle River and Wildcat Brook (0-6 map-area), northern New Brunswick: Mineral Resources Branch, Dept. Natural Resources, New Brunswick, Map Series 70-7.

_____, 1971, Structural geology of Portage Lakes area, Bathurst-Newcastle district, New Brunswick: Geol. Survey Canada Paper 70-28.

_____, 1973a, Structural geology of the Bathurst-Newcastle district: in Rast, N., ed., Geology of New Brunswick: New England Intercollegiate Geological Conference Field Guide, p. 34-46.

_____, 1973b, Deformational history of the Caribou stratabound deposits, Bathurst, New Brunswick, Canada – a discussion: Econ. Geol., v. 68, p. 571-572.

Hynes, A. 1976, Magmatic affinity of Ordovician volcanic rocks in northern Maine and their tectonic significance: American Jour. Science, v. 276, p. 1208-1224.

Johnson, J.G. and Dasch, E.J. 1972, Origin of the Appalachian Faunal Province of the Devonian: Nature Physical Science, v. 236, p. 125-126.

Kennedy, D.J., Barnes, C.R. and Uyeno, T.T., 1979, A Middle Ordovician conodont faunule from the Tetagouche Group, Camel Back Mountain, New Brunswick: Canadian Jour. Earth Sci., v. 16, p. 540-551.

Lutes, G., 1979, Geology of Fosterville-North and Eel Lakes (map-area G-23) and Canterbury-Skiff Lake (map-area H-23): Mineral Resources Branch, Dept. Natural Resources, New Brunswick, Map Rept. 79-3.

McBride, D.E., 1976, Tectonic setting of the Tetagouche Group, host to the New Brunswick polymetallic massive sulphide deposits: in Strong, D.F., ed., Metallogeny and Plate Tectonics: Geol. Assoc. Canada Spec. Paper 14, p. 473-485.

McKerrow, W.S. and Ziegler, A.M., 1971, The Lower Silurian palaeography of New Brunswick and adjacent areas: Jour. Geol., v. 71, p. 635-646.

Neuman, R.B., 1968, Palaeographic implications of Ordovician shelly fossils in the Magog belt of the northern Appalachian region: in E-an Zen, White, W.S., Hadley, J.B. and Thompson, J.B., eds., Studies of Appalachian Geology: Northern and Maritime: New York, Interscience Publishers, p. 35-48.

Noble, J.P.A., 1976, Silurian stratigraphy and palaeogeography, Point Verte area, New Brunswick, Canada: Canadian Jour. Earth Sci., v. 13, p. 537-545.

Nowlan, G.S., 1979, Report on thirteen samples collected for conodonts from Lower Palaeozoic strata in New Brunswick: Geol. Survey Canada, Eastern Palaeontology Section, Rept. No. 08-GSN-1979.

O'Brien, B.H., 1977, Pre-Acadian deformation, metamorphism, and intrusion in the vicinity of the Pokiok Pluton, west-central New Brunswick and its regional implications: Canadian Jour. Earth Sci., v. 14, p. 1796-1808.

Pajari, G.E., Rast, N. and Stringer, 1977, Palaeozoic volcanicity along the Bathurst-Dalhousie geotraverse, New Brunswick, and its relations to structure: in Baragar, W.R.A., Coleman, L.C. and Hall, J.M., eds., Volcanic Regimes in Canada: Geol. Assoc. Canada Spec. Paper 16, p. 111-124.

Pickerill, R.K., 1976, Significance of a new fossil locality containing a Salopina community in the Waweig Formation: Canadian Jour. Earth Sci., v. 13, p. 1328-1331.

Poole, W.H., 1963, Hayesville, New Brunswick: Geol. Survey Canada Map 6-1963.

Potter, R.R., 1969, The geology of the Burnt Hill area and ore controls of the Burnt Hill tungsten deposit: Ph. D. Thesis, Carleton University, Ottawa, Ontario.

Rast, N., Kennedy, M.J. and Blackwood, R.R., 1976, Comparison of some tectonostratigraphic zones in the Appalachians of Newfoundland and New Brunswick: Canadian Jour. Earth Sci., v. 13, p. 868-875.

Ruitenberg, A.A., Fyffe, L.R., McCutcheon, S.R., St. Peter, C.J., Irrinki, R.R. and Venugopal, D.V., 1977, Evolution of pre-Carboniferous tectonostratigraphic zones in the New Brunswick Appalachians: Geosci. Canada, v. 4, p. 171-181.

Skinner, R., 1974, Geology of Tetagouche Lakes, Bathurst and Nepisiquit Falls map-areas, New Brunswick: Geol. Survey Canada Memoir 371, 133 p.

St.-Julien, P. and Hubert, C., 1975, Evolution of the Taconian Orogen in the Quebec Appalachians: American Jour. Sci., v. 275-A, p. 337-362.

St. Peter, C., 1978, Geology of parts of Restigouche, Victoria, and Madawaska Counties, Northwestern New Brunswick: Mineral Resources Branch, Dept. Natural Resources, New Brunswick, Rept. of Investigation No. 17.

——————————, 1979, Geology of Wapske-Odell River-Arthurette region, New Brunswick (map-areas I-13, I-14, H-14): Mineral Resources Branch, Dept. Natural Resources, New Brunswick, Map Report 79-2.

Strong, D.F., 1977, Volcanic regimes of the Newfoundland Appalachians: in Barager, W.R.A., Coleman, L.C. and Hall, J.M., eds., Volcanic Regimes in Canada: Geol. Assoc. Canada Special Paper 16, p. 61-90.

Whitehead, R.E.S. and Goodfellow, W.D., 1978, Geochemistry of volcanic rocks from the Tetagouche Group, Bathurst, New Brunswick, Canada: Canadian Jour. Earth Sci., v. 15, p. 207-219.

Williams, H., 1979, Appalachian Orogen in Canada: Canadian Jour. Earth Sci., v. 16, p. 792-807.

Manuscript Received December 15, 1979
Revised Manuscript Received April 22, 1980

Major Structural Zones and Faults of the Northern Appalachians, edited by
P. St-Julien and J. Béland, Geological Association of Canada Special Paper 24, 1982

ACADIAN AND HERCYNIAN STRUCTURAL EVOLUTION
OF SOUTHERN NEW BRUNSWICK

A.A. Ruitenberg and S.R. McCutcheon
Department of Natural Resources Geological Surveys Branch
P.O. Box 1519, Sussex, N.B. E0E 1P0

ABSTRACT

Structural trends in southern New Brunswick mainly reflect effects of the Acadian and Hercynian orogenies. Pre-Acadian deformation effects occur in the Miramichi Zone and parts of the Caledonia Zone. The Acadian and Hercynian orogenic effects coincided with extensive right lateral movements between the Avalon and Meguma microcontinental blocks along the Cobequid-Chedebucto Fault.

The early Middle Devonian Acadian orogeny was responsible for the northeast trending, steep penetrative fabric and faults that characterize the Lower Paleozoic rocks in the St. Croix and Magaguadavic Zones of southwestern New Brunswick and probably the gently dipping fabric that occurs in the northeastern Fundy Cataclastic Subzone. Ordovician and Lower Silurian rocks are generally more intensely deformed than the Upper Silurian strata except in the southwestern Mascarene-Nerepis Belt (immediately north of the contact zone of the Avalon and Meguma blocks), where Upper Silurian strata are intensely deformed.

The final episode of the Acadian orogeny was characterized by northward and locally southward directed thrusting. The major granitoid intrusions were emplaced immediately after this episode, although a few deformed granitoid intrusions are earlier.

In the Cumberland and Moncton subbasins, rift tectonics (reflected by block faulting), controlled sedimentation from late Devonian to Westphalian time. Block faulting resulted at least in part from reactivation of older faults that originated during the Acadian orogeny or earlier. The rifting episode was interrupted by two compressive deformation episodes that appear to have occurred penecontemporaneously with two Hercynian orogenic events in Europe. The only penetrative fabric formed during this time is found in the southwestern part of the Fundy Cataclastic Subzone east and west of Saint John.

IGCP Project 27   Canadian
IUGS
UNESCO   Caledonide   Contribution
Orogen   No. 26

RÉSUMÉ

Dans la partie sud du Nouveau Brunswick les directions structurales résultent surtout des orogénèses acadienne et hercynienne. Les effets de la déformation pré-acadienne s'observent dans la zone de Miramichi et dans une partie de la zone de Calédonia. Les déformations reliées aux orogénèses acadienne et hercynienne sont associées aux importants mouvements de décrochement dextres de la faille de Cobequid-Chedebucto laquelle sépare les microcontinents d'Avalon et de Méguma.

L'orogénèse acadienne, du début du Dévonien moyen, est caractérisée dans les roches du Paléozoïque Inférieur des zones de Ste-Croix et de Magaguadavic du sud-ouest du Nouveau Brunswick, par une fabrique pénétrative et des failles de direction nord-est fortement pentées. Elle est en outre probablement responsable de la fabrique à faible pendage observée dans la partie nord-est de la sous-zone cataclastique de Fundy. Les roches de l'Ordovicien et du Silurien inférieur sont généralement plus déformées que les strates du Silurien supérieur à l'exception de celles de la partie sud-ouest de la zone de Mascarene-Nerepis (au nord immédiat de la zone de contact entre les compartiments tectoniques d'Avalon et de Méguma) où les strates du Silurien supérieur sont intensément déformées.

La fin de l'orogénèse acadienne fut caractérisée par des chevauchements vers le nord et localement vers le sud. La mise en place des intrusions granitoïdes majeures a suivi immédiatement la fin de cette orogénèse quoique quelques intrusions granitoïdes déformées soient antérieures.

Dans les sous-bassins de Cumberland et de Moncton, la sédimentation du Dévonien terminal-Westphalien s'effectua dans un système de rift. Ce réseau de failles résulte, en partie, d'une réactivation de failles anciennes qui se sont développées durant l'orogénèse acadienne ou antérieurement. Cette période d'extension (rifting) fut interrompue par deux périodes de compression lesquelles semblent contemporaines à deux évènements de l'orogénèse hercynienne d'Europe. La seule fabrique pénétrative développée durant cette période est observée dans la partie sud-ouest de la sous-zone cataclastique de Fundy à l'est et à l'ouest de Saint-Jean.

ACADIAN AND HERCYNIAN
STRUCTURAL EVOLUTION OF SOUTHERN NEW BRUNSWICK

The structural geology of southern New Brunswick shows effects of both the Acadian and Hercynian orogenies that were, in part, superimposed upon older pre-Acadian (Precambrian?) structures. Effects of Acadian polyphase deformation in parts of southwestern New Brunswick were described previously by Ruitenberg (1967, 1968), Helmstaedt (1968), Brown and Helmstaedt (1970), Garnett (1973), Garnett and Brown (1973), and Donohoe (1978). Hercynian and possibly older deformational structures were described by Ruitenberg *et al.* (1973) in the "Fundy Cataclastic Subzone" along the Bay of Fundy coast and by Rast and Grant (1973) in the southwestern part of this sub-zone. Carboniferous (Hercynian) deformational structures in the Moncton Subbasin of southeastern New Brunswick, were described by Gussow (1953), and more recently by McCutcheon (1978), McLeod and Ruitenberg (1978) and McLeod (1979b).

The purpose of this paper is to summarize and compare these deformational effects in different parts of the region. The possible reasons for the variations in structural style in the region are also briefly discussed.

GENERAL GEOLOGY

Several well-defined tectonostratigraphic zones can be distinguished in southern New Brunswick (Ruitenberg *et al.*, 1977a; Fig. 1). Each of these zones can be further sub-divided, mainly on the basis of structural style (see next section).

The *Caledonia Zone* (Fig. 1, zone 1; part of Avalon Zone, Williams, 1978) contains the oldest rocks in southern New Brunswick. Precambrian (Helikian-Hadrynian) carbonate and clastic sedimentary rocks of the Green Head Group are found in subzone 1d (Alcock, 1938; Leavitt, 1963; Wardle, 1978). These are overlain by Precambrian (Hadrynian) felsic and mafic volcanic rocks of the Coldbrook Group (Alcock, 1938; Giles and Ruitenberg, 1977; Ruitenberg *et al.*, 1979), which constitute most of subzones 1b, c, and e. Subzone 1e is atypical of the Coldbrook Group in that it is dominated by a mylonite zone and dyke swarm.

Coldbrook Group rocks in zone 1 are unconformably overlain by Cambrian fluviatile sedimentary rocks, succeeded by shallow marine clastic and carbonate rocks (Alcock, 1938; McLeod and McCutcheon, 1981). Mississippian and Pennsylvanian fluvatile sedimentary and volcanic rocks also lie upon Precambrian rocks in part of the zone. The youngest rocks in the zone are Triassic red beds and volcanic rocks, which occupy a few small areas along the Bay of Fundy coast.

The *St. Croix Zone* (zone 2) comprises three lithologic divisions that are separated by major faults or intrusions. The Rolling Dam Belt (subzone 2c) is composed mainly of Silurian and Lower Devonian wackes and slates that unconformably overlie Ordovician graphitic and pyritiferous slate, siltstone, quartzite and greywacke (Ruitenberg, 1967; Ruitenberg and Ludman, 1978). The Mascarene-Nerepis Belt (subzone 2a) is composed of Silurian marine felsic and mafic volcanic and shallow water clastic sedimentary rocks (Ruitenberg, 1968; Donohoe, 1978). The Ovenhead Belt (subzone 2b) consists mainly of Lower Devonian mafic and felsic terrestrial volcanics and intercalated fluviatile sedimentary rocks (Ruitenberg, 1968; Pickerell and Pajari, 1976; Ruitenberg and McCutcheon, 1978). Major granitoid intrusions were emplaced in the St. Croix Zone during Middle Devonian and Early Carboniferous time (Ruitenberg 1967, 1968, 1969).

The *Magaguadavic Zone* (zone 3) is characterized by a thick sequence of Silurian greywackes and slates. It is separated from the St. Croix Zone by the Fredericton Fault.

The *Miramichi Zone* (zone 4) is composed of a thick succession of Ordovician slate, quartzite, felsic and mafic volcanic rocks and locally iron formation (Tetagouche Group; Ruitenberg *et al.*, 1977b). The Tetagouche rocks are, in places, overlain unconformably by Silurian clastic sedimentary rocks.

The *Central Basin* (zone 5) chiefly contains Carboniferous clastic sedimentary rocks, but also a volcanic sequence (van de Poll, 1967; Ruitenberg, 1967; Gemmell, 1975) and a thin carbonate unit.

The *Moncton and Cumberland subbasins* (subzones 6a and b) are composed chiefly of Carboniferous fluviatile and deltaic clastic sedimentary rocks with the exception of one major unit, the Windsor Group, which is composed mainly of carbonates and evaporites. Volcanic rocks occur locally in the sequence (Gussow, 1953; McCutcheon, 1978; McLeod, 1979b).

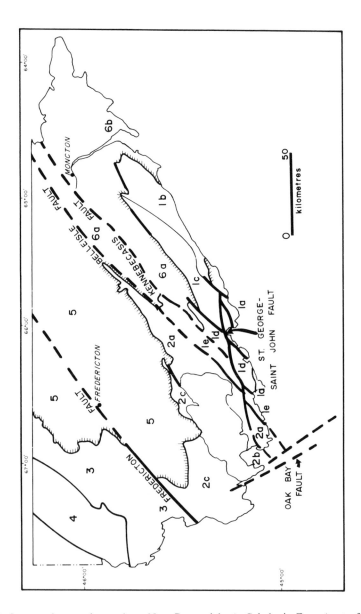

Figure 1. Structural zones in southern New Brunswick; 1. Caledonia Zone (part of Avalon Zone, Williams 1978); 1a, southwestern and 1b, northeastern "Fundy Cataclastic Subzones"; 1c, Loch Lomond Subzone; 1d, Saint John Subzone; 1e, Kingston Subzone; 2, St. Croix Zone; 2a, Mascarene-Nerepis Belt (Subzone); 2b, Oven Head Belt (Subzone) and 2c, Rolling Dam Belt (Subzone); 3, Magaguadavic Zone; 4, Miramichi Zone; 5, Central Carboniferous Basin; 6a, Moncton Subbasin; 6b, Cumberland Subbasin.

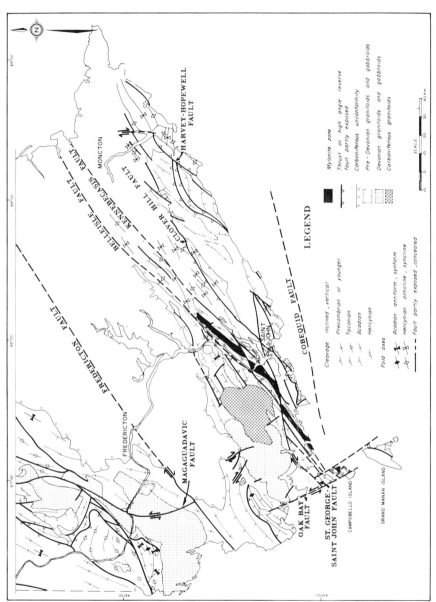

Figure 2. Structural geology map of southern New Brunswick.

PRE-ACADIAN DEFORMATION

Pre-Acadian deformation can be demonstrated with certainty only in zones 1 and 4.

Precambrian deformation is indicated in zone 1 by the local presence of deformed pebbles in basal Cambrian conglomerate (Ruitenberg et al., 1979, p. 83; McLeod and McCutcheon, 1981), but it does not appear that all Precambrian rocks were affected by this orogeny. Subzone 1d was deformed during Precambrian time (Ruitenberg et al., 1979). This resulted in a well-developed, northeast striking, steeply dipping, penetrative fabric. A broad, northeast striking, mylonite zone in subzone 1e (Fig. 2), which is flanked by and grades into less deformed Coldbrook Group rocks, is probably also Precambrian in age. This mylonite zone is intruded by a mafic dyke swarm that is at least in part contemporaneous with the deformation. McLeod (1979a) demonstrated that Silurian rocks unconformably overlie this dyke swarm on Campobello Island in the southwestern part of subzone 1e. Wardle (1978) and Rast (1979) suggest that the dykes are Precambrian because of their absence in Cambrian and Ordovician outliers. A K-Ar date of 369 ± 12 Ma reported by Helmstaedt (1968) from an amphibolitized dyke, however, indicates that the mylonites were recrystallized during the Acadian orogeny. It is notable that the intense deformational effects found in the Coldbrook rocks of subzone 1e are not reflected in subzone 1c, where rocks of the same age generally exhibit no tectonic fabric.

Ordovician rocks in the southern Miramichi Zone (zone 4) display a composite cleavage (S_1, S_2). S_2 cuts S_1 in F_2 fold hinges, but along the limbs they are nearly parallel. F_1 fold closures are not common, but they were found in some localities (Venugopal, 1978). The S_1, S_2 fabric defines a regional northerly closing fold in this area. The penetrative fabric (S_1, S_2) probably formed during the Taconian orogeny (Fyffe, this volume). In the southwestern part of subzone 2c (northeast of St. Stephen), Silurian rocks unconformably overlie Ordovician strata, but no convincing evidence for a Taconian penetrative fabric has been found (Ruitenberg, 1967; Brown and Helmstaedt, 1970).

ACADIAN DEFORMATION

Ordovician and Silurian rocks of the St. Croix (zone 2), Magaguadavic (zone 3) and Miramichi (zone 4) zones, Upper Precambrian rocks of the Kingston Peninsula (subzone 1e) and probably the northeastern Fundy Cataclastic Subzone (1b) were deformed during the Acadian orogeny (Ruitenberg, 1967, 1968; Helmstaedt, 1968; Brown and Helmstaedt, 1970; Garnett, 1973; Garnett and Brown, 1973; Donohoe, 1978; Ruitenberg et al., 1973, 1979). The main compressive deformation phases (D1 and D2) attributed to this orogeny occurred during early Middle Devonian time (Ruitenberg, 1967, 1968; Pajari et al., 1974), but subsequent deformations (D3 and D4) are younger.

Folding

Most of the Acadian cleavage trends reflect the effects of the first or main deformation phase (D1). Various subzones or belts in the region that were deformed by the Acadian orogeny exhibit distinct differences in both style and intensity of folding. These differences are schematically depicted in Table I.

TABLE I

TABLE SHOWING DEFORMATION EPISODES IN SOUTHERN NEW BRUNSWICK

		PERMIAN	ROLLINGDAM BELT	MASCARENE-NEREPIS BELT	FUNDY CATACLASTIC SUBZONE	MONCTON AND CUMBERLAND SUBBASINS
EARLY HERCYNIAN LATE P.	PENNSYLVANIAN	STEPHANIAN		NE		NE
		WESTPHALIAN			NNW S₈,S₇ W SSE S₈,S₇ SW part of zone	NNW SSE NW SE S₈ S₈
	MISSISSIPPIAN	NAMURIAN				SSW ENE S₈
		VISEAN	NNW P Sn,Zn SSE	V W,Mo,Sn Cu,Pb,Zn Bi,U	M P Zn,Pb SSE	NNW P Cu NNW S₁ SSE SSE NE part of zone	NNW SSE S₈ V
LATE & POST ACADIAN		TOURNASIAN	NNW SSE Sb,U,Au	NNW SSE W,Sn,Mo,Cu	WSW ENE W,Sn,Mo,Cu	WSW ENE Pb,Zn,Cu,Ag	
		LATE DEVONIAN					
	LATE DEVONIAN	LATE	S₃ WSW ENE S₁,S₈ P Au	S₃ WSW ENE S₁,S₈ SW P	NE		
ACADIAN	MIDDLE DEVONIAN	EARLY	NW time uncertain SE S₁,S₈ NW SE S₂ NW SE S₈ S₁ S₂ S₁b S₂,S₁a	NW S₁ SE S₁	NW S₁ SE S₁	S₈ SE NW SE ? SW S₁ NW SE NE part of zone	

LEGEND

Fold with axial surface cleavage (S₁)
Fold with fanning cleavage (S₁)
Overturned fold in S₁
Upwarp (bending fold)
Fold with concentric (S₁a) and axial surface (S₁b) cleavage
Close fold without or with poorly developed axial surface cleavage
Steeply plunging cross-fold in S₁
Gently plunging kink band in S₁
Open fold
Thrust fault
Wrench fault
Block fault
Mineralized fracture
P Plutonism
V Volcanism

Ag – Silver
Au – Gold
Bi – Bismuth
Cu – Copper
Mo – Molybdenum
Pb – Lead
U – Uranium
Zn – Zinc

Deformed Upper Silurian rocks in the southwestern part of the Mascarene-Nerepis Belt (subzone 2a) show that the first or main phase folds (F_1) are isoclinal and steeply plunging, and exhibit a well-defined axial surface cleavage (S_1). These folds resulted from intense northwest-southeast compression (Ruitenberg, 1968; Brown and Helmstaedt, 1970; Donohoe, 1978). Similar rocks in the northeastern part of this belt (McCutcheon and Ruitenberg, in press) are deformed by more open, gently plunging folds (F_1) that are slightly overturned to the south. Lower Silurian rocks on Campobello Island are deformed by open, gently plunging folds (McLeod, 1979a). F_1 folds in the Rolling Dam Belt (subzone 2c) are nearly isoclinal, gently plunging and slightly overturned to the north or south (Ruitenberg, 1967, 1968). A penetrative cleavage approximately parallel to the axial surfaces is usually well developed.

Some parts of the Rolling Dam and Mascarene-Nerepis belts (subzones 2a and 2c), exhibit F_1 folds (with associated axial surface cleavage) that deform an earlier (pre-F_1) cleavage that is parallel to bedding (concentric cleavage S_{1a}, Table I; Ruitenberg, 1967; Ruitenberg and McCutcheon, 1978; McCutcheon and Ruitenberg, in press). Detailed investigation did not reveal any evidence for mesoscopic axial surface cleavage folds associated with the pre-F_1 cleavage nor is there any evidence for such folds in the regional map pattern. Therefore, Ruitenberg and McCutcheon (1978) suggested that this cleavage, parallel to bedding (S_{1a}), may reflect flattening produced by arching or bending during an early stage of the main deformation episode. Ramberg (1967) produced cleavage of this type in scale models by using a similar mechanism. McLelland and Isachson (1980) observed similar pre-F_1 cleavage parallel to bedding in the Adirondacks.

In the southwestern Mascarene-Nerepis Belt, a second steeply dipping cleavage (S_2) cuts S_1 at a small angle. A second, gently southward dipping fabric (S_2) locally cuts S_1 in the northeastern part of the belt. In part of the Rolling Dam Belt, the penetrative cleavage (S_1) is deformed by gently plunging recumbent folds F_2; overturned to the south). A gently northeastward dipping fabric is locally associated with these F_2 folds (Ruitenberg, 1967). These second folds probably formed during the waning stages of the Acadian compression.

Regional reconnaissance mapping of the Magaguadavic Zone (zone 3) indicates that this zone is characterized by gently plunging, sub isoclinal F_1 folds. A steeply dipping axial surface cleavage is associated with these folds and strikes north-northeast rather than northeast as in the St. Croix Zone (zone 2).

It was mentioned earlier that the composite fabric S_1, S_2 in the southern Miramichi Zone (zone 4) defines a northward closing fold. It is possible that this post-S_2 folding occurred during or later than the Acadian orogeny.

Effects of the Acadian orogeny in parts of the Caledonia Zone (zone 1; mainly Upper Precambrian) are uncertain. In the Kingston Subzone (subzone 1e), effects of the Acadian orogeny are difficult to differentiate from pre-Acadian deformation. Tight, steeply plunging folds with a well developed axial surface cleavage characterize this subzone (McCutcheon and Ruitenberg, in press). This style of folding resembles that in the southwestern Mascarene-Nerepis Belt, but the fact that these intensely folded rocks appear to grade into a pre-Acadian (probably Precambrian) mylonite zone (Fig. 2; pre-Acadian deformation) suggests that the penetrative deformation in this area mainly predates the Acadian orogeny. Acadian effects are indicated, however, by a K-Ar date of 369 ± 21 Ma from an amphibolitized dyke (Helmstaedt, 1968).

Subzone 1b (northeastern Fundy Cataclastic Subzone) was probably deformed during the Acadian orogeny. This is a complex belt containing numerous fault blocks. A penetrative cataclastic fabric (S_1) appears to be mostly the result of flattening associated with arching (or bending; Ruitenberg et al., 1973, 1979). Gently northeast or southwest plunging shear type folds (F_1) are locally associated with the S_1 fabric (mainly along the margins of uplifted blocks), but there is no evidence for regional folds of this type (Ruitenberg et al., 1973, 1979). Recent detailed mapping by McLeod (in prep.) has demonstrated the local presence of a second penetrative cleavage S_2. F_2 folds associated with S_2, are overturned to the north and south.

Penetrative deformation in the northeastern Fundy Cataclastic Subzone (sub-zone 1b) occurred prior to the deposition of the Upper Devonian (Famennian) Mem-ramcook Formation (Ruitenberg *et al.*, 1973, 1979), and after deposition of the Cam-brian Saint John Group (McLeod and McCutcheon, 1981). This deformed zone is confined to an area immediately north of the Cobequid Fault Zone (Fig. 2), along which the Meguma and Avalon blocks (Williams, 1978) collided during the Acadian orogeny (Keppie, this volume). This suggests that this deformation occurred during the Acadian rather than during the Taconian orogeny.

Faulting

The final episode of the Acadian orogeny in southern New Brunswick was characterized by prominent northeasterly striking thrusts. These faults commonly terminate against northwesterly striking wrench faults.

In the Mascarene-Nerepis Belt (subzone 2a), there are several thrust- or high angle reverse faults that dip to the south (McCutcheon, 1981; McCutcheon and Ruitenberg, in press). These faults are commonly terminated by northwesterly strik-ing wrench faults. All the faults postdate the oldest of the Devonian granites (de-formed) and are truncated by younger Devonian and Carboniferous intrusions (un-deformed).

A southward dipping thrust also separates Ordovician and Silurian strata in the Rolling Dam Belt (subzone 2c). The most prominent wrench fault in the region is the sinistral Oak Bay Fault, which offsets the Ordovician-Silurian unconformity nearly 4 km (Ruitenberg, 1967, 1968).

The southern boundary of the St. Croix Zone (zone 2) is the Belleisle Fault. Garnett (1973) concluded that only reverse dip-slip movement took place during the Acadian orogeny. The similarity of the Silurian Long Reach Formation on both sides of the fault precludes extensive post-Silurian strike-slip displacement. A Silurian or Lower Devonian granite northeast of Loch Alva (northeastern Mascarene-Nerepis Belt; Giles *et al.*, 1974) is also not significantly displaced by the fault. On the other hand, dissimilar Cambrian rocks north and south of the fault indicate that major movement (strike-slip) occurred prior to the Acadian Orogeny.

A major southward-directed thrust was delineated recently in the northeastern Fundy Cataclastic Subzone 1b (McLeod and McCutcheon, 1981). This thrust is pos-sibly related to a late phase of the Acadian orogeny.

Major strike slip movement is postulated along the Fredericton Fault, which constitutes the boundary between the St. Croix and Magaguadavic Zones (Fig. 2). The difference in the trend of the Acadian fabric across this fault (Fig. 2) indicates that the movement postdates the penetrative deformation. Gravity and magnetic maps compiled by Haworth (1975) also support a major change in structural trend across the Fredericton Fault.

One of the most important structures in the Bay of Fundy coastal area is the Cobequid Fault (Fig. 2). The Cambro-Ordovician Meguma Group of southern Nova Scotia was juxtaposed, along this fault, against the Upper Precambrian rocks of northern Nova Scotia and southern New Brunswick (Avalon Microcontinent).

Aeromagnetic maps (Geological Survey of Canada, Maps 7037G and 7036G, 1958) indicate that the Cobequid Fault extends westward from the Cobequid Highlands in Nova Scotia under the Bay of Fundy, to terminate against a north-northwesterly lineament, which bounds the eastern margin of the Upper Precambrian basement exposed on Grand Manan Island (Fig. 2). Juxtaposition of the Meguma and Avalon blocks probably took place prior to late-Early Devonian time, because the oldest rocks that can be correlated across the Cobequid Fault are grey siltstones and quartzites of Emsian age (Donohoe, 1979, oral commun.). Detailed mapping by Donohoe (1979) and Keppie (this volume) shows that extensive dextral offset has taken place along the Cobequid Fault during a late phase of the Acadian Orogeny.

Movements along major thrust and wrench faults undoubtedly did occur during the final stages of the Acadian orogeny, but it is possible that some of these faults were initiated earlier. The northeasterly trending mylonitized belt in the Kingston Subzone (subzone 1e) indicates that northeasterly trending zones of structural weakness existed in the basement rocks prior to the Acadian orogeny and were reactivated subsequently. Geophysical evidence (Haworth, 1975) also suggests that some major northwesterly trending wrench faults may have existed prior to the Acadian orogeny.

POST ACADIAN DEFORMATION

The major Acadian and Hercynian orogenic events were separated by a long transitional episode (Table I) mainly characterized by distension. Major intrusions were emplaced in the St. Croix Zone and locally the Caledonia Zone. Rift-controlled sedimentation occurred during this episode in the Moncton and Cumberland sub-basins (Gussow, 1953; McCutcheon, 1978; McLeod and Ruitenberg, 1978).

Folding

In the Mascarene-Nerepis and Rolling Dam belts, the Acadian S_1 fabric (and S_2 where present) is commonly deformed by S- and Z- shaped and locally conjugate sets of steeply to moderately plunging chevron-type crossfolds (F_3 folds, Ruitenberg, 1967, 1968; Brown and Helmstaedt, 1970; Donohoe, 1978; Table I).

The attitudes and geometry of F_3 folds are consistent with sub-horizontal slip along S_1 surfaces. These structures appear to be most abundant in broad northwesterly trending fault zones that bound major fault blocks. Ruitenberg (1967, 1968) demonstrated that Devonian granitoid intrusions in southwestern New Brunswick were emplaced contemporaneously with this episode of cross folding. Gold-bearing quartz veins were emplaced in the southwestern Rolling Dam Belt (Ruitenberg, 1967) and copper deposits in the northeastern Mascarene-Nerepis Belt were remobilized as a result of this deformation (Ruitenberg, 1972, 1976).

Gently plunging kink bands in S_1 represent the final folding phase (F_4) in the Mascarene-Nerepis and Rolling Dam Belts (Ruitenberg 1967, 1968; Brown and Helmstaedt, 1970; Donohoe, 1978; Table I). These structures are consistent with shortening approximately parallel to the dip direction of the cleavage and a north-northwest – south-southeast sub-horizontal extension (Ruitenberg, 1967, 1968). These kink bands postdate all other folds and could have formed during the episode of rifting that controlled sedimentation in the Moncton and Cumberland subbasins.

Although no detailed structural studies were conducted in the Magaguadavic Zone, it appears that refolding of Acadian cleavage (S_1) is less common than in the Rolling Dam and Mascarene-Nerepis belts of the St. Croix Zone. Nevertheless, structures similar to F_3 folds in the St. Croix Zone were observed at a few localities in the Magaguadavic Zone.

As mentioned earlier, refolding of the composite S_1, S_2 fabric in the southern Miramichi zone could have occurred during the Acadian orogeny or later.

Penetrative cleavage (S_1 and locally S_2) in the northeastern Fundy Cataclastic Subzone 1b, is deformed by open to close, mostly gently plunging folds (F_3; F_2 folds, Ruitenberg et al., 1973) that vary in size from small crenulations in S_1 to regional folds. These folds appear to have formed as a result of sliding of the upper layers over the lower layers down the regional dip of S_1, although in some localities the reverse appears to be the case. The cleavage locally is deformed by chevron type folds (F_4; F_3 folds, Ruitenberg et al., 1973) that plunge down the dip of S_1. The folds range from small kinks to folds with limbs several kilometres across (Ruitenberg et al., 1973, 1979).

Faulting

Stratigraphic and structural studies of the Moncton and Cumberland subbasins (Gussow, 1953; McCutcheon, 1978; McLeod and Ruitenberg, 1978; McLeod, 1979b) have demonstrated that northeasterly trending faults (associated with rifting) controlled the deposition of coarse- and fine-grained sediments during Late Devonian and Early Carboniferous time (McCutcheon, 1978; McLeod and Ruitenberg, 1978; McLeod, 1979b).

In the St. Croix Zone, there is evidence for strike slip and vertical movements along northwesterly trending faults that postdate F_3 folds, but the exact time of these fault movements is not known (McCutcheon and Ruitenberg, in press).

HERCYNIAN OROGENIC EVENTS

Effects of two major folding events were distinguished in the Moncton and Cumberland subbasins (McCutcheon, 1978; McLeod and Ruitenberg, 1978; McLeod, 1979b). It is notable that these folding events occurred approximately synchronously with the early (Sudetian) and main (Asturian) episodes of the Hercynian orogeny in Europe, although it is difficult to make a direct comparison. Intense penetrative deformation in the southwestern part of the Fundy Cataclastic Subzone occurred during the latest of these deformation episodes (Rast and Grant, 1973).

Folding

The Memramcook, Albert and Weldon Formations (Famennian to Tournaisian), were deformed by east-northeasterly trending, gently plunging, close folds in the Dorchester-Hillsborough area of the Moncton Subbasin (Figs. 1 and 2; McLeod and Ruitenberg, 1978; McLeod, 1979b). Locally, broad north-northwesterly trending folds, superimposed upon the east-northeasterly ones, constitute the final phase of this deformation episode (McLeod and Ruitenberg, 1978; McLeod, 1979b). These late folds occur in an area with small thrusts along the contact of the Upper Precambrian rocks in the Hillsborough area (McLeod, 1979b). It is possible that the folding and thrusting in this area reflects lateral movements of the Precambrian faulted block. In the Apohaqui-Markhamville area (McCutcheon, 1978), farther west in the Moncton Subbasin, this deformation episode is represented · only by north-northwesterly trending folds.

The lower time limit of the Early Carboniferous deformation episode depends on the age of the *Vallatisporites Vallatus* miospore zone of the Albert Formation which has been assigned to the Late Tournaisian or Early Viséan (Barss, written commun. 1977), but could be as old as Lower or Middle Tournaisian (George *et al.*, 1976). The upper time limit is determined by the fact that the Windsor Group (Middle Viséan) was not affected by this deformation. Eysinga (1975) and Autran (1978) stated that the earliest Hercynian orogenic event (Sudetian) in western Europe commenced during the Dinantian. This was penecontemporaneous with the earliest deformation in the Moncton and Cumberland subbasins. The folds associated with this deforma-tion episode in the Moncton and Cumberland subbasins represent the first major compressive event after the Acadian orogeny. This Early Carboniferous deformation episode is here referred to as early Hercynian (Table I).

The second major deformation episode in the Moncton Subbasin produced broad, open east-northeasterly trending folds that affected rocks ranging in age from Middle Viséan (Windsor Group) to Westphalian B (McLeod and Ruitenberg, 1978; McCutcheon, 1978; McLeod, 1979b). This deformation appears to have occurred about the same time as the penetrative deformation in the southwestern Fundy Cataclastic Subzone 1a.

Both the Coldbrook (Hadrynian) and Mispeck (probably Namurian, McCut-cheon, 1979) rocks exhibit a well-developed, south-dipping penetrative fabric in the southwestern Fundy Cataclastic Subzone 1a (Ruitenberg *et al.*, 1973; Rast and Grant, 1973). The fabric is similar to that in the northeastern Fundy Cataclastic Subzone 1b, although these rocks were penetratively deformed previously, probably during the Acadian orogeny. As in Subzone 1b, overturned folds (F_1) associated with S_1 occur only locally. Possible correlatives of the Pictou Group (Westphalian C to Lower Permian) at McCoy Head in this area, exhibit no effects of penetrative deformation. This supports the suggestion by Rast and Grant (1973) that the penetra-tive deformation in these rocks occurred penecontemporaneously with the main episode of the Hercynian orogeny in Europe (Westphalian, Autran, 1978; Westpha-lian B, Zwart, 1963; or as late as Stephanian, Owen, 1976).

Faulting

High angle thrusting and wrench faulting accompanied the Hercynian orogenic events. These movements in part occurred along pre-existing structures.

In the southwestern Fundy Cataclastic Subzone 1a (Ruitenberg *et al.*, 1973, 1979; Rast and Grant, 1973), an east-northeasterly trending thrust zone separates intensely deformed rocks of the Carboniferous Mispeck and Upper Precambrian Coldbrook groups from less deformed rocks to the north. The thrusting deformed the penetrative fabric. The youngest rocks affected by this deformation are Middle Westphalian in age (Alcock, 1938).

The thrust zone in the southwestern Fundy Cataclastic Subzone 1a occurs about 10 km north of the western end of the Cobequid Fault. The thrust zone and Cobequid Fault probably are related. Thrusting in this part of the Fundy Cataclastic Subzone is probably reflected, in the Cobequid Fault Zone, by right lateral Carboniferous wrench movements inferred from mapping in Nova Scotia (Arthaud and Matte, 1977; Donohoe and Wallace, 1978; Donohoe, 1979; Keppie, this volume).

The Harvey-Hopewell Fault (Fig. 2) in the Cumberland Subbasin is possibly of similar age as the thrust zone in the southwestern Fundy Cataclastic Subzone. Movements along the Harvey-Hopewell Fault have placed red beds of the Mississippian Hopewell Group on grey fluviatile sedimentary rocks of the Pennsylvanian Boss Point Formation. Extensive reverse movement along the Clover Hill Fault (Gussow, 1953; McCutcheon, 1978) may also have occurred about this time.

The east-northeasterly trending St. George-Saint John Fault (Fig. 2) first delineated by McCutcheon (1976) produced a 3 km dextral offset on the northeasterly trending Lubec-Belleisle Fault. The St. George-Saint John Fault in part follows the trace of an older fault that originated during the Acadian orogeny. Latest movements along this fault offset Upper Carboniferous strata and therefore occurred during the Stephanian or later.

The northwesterly and northeasterly trending faults in the Lower Paleozoic rocks of the St. Croix Zone were reactivated during the Hercynian orogenic events. Reactivation of these faults was in places accompanied by terrestrial volcanism and emplacement of granitoid stocks (Ruitenberg, 1967, 1968). In part of the Moncton Subbasin, felsic volcanism occurred during latest Tournaisian or earliest Viséan (McLeod, 1979b). This igneous activity coincided approximately with the onset of the early Hercynian deformation episode. The large Magaguadavic Fault, in the Magaguadavic Zone, also was reactivated during late Carboniferous time (Ruitenberg and Fyffe, in prep.).

DISCUSSION

During Late Silurian to Early Devonian time, the Acadian orogeny was preceded by the closing of a marine basin in the St. Croix and Magaguadavic Zones (Ruitenberg *et al.*, 1977b). Closing of this basin was accompanied initially by submarine volcanism (Mascarene-Nerepis Belt) and later by terrestrial (Ovenhead Belt) volcanism. Poole (1976) suggested that this ensialic volcanism was related to a north-dipping subduction zone, but supporting evidence for this is sparce (Ruitenberg *et al.*, 1977b).

Closing of the Siluro-Devonian basin was followed by juxtaposition and subsequent right lateral offset, along the Cobequid-Chedebucto Fault, of the Meguma and Avalon microcontinental blocks. The most extensive dextral movement apparently occurred during a late phase of the Acadian orogeny.

The effects of the Acadian orogeny varied considerably throughout southern New Brunswick. Ordovician and Lower Silurian rocks are more intensely deformed than Upper Silurian strata in most of the region. The tight, steeply plunging main phase folds (F_1) in the Upper Silurian rocks of the southwestern Mascarene-Nerepis Belt (compared with the more open, gently plunging folds in rocks of the same age in the northeastern part) reflect intense northwest-southeast lateral compression immediately north of the impact zone of the Meguma and Avalon blocks along the northwestern end of the Cobequid Fault.

The gently plunging F_1 folds in the Rolling Dam Belt reflect northwest-southeast lateral compression as in the Mascarene-Nerepis Belt. There is no evidence for axial surface cleavage folds associated with the pre-F_1 cleavage that locally occurs parallel to bedding in both Ordovician and Silurian rocks of this belt. The pre-F_1 cleavage was probably formed as a result of arching or bending that occurred during the onset of the Acadian orogeny. On the other hand, the gently plunging, overturned F_2 folds in this zone probably formed during the waning stages of the Acadian compression.

In the northeastern Fundy Cataclastic Subzone 1b, deformation of probable Acadian origin was marked predominantly by arching or bending that produced a penetrative fabric (S_1). Shear folds (F_1) are locally associated with this penetrative fabric mainly along the margins of uplifted blocks. The S_1 fabric is cut by a second penetrative fabric (S_2) in some localities. F_2 folds associated with S_2 are overturned to north or south. The reason for extensive arching or bending along this southern margin of the Caledonia Zone in southern New Brunswick is not well understood. It is possible that arching was produced by northward directed underthrusting. This would be consistent with major southerly directed overthrusts or reverse faults of possible late Acadian age that occur in this area.

The final episode of Acadian compression in the Mascarene-Nerepis and Rolling Dam Belts of the St. Croix Zone is marked by northwesterly directed thrusts and high angle reverse faults. These structures commonly terminate against northwesterly striking wrench faults which also probably were initiated during this time. The major nondeformed granitoid intrusions in the region were emplaced immediately after this episode of thrusting.

Evidence for extensive Acadian or younger strike-slip movement along the northeasterly trending faults as postulated by Webb (1963), is generally lacking in southern New Brunswick, with one notable exception. The marked change in trend of the Acadian penetrative fabric and contrasting magnetic and gravity signatures on opposite sides of the Fredericton Fault suggest that major post-folding movement (probably strike slip displacement), occurred. Confirmation awaits detailed structural studies.

Rift tectonics reflected by block faulting dominated the region and controlled fluviatile and shallow marine sedimentation from Late Devonian to about Westphalian time (McCutcheon, 1978; McLeod and Ruitenberg, 1978; McLeod, 1979b). The block faulting resulted at least in part from reactivation of older faults that originated during the Acadian orogeny or earlier. Unlike the Hercynian belt in Europe, this episode of rift tectonics did not result in the opening of oceanic basins in the Canadian Appalachians.

The episode of rifting was interrupted by two episodes of compressive deformation. The earliest of these deformations occurred during the lowermost Viséan or Late Tournaisian, which is penecontemporaneous with the onset of the early Hercynian (Sudetian) orogenic event (Dinantian of Eysinga, 1975; Autran, 1978) in Europe. Activation (or reactivation) of major fault zones during this deformation episode was accompanied by volcanism and emplacement of granitoid stocks in the

St. Croix Zone (Ruitenberg, 1967; Ruitenberg and Ludman, 1978) and volcanism in part of the northeastern Moncton Subbasin (McLeod, 1979b). Important mineral deposits formed during late phases of this igneous activity (Table I; Ruitenberg *et al.*, 1977a) in the Rolling Dam Belt. Prominent east-northeast trending folds resulted from the early Hercynian deformation episode in the northeastern part of the Moncton-Cumberland Subbasin (McLeod and Ruitenberg, 1978; McLeod, 1979b). North-northwest trending open folds locally superimposed upon major east-northeasterly folds, are consistent with lateral movement of the Caledonia Block. North-northwesterly trending folds represent the earliest Carboniferous deformation in the southwestern part of the Moncton Subbasin (McCutcheon, 1978). The second (Carboniferous) deformation event postdated Westphalian B and probably predated the Stephanian and hence occurred about the same time as the main Hercynian deformation of Westphalian to Stephanian age in Europe (Zwart, 1963; Owen, 1976; Autran, 1978). This deformation produced the penetrative fabric in the southwestern Fundy Catalastic Subzone 1a and east-northeast trending open folds in the Moncton-Cumberland Subbasin.

Northerly directed thrusting in the southwestern Fundy Cataclastic Subzone deformed the gently south-dipping penetrative fabric. This fabric is similar to an earlier fabric in the northeastern Fundy Cataclastic Zone 1b which defines a complex arch (bending fold). This suggests that the penetrative fabric in the southwestern Fundy Cataclastic Zone resulted from renewed arching, but subsequently the southern limb of this arch was thrust northward giving rise to the structure presently exposed (Ruitenberg *et al.*, 1973). The thrusting in the southwestern Fundy Cataclastic Zone possibly coincided with reactivation of dextral movements along the Cobequid Fault in Nova Scotia. Thrusting at Harvey Hopewell and extensive reverse movements along the Clover Hill Fault (Gussow, 1953; McCutcheon, 1978) may have occurred also during the late Hercynian deformation episode.

ACKNOWLEDGEMENTS

Constructive criticism of this paper by Drs. J. Béland, R. Béland, T. Fenninger, J.B. Hamilton, R.R. Potter, P. St-Julien and an anonymous reviewer is greatly appreciated. The writers would like to thank Mr. Maurice R. Mazerolle for drafting the figures.

Field trips in the Cobequid Highlands and Meguma Zone of Nova Scotia guided by Drs. Howard Donohoe and Duncan Keppie (Nova Scotia Department of Mines) were very helpful. The field work incorporated in this study was supported over the years by the New Brunswick Department of Natural Resources, the Geological Survey of Canada and the Department of Regional Economic Expansion, Ottawa.

REFERENCES

Alcock, F.J., 1938, Geology of Saint John region New Brunswick: Geol. Survey Canada Memoir 216.

Arthaud, F. and Matte, P., 1977, Late paleozoic strike-slip faulting in southern Europe and northern Africa. Result of a right lateral shear zone between the Appalachians and the Urals: Geol. Soc. America Bull., v. 88, p. 1305-1320.

Autran, A., 1978, Synthèse provisoire des événements orogéniques calédoniens en France: in Caledonian-Appalachian Orogen of the North Atlantic Region, Project 27, Internatl. Geol. Correlation Program, Geol. Survey Canada Paper 78-13.

Brown, R.L. and Helmstaedt, H., 1970, Deformation history in part of the Lubec-Belleisle Zone of southern New Brunswick: Canadian Jour. Earth Sciences, v. 7, p. 748-767.

Donohoe, H.V., 1978, Analysis of structures in the St. George area, Charlotte County: Ph.D. Thesis, University of New Brunswick.

_____, 1979, Poster session on major faults in northern Nova Scotia and southern New Brunswick: Atlantic Geoscience Annual Meeting, Amherst, Nova Scotia.

Donohoe, H.V. and Wallace, P.E., 1978, Geology Map of the Cobequid Highlands, Preliminary Map 78-1: Nova Scotia Dept. Mines and Dept. Regional Economic Expansion, Ottawa.

Eysinga, F.W.B., 1975, Geological Time Table: Amsterdam, Elsevier Scientific Publishing Company.

Garnett, J.A., 1973, Structural analysis of part of the Lubec-Belleisle Fault zone, southwestern New Brunswick: Ph.D. Thesis, University of New Brunswick.

Garnett, J.A. and Brown, R.L., 1973, Fabric variation in the Lubec-Belleisle zone of southern New Brunswick: Canadian Jour. Earth Sci., v. 10, p. 1591-1599.

Gemmell, D.E., 1975, Carboniferous volcanic and sedimentary rocks of the Mount Pleasant Caldera and Hoyt Appendage, New Brunswick: M.Sc. Thesis, University of New Brunswick.

George, T.N., Johnson, G.A.L., Mitchell, M., Prentice, J.E., Ramsbottom, W.H.C., Sevastopulo, G.D. and Wilson, R.B., 1976, A correlation of Dinantian rocks: Geol. Soc. London, Special Report No. 7.

Giles, P.S. and Ruitenberg, A.A., 1977, Stratigraphy, paleogeography, and tectonic setting of the Coldbrook Group in the Caledonia Highlands of southern New Brunswick: Canadian Jour. Earth Sci., v. 14. p. 1263-1275.

Giles, P.S., Howells, K.D.M., McCutcheon, S.R., Ruitenberg, A.A. and Venugupal, D.V., 1974, Geology of Browns Flats-Long Reach-Moss Glen, Plate 74-78: Mineral Resources Branch, New Brunswick Dept. Natural Resources.

Gussow, W.C., 1953, Carboniferous stratigraphy and structural geology of New Brunswick, Canada: Bull. American Assoc. Petroleum Geol., v. 37, p. 1713-1816.

Haworth, R.T., 1975, Paleozoic continental collision in the northern Appalachians in light of gravity and magnetic data in the Gulf of St. Lawrence: in van der Linden W.J.M. and Wade, J.A. eds., Offshore of Eastern Canada: Geol. Survey Canada Paper 74-30.

Helmstaedt, H., 1968, Structural analysis of the Beaver Harbour area, Charlotte County, New Brunswick: Ph.D. Thesis, University of New Brunswick.

Leavitt, E.M., 1963, Geology of the Precambrian Green Head Group in the Saint John area, New Brunswick: M.Sc. Thesis, University of New Brunswick.

McCutcheon, S.R., 1976, Geology of Map Area M-29, Parts of New Lepreau and Musquash Rivers (21G/1W, 2E, 7E, 8W). Plate 76-56: Mineral Resources Branch, New Brunswick Dept. Natural Resources.

_____, 1978, Geology of the Apohaqui-Markhamville area. Map Report 78-5: Mineral Resources Branch, New Brunswick Dept. Natural Resources.

_____, 1979, Stratigraphy of the Mispeck Group appendix: in Ruitenberg, A.A., Giles, P.S., Venugopal, D.V., Buttimer, S.M., McCutcheon, S.R. and Chandra, J., Geology and mineral deposits of the Caledonia area. New Brunswick Dept. Natural Resources and Dept. Regional Economic Expansion Ottawa, Memoir I.

_____, 1980, Revised stratigraphic interpretation of the Long Reach Area, southern New Brunswick: Evidence for major northwest directed Acadian thrusting: Geol. Assoc. Canada, Annual Meeting Abstracts, Halifax.

_____, 1981, Revised stratigraphy of the Long Reach area, southern New Brunswick: evidence for major northwestward-directed Acadian thrusting: Canadian Jour. Earth Sci., v. 18, p. 646-656.

McCutcheon, S.R. and Ruitenberg, A.A., in press, Geology and mineral deposits of the Annidale-Nerepis area: Mineral Resources Branch, New Brunswick Dept. Natural Resources.

McLelland, J. and Isachson, I., 1980, Structural synthesis of the southern and central Adirondacks. A model for the Adirondacks as a whole and plate tectonic interpretation: Geol. Soc. America Bull., Part 1, v. 91, p. 68-72.

McLeod, M.J., 1979a, Geology of Campobello Island. M.Sc. Thesis, University of New Brunswick.

_____, 1979b, Geology and mineral deposits of the Hillsborough Area: Mineral Resources Branch, New Brunswick Dept. Natural Resources, Map Report 79-6.

McLeod, M.J. and Ruitenberg, A.A., 1978, Geology and mineral deposits of the Dorchester Area, Map Report 78-4: Mineral Resources Branch, New Brunswick Dept. Natural Resources and Dept. Regional Economic Expansion, Ottawa.

McLeod, M.J. and McCutcheon, S.R., 1981, A newly recognized sequence of possible Early Cambrian age in southern New Brunswick: evidence for major southward directed thrusting: Canadian Jour. Earth Sci., v. 18, p. 1012-1017.

Owen, T.R., 1976, The geological evolution of the British Isles: Pergamon International Library of Science, Technology, Engineering and Social Studies, William Clowes and Sons Limited, London, U.K., 167 p.

Pajari, G.E., Jr., Trembath, L.T., Cormier, R.F. and Fyffe, L.R., 1974, The age of the Acadian deformation in southwestern New Brunswick: Canadian Jour. Earth Sci., v. 11, p. 1309-1313.

Pickerill, R.K. and Pajari, G.E., 1976, The Eastport Formation (Lower Devonian) in the northern Passamaquoddy Bay area, southwest New Brunswick: Canadian Jour. Earth Sci., v. 13, p. 266-270.

Poole, W.H., 1976, Plate tectonic evolution of the Canadian Appalachian region: Geol. Survey Canada Report of Activities, Part B, p. 113-126.

Ramberg, H., 1967, Gravity Deformation and the Earth's Crust: New York, Academic Press, 214 p.

Rast, N., 1979, Precambrian metadiabases of southern New Brunswick – the opening of the Iapetus Ocean?: Tectonophysics, v. 59, p. 127-137.

Rast, N. and Grant, R., 1973, Transatlantic correlation of the Variscan-Appalachian orogeny: American Jour. Sci., v. 273, p. 572-579.

Ruitenberg, A.A., 1967, Stratigraphy, structure and metallization, Piskahegan-Rolling Dam area: Leidse Geologische Mededelingen, 40, p. 79-120.

_____, 1968, Geology and mineral deposits, Passamaquoddy Bay area: Mineral Resources Branch, New Brunswick Dept. Natural Resources, Rept. Investigation 7, 47 p.

_____, 1969, Mineral deposits in granitic intrusions and related metamorphic aureoles in parts of Welsford, Loch Alva, Musquash and Pennfield areas: Mineral Resources Branch, New Brunswick Dept. Natural Resources, Rept. Investigation 9, 24 p.

_____, 1972, Metallization episodes related to tectonic evolution, Rolling Dam and Mascarene-Nerepis Belts, New Brunswick: Economic Geology, v. 67, p. 434-444.

_____, 1976, Comparison of volcanogenic mineral deposits in the northern Appalachians and their relationship to tectonic evolution: in Wolf, K.H., ed., Handbook of Strata-Bound and Stratiform Ore Deposits, v. 5: Amsterdam, Elsevier Scientific Publishing Co., p. 109-159.

Ruitenberg, A.A. and Ludman, Allan, 1978, Stratigraphy and tectonic setting of early Paleozoic sedimentary rocks of the Wirral-Big Lake area, southwestern New Brunswick and southeastern Maine: Canadian Jour. Earth Sci., v. 15, p. 22-32.

Ruitenberg, A.A. and McCutcheon, S.R., 1978, Field guide to lower Paleozoic sedimentary and volcanic rocks of southwestern New Brunswick: in Guide Book for Field Trips, New England Intercollegiate Geological Conference, 70th Annual Meeting.

Ruitenberg, A.A., Venugopal, D.V. and Giles, P.S., 1973, Fundy Cataclastic Zone, New Brunswick: evidence for post-Acadian penetrative deformation: Geol. Soc. America Bull., v. 84, p. 3029-3044.

Ruitenberg, A.A., Giles, P.S., Venugopal, D.V., Buttimer, S.M. and Chandra, J., 1979, Geology and mineral deposits, Caledonia area: Mineral Resources Branch, New Brunswick Dept. Natural Resources, Memoir I, 213 p.

Ruitenberg, A.A., McCutcheon, S.R., Venugopal, D.V. and Pierce, G.A., 1977a, Mineralization related to post-Acadian tectonism in southern New Brunswick: Geosci. Canada, v. 4, p. 13-22.

Ruitenberg, A.A., Fyffe, L.R., McCutcheon, S.R., St. Peter, C.J., Irrinki, R.R. and Venugopal, D.V., 1977b, Evolution of pre-Carboniferous tectonostratigraphic zones in the New Brunswick Appalachians: Geosci. Canada, v. 4, p. 171-181.

Van de Poll, H.W., 1967, Carboniferous volcanic and sedimentary rocks of the Mount Pleasant area, New Brunswick: Mineral Resources Branch, New Brunswick Dept. Natural Resources, Rept. Investigation No. 3.

Venugopal, D.V., 1978, Geology of Benton-Kirkland-Upper Eel River Bend: Mineral Resources Branch, New Brunswick Dept. Natural Resources, Map Report 78-3, 16 p.

Wardle, R.J., 1978, The stratigraphy and tectonics of the Green Head Group: in relationship to Hadrynian and Paleozoic rocks, southern New Brunswick: Ph.D. Thesis, University of New Brunswick.

Webb, G.W., 1963, Occurrence and exploration significance of strike-slip faults in southern New Brunswick, Canada: American Assoc. Petroleum Geol. Bull., v. 57, p. 1904-1927.

Williams, H., 1978, Tectonic lithofacies map of the Appalachian Orogen: Memorial University of Newfoundland, Map No. 1.

Zwart, H.J., 1963, The structural evolution of the Paleozoic rocks of the Pyrenees: Geologischen Rundschau, Bd. 53, p. 170-205.

Manuscript Received July 5, 1979
Revised Manuscript Received December 8, 1980

Major Structural Zones and Faults of the Northern Appalachians, edited by
P. St-Julien and J. Béland, Geological Association of Canada Special Paper 24, 1982

THE STRUCTURE OF PALEOZOIC OCEANIC ROCKS BENEATH NOTRE DAME BAY, NEWFOUNDLAND

R. T. Haworth
Atlantic Geoscience Centre, Geological Survey of Canada, Bedford
Institute of Oceanography, P.O. Box 1006, Dartmouth, N.S. B2Y 4A2

H.G. Miller
Department of Physics, Memorial University, St. John's,
Newfoundland A1C 5S7

ABSTRACT

The ophiolitic rocks at Betts Cove, western Notre Dame Bay, Newfoundland, are associated with a belt of high magnetic and gravity anomalies that fringe the Bay. The offshore extension of these anomalies and their relationship to the general anomaly pattern of Newfoundland shows that oceanic crustal rocks consistently dip seaward from the margins of the Bay, and form a generally synclinal structure having a northeastward trending axis. On the western side of Notre Dame Bay, the ophiolite units dip steeply eastwards in imbricate thrust sheets. The dip decreases seawards and the ultramafic units reach a probable maximum depth of 7 km. On the southeastern side of Notre Dame Bay, the ophiolite sequence is truncated by the Chanceport-Lobster Cove fault. The ultramafic section appears to shallow northeastwards along the fault contact, but it is nowhere exposed.

Southeast of the Chanceport-Lobster Cove Fault lies a zone of granite bodies that extends northeastwards from Bay of Exploits to approximately 50°N, 53.5°W. The northeasternmost granites appear to be bordered on their southeastern side by diorite. A major decrease in gravity southeast of the Gander River Ultrabasic Belt, if due entirely to the granites of the Gander Zone, indicates that the granites must extend to a depth of 10 km or more.

RÉSUMÉ

A l'anse Betts dans la partie ouest de la baie de Notre-Dame, les roches ophiolitiques coincident avec une ceinture de fortes anomalies magnétiques et gravimétriques encerclant la baie. L'extension au large des côtes de ces anomalies et leurs relations avec l'arrangement

IGCP Project 27
Caledonide
Orogen

Canadian
Contribution
No. 30

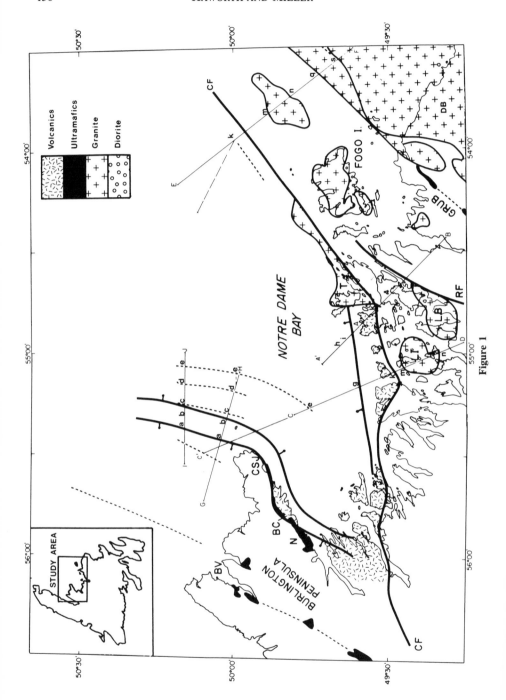

Figure 1

général des anomalies à Terre-Neuve montre que la croûte océanique est pentée du rivage vers l'océan, et, ainsi forme un synclinal dont l'axe a une direction nord-est. Du côté ouest de la baie de Notre-Dame l'ophiolite est imbriquée, par des failles de chevauchement fortement pentées. Le pendage diminue en allant du rivage vers l'océan et les unités de roches ultramafiques atteignent une profondeur maximum probable de 7 km. Du côté sud-est de la baie de Notre Dame la séquence ophiolitique est interrompue par la faille de Chanceport-Lobster Cove. La séquence ultramafique semble remonter vers le nord est, le long du contact de faille, mais sans toutefois qu'elle n'affleure.

Au sud-est de la faille de Chanceport-Lobster Cove une zone de massifs granitiques s'étend de la baie d'Exploits, vers le nord-est, jusqu'à approximativement la latitude 50°N et la longitude 53.5°W. Les granites les plus au nord-est semblent être limités, du côté sud-est, par des diorites. Si la forte diminution de la gravité du côté sud-est de la zone de roches ultrabasiques de la rivière Gander est entièrement attribuée aux granites de la zone de Gander, cela signifie que les granites doivent se prolonger jusqu'à une profondeur de plus de 10 km.

INTRODUCTION

The plate tectonic hypothesis, based on geophysical observations in the ocean, has brought about a rennaissance in geological evolutionary models proposed for continental structures. In Newfoundland, tectonostratigraphic zones have been defined and regionally interpreted in terms of the results of a collision between two opposing continental margins in early Paleozoic time that caused deformation and metamorphism, involving subduction and obduction of the intervening continental margin and oceanic sequences. Oceanic crustal units have been recognized along the margin of Notre Dame Bay (Fig. 1) and on the basis of their structure as deduced from surface exposure, localized crustal cross sections within the collision zone have been proposed. Conflict between different interpretative plate tectonic models and crustal structures proposed on the basis of surface geology can be resolved only through the use of independent information to add the dimension of depth to these interpretations. Geophysical data collected adjacent to the coast of Notre Dame Bay are examined here in an attempt to determine the subsurface extension of structures partially exposed on land and add that extra dimension.

Ophiolitic rocks of the Burlington Peninsula have been interpreted as representing the crustal rocks of the Iapetus Ocean (Harland and Gayer, 1972) which developed in late Precambrian time and closed in early Paleozoic time (Bird and

Figure 1. Summary map of geological features and their offshore and subsurface extension as sensed geophysically on the margins of Notre Dame Bay. Black areas are ultramafic and associated rocks (high density, high magnetization). Random stipple denotes volcanic rocks (high density, low magnetization). + pattern indicates the extent of granite plutons at or close to the surface (low density, moderate magnetization). T = Twillingate, LI = Long Island, LB = Loon Bay, and DB = Deadman's Bay plutons. BV = Baie Verte, BC = Betts Cove, N = Nippers Harbour, CSJ = Cape St. John, CF = Chanceport-Lobster Cove Fault, RF = Reach Fault, GRUB = Gander River Ultrabasic Belt. In the vicinity of Betts Cove and Twillingate the heavy lines indicate the approximate faulted limits of the high density rocks close to the surface. The dashed lines parallel to these faults represent the trend of ophiolite slices correlative between the models. The circle pattern on the southeast side of Fogo Island indicates the outcrop of diorite adjacent to the Fogo granite (see analysis of profile EF). The sections analyzed in this paper are indicated by straight solid lines, adjacent to which the letters indicate features discussed individually.

Dewey, 1970; Church and Stevens, 1971; Dewey and Bird, 1971; Upadhyay *et al.*, 1971; Kennedy, 1975; Williams, 1977). However, these authors were not unanimous in their interpretation of the direction of subduction leading to preservation of the ophiolitic sequences. Bird and Dewey (1970) were followed initially by most other workers in invoking a northwestward dip whereas Church and Stevens (1971) suggested that the dip was southeastward. There followed a time during which several subduction zones of opposite polarity within numerous small ocean basins were invoked to thrust, assemble and transport the different ophiolites of western Newfoundland (e.g., Dewey and Bird, 1971; Williams, 1975; Kidd, 1977). However, there are supporters of the view that all the ophiolites are the product of a single ocean (Strong *et al.*, 1974a; Williams, 1979).

Along the south coast of Notre Dame Bay, Strong (1973) and Strong and Payne (1973) classified the exposed volcanic rocks according to the environment in which they were formed and deduced their plate tectonic setting as correlative with that of island arcs in the southwest Pacific (Mitchell and Reading, 1971). This interpretation was maintained by Kean and Strong (1975) in the context of subduction to the southeast. Such a direction of subduction had been deduced by Strong *et al.* (1974a) in an attempt to use regional geochemical data as the primary basis for interpretation of the plate tectonic regime. However, the ages of the plutons whose geochemistry was investigated were inconsistent, thereby invalidating the conclusion (Bell *et al.*, 1977). Dean (1978) provides a complete summary of the geology of this area.

Many of the conflicts between the different views of the plate tectonic environment of Newfoundland in the early Paleozoic seem to arise because of a preoccupation with a multitude of local geological situations rather than with a view to their regional correlation. The essentially regional techniques of geophysics can perhaps suggest such correlations. For example, the ophiolites of Betts Cove and Baie Verte have rather limited outcrop compared with the extent of the magnetic anomalies which are the most obvious indication of their subsurface extension.

Miller and Deutsch (1973, 1976) examined high gravity anomalies along the margins of Notre Dame Bay and deduced that the coastal zone is underlain by dense material that correlates with rocks interpreted to be oceanic in origin. Jacobi and Kristoffersen (1976) collected gravity and magnetic data on one ship's track into and out of the Bay and showed that the gravity high associated with the mafic rocks appeared to be confined to the margins of the Bay and continued in a northeasterly direction from both margins. Two field operations were therefore deployed to collect data which might adequately describe the structure beneath the Bay: (1) The gravity program of Memorial University was extended to increase the density of stations along the western margin and to collect rock samples for density and susceptibility measurements. These data were compiled with those of Weaver (1968), Weir (1970), and Miller and Deutsch (1973, 1976) to provide a gravity map of the margins of Notre Dame Bay having a mean station spacing of 2.5 km (Miller and Deutsch, 1978). (2) Gravity and magnetic data were collected on a marine survey of Notre Dame Bay (Haworth *et al.*, 1976; Folinsbee *et al.*, 1978) to provide continuity between (and northeastward extension of) the data on its margins. Seismic work of Haworth *et al.* (1976) and Jacobi and Kristoffersen (1976) provided some information on the sedimentary sequence overlying the 'basement' structures under investigation.

Figure 2. Bathymetric map of the Notre Dame Bay area. Contours in metres, at 20 m intervals, except inshore of the 100 m contour and where gradients are large. Data are from Haworth *et al.* (1976) and Dale and Haworth (1979).

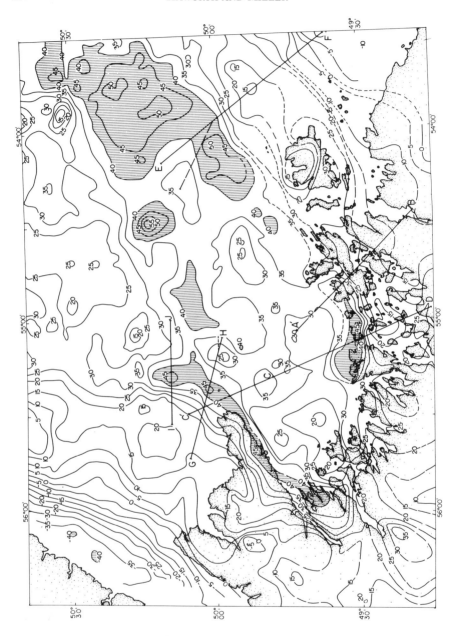

Figure 3. Bouguer gravity anomaly map of the Notre Dame Bay area. Contours in milligals at 5 mgal intervals. The area with Bouguer anomalies higher than 40 mgal is shaded. Data are from Haworth *et al*. (1976) and Miller and Deutsch (1978).

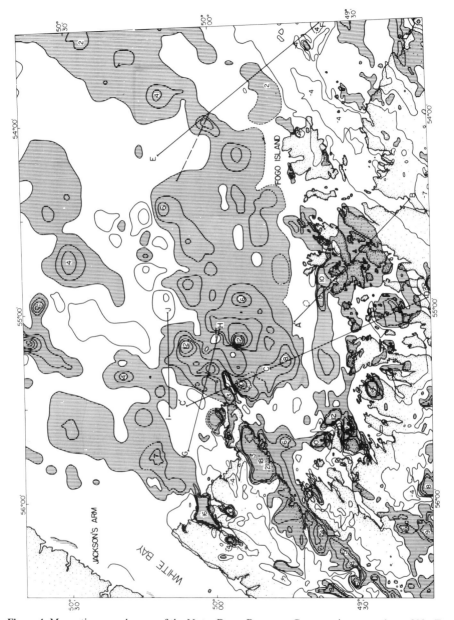

Figure 4. Magnetic anomaly map of the Notre Dame Bay area. Contours in nanoteslas at 200 nT intervals except where the contours cannot be adequately shown, in which case the maximum and minimum values are shown in hundreds of nanoteslas. The area of positive anomalies is shaded. Data offshore are from Haworth *et al.* (1976). Data onshore are from the Geological Survey of Canada federal/provincial aeromagnetic survey program and Hood and Reveler (1977).

TABLE I

DENSITIES AND MAGNETIZATIONS USED IN MODELLING

Geophysical Property Map Unit	Density (g cm^{-3})	Density Contrast* (g cm^{-3})	Magnetiz- ation (10^{-3} A m^{-1})	Model Pattern in Figures
Matrix of sedimentary and volcanic rocks (e.g. Cape St. John Group, Dunnage Mélange)	2.75	+0.04	0	Unshaded
Ophiolite Ultramafics Volcanics	 2.97 2.97	 +0.26 +0.26	 25 0	 Black Random stipple
Granite	2.65	−0.06	10	+pattern

*with respect to a mean regional density of 2.71 g cm^{-3}

Compiled maps of bathymetry and gravity and magnetic anomalies for the Notre Dame Bay region are presented in Figures 2, 3, and 4. In the following sections, profiles of the data will be discussed with reference to the known surface geology and interpreted to give structural sections which will be synthesized to provide an interpretation of the regional structure of the area.

DATA DISPLAY AND INTERPRETATION METHODS

A few general words of caution should be made before examining the interpreted profiles. A contour map can only accurately depict features that have a wavelength of the order of separation of the survey lines (in this case, approximately 10 km), whereas the survey lines themselves may contain much shorter wavelength information. The same is true on land, but whereas there is little to restrict the location of ship's tracks, the land gravity data are primarily confined to measurements along roads and the coast. The gravity and magnetic anomalies have been interpreted by using an interactive graphics modelling system (Wells, 1979; Haworth and Wells, 1980) that employs two-dimensional calculation of magnetic (Talwani and Heirtzler, 1964) and gravity anomalies (Talwani et al., 1959). Such a two-dimensional approximation is valid only where the profiles are perpendicular to an elongate structure. Although the anomalies to be interpreted (Figs. 3 and 4) have a length-to-width ratio of 6:1, sufficient to satisfy the elongation requirements, it has been necessary to effect compromises between the availability of higher fidelity data along ships' tracks and roads and the orientation of the lines with respect to the structure.

Profiles are presented in a common format to facilitate their comparison. Each profile and section has the same width, depth and anomaly scales. Bouguer anomalies with reference to IGSN '71 (Morelli et al., 1971) and the 1967 gravity formula (Geodetic Reference System, 1967) and magnetic anomalies with respect to IGRF '65 (IAGA, 1969) for epoch 1975 have been used. The locations of the profiles with respect to the exposures of geological units that may be sensed geophysically are shown in Figure 1, which also serves as a compilation diagram for the results of each section's interpretation.

The density contrasts used in the modelling (Table I) are quoted with respect to a mean regional density of 2.71 g cm^{-3}, compatible with the values of 2.70 to 2.72 g cm^{-3} used by Miller and Deutsch (1973, 1976). Little evidence was seen from the gravity data for a consistent difference in density between the rocks west of the Betts Cove ophiolites, the primarily Ordovician sedimentary rocks overlying the volcanic rocks within Notre Dame Bay and the rocks south of the Chanceport-Lobster Cove Fault. A mean density of 2.75 g cm^{-3} was therefore used in all models as the matrix for structures to a depth of 10 km, compatible with the mean densities of 2.79 ±0.16 g cm^{-3} and 2.74 ±0.10 g cm^{-3} measured by Miller and Deutsch (1976). Granite bodies were assigned a density of 2.65 g cm^{-3}, which is slightly less than the 2.66 to 2.68 mean values measured by Miller and Deutsch (1973) for the Twillingate, Long Island, and Loon Bay plutons (Fig. 1), so that the calculated thicknesses of granite bodies are perhaps minima.

Measurements of the magnetic properties of Newfoundland ophiolites (E.R. Deutsch, pers. commun.) yield arithmetic means of the remnant magnetization at three ultramafic sites of 3.5 to 50·10^{-3} A m^{-1} and Koenigsberger ratios of 0.5 to 0.9. At one site pillow lavas had a remnant magnetization of 13·10^{-3} A m^{-1} but were generally less than 5·10^{-3} A m^{-1}. The scatter in the measured values of these magnetic properties is so large, that with little possibility of being able to be compatible with the measured values everywhere, a constant value of 25·10^{-3} A m^{-1} was adopted for the induced magnetization of the ultramafics in determining the general shapes and attitudes of the causative bodies rather than trying to fit each magnetic anomaly exactly. A magnetization of 10·10^{-3} A m^{-1} for the granites was found to be necessary in reconciling the gravity and magnetic anomalies with the areal outcrop of the Long Island granite (see Section CD, Fig. 8).

NORTHWEST NOTRE DAME BAY (SOUTHEAST BURLINGTON PENINSULA)

The western margin of Notre Dame Bay is steep sided (Fig. 2), and is interpreted as one margin of a graben containing Carboniferous sedimentary rocks (Haworth *et al.*, 1976; Jacobi and Kristoffersen, 1976). Although the ship's tracks ventured as close to shore as possible, only the extremities of the gravity high encircling Notre Dame Bay were observed. Consequently the contouring along the western margin of the Bay is somewhat interpretative between the maximum values reached at sea and the most easterly values recorded on land. The contours of the land gravity data continue uninterrupted northeastwards as shown by the marine gravity data northeast of Cape St. John. This 'Cape St. John trend' can be clearly seen on the basis of east-west profiles as far north as 50°25'N. Since the seismic reflection information also indicates a northeastward extension offshore of the structures of the Burlington Peninsula, the marine profiles immediately northeast of Cape St. John provide an accurate indication of the gravity variations over the structures partially exposed on the eastern side of the peninsula.

Section IJ

Section IJ (Fig. 5) follows a ship's track orthogonal to the magnetic and gravity anomalies trending northwards from Cape St. John (Haworth *et al.*, 1976), so that two-dimensional modelling is valid. Magnetic anomaly values are available along this track at intervals of approximately 35 m, but since the modelling program is limited to a maximum of 100 observations, an observation interval of 0.5 km was used. Maximum and minimum magnetic observations were assigned to the closest point on the digitized observation profile. Since the magnetic variation is the key factor in defining the surface contact between geophysically sensed units, the geological model deduced has a limiting accuracy along profile of 0.5 km. The gravity profile was digitized at 1 km intervals.

The gravity high in the centre of profile IJ is the northward extension of the gravity high associated with the Betts Cove ophiolite (Fig. 3). The Betts Cove ophiolitic rocks on the eastern side of Burlington Peninsula appear to be dipping steeply to the southeast and are in fault contact with the surrounding rocks at most places (DeGrace et al., 1976). If the high in the gravity anomaly profile were an expression of the upper edge of a dipping high density layer, the skewness of the profile would indicate that the overall dip of such a layer must be to the east. The steep gravity gradient on the western side of the anomaly would be controlled by the surface truncation of the high density layer, whereas the gentler gravity gradient on the eastern side would be controlled by its dip.

In the initial stages of modelling, the gravity profile could be fitted reasonably well by the effect of a slab with a mean thickness of approximately 4 km dipping eastwards at about 20°. Only broad limits on the outcrop of that slab were imposed by the gravity data, but they suggested that the outcrop was coincident with the location of the bathymetric high trending north from Cape St. John (Fig. 2). The simplicity of that initial approach was considerably modified by examination of the magnetic data.

In the magnetic profile (Fig. 5), features having a wavelength of 1 to 2 km are superimposed upon others that have a wavelength of 10 to 20 km. A uniform dipping slab, as suggested by the gravity data, but having a constant magnetization, could not explain the observed anomalies. An acceptable fit of the magnetic anomalies was possible only by defining two distinct zones within the high density slab, each having a different magnetization. This is compatible with the geological situation of Burlington Peninsula where the high density, eastdipping ophiolites are composed of highly magnetic, ultramafic units and minimally magnetic volcanic and sedimentary rocks.

Similar compromise between the control of the magnetic anomalies over the location of the surface contacts and that of the gravity anomalies over the depth reached by bodies between such contacts governed the iterative approach towards the final model (Fig. 5). The following aspects of that model should be noted:

(a) Each short wavelength magnetic anomaly necessitates the introduction of a separate magnetic sheet which, because of the extrapolation of some of them towards the location of the ultramafic units on Burlington Peninsula, is also assigned a high density.

(b) Although the existence of the magnetic sheets is well supported, their individual sizes and shapes are subject to fairly wide tolerances. Primary control is on their location and inclination. The location of the magnetic peaks gives precise location of the upper edge of the causative sheets while the asymmetry of most of the short wavelength magnetic anomalies indicates that they dip eastwards (the only exception being the most easterly sheet).

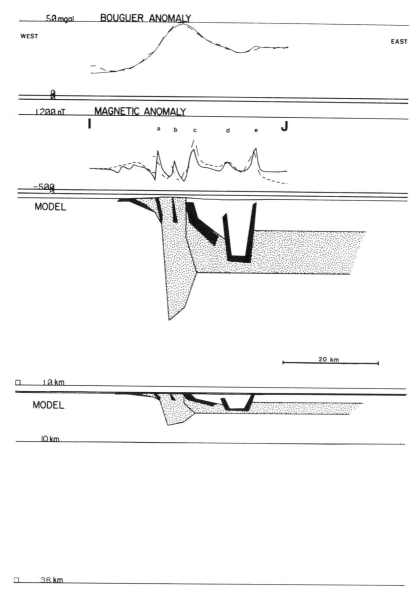

Figure 5. Section IJ. Bouguer gravity anomaly scale 0 to 50 mgal, magnetic anomaly scale −500 to 1200 nT, model scale 0 to 10 km depth, and 81 km width, a vertical exaggeration of approximately 3.8. The model with no vertical exaggeration is shown at the bottom of the diagram. Observed anomalies are indicated by the dashed line joining successive pairs of observations. The anomalies calculated from the postulated body are shown by the solid line. The patterns shading the model structures are as used in Figure 1. The densities and magnetizations are given in the text.

(c) The wavelength of the magnetic anomalies gives some indication of the depth to the upper edge of the causative body. Only the sheet at location 'b' on the profile correlates with any bathymetric feature which might indicate that it outcrops at the sea floor. The other sheets at locations 'a', 'c', and 'e' could outcrop according to the magnetic evidence, but they have no bathymetric expression and are therefore assumed to be covered. The depth extent of the sheets is poorly controlled, so that the joining of the two easterly sheets ('d' and 'e') at depth is *compatible* with the geophysical data rather than being *required* by the data.

(d) The constraint imposed by the magnetic data on the eastern limit in 'outcrop' of the main dipping slab necessitates the presence of an approximately 6 km deep 'root' to explain the gravity high. If a higher density contrast (up to 0.4 g cm^{-3}) were introduced in the model directly beneath the main gravity high, as might be argued on the basis of samples in the equivalent location onshore (Miller and Deutsch, 1976), the depth of this root would be reduced to give a more regularly east-dipping body. However, without more complete density data there seems little justification for changing the density of the slab down dip, in which case a discontinuity at locations 'a' and 'c' is still required.

(e) Much difficulty was encountered in fitting the magnetic anomalies west of location 'a' while assuming that the high magnetization and high density body subsurface was still responsible for them. They can only approximate the magnetic profile and still be compatible with the gravity data if such a body has only a few degrees dip extending as far as the steep contact at location 'a'.

All the constraints on the shape of the subsurface structure were deduced solely by compromise between the gravity, magnetic and bathymetry data. The remarkable correlation between Figure 5 and the surface geology in the Nippers Harbour-Betts Cove area along strike to the southwest (Fig. 1), as described by DeGrace *et al.* (1976) is therefore very reassuring. They state that "the ophiolitic rocks around Nippers Harbour, however, have been emplaced differently than those between Pittman Bight and Tilt Cove; since the former are more or less flat-lying and have intrusive contacts with the enclosing rocks, whereas the latter dip steeply to the southeast and are in fault contact with the surrounding rocks in most places." This is precisely the contrast deduced for the subsurface model, as described in preceding paragraphs (d) and (e). The gently dipping section of the model west of location 'a' (Fig. 5) corresponds with the Nippers Harbour ophiolitic rocks, in fault contact at location 'a' with the steeply dipping Betts Cove ophiolites. On the basis of geological-geophysical correlations on eastern Burlington Peninsula, it is apparent that the magnetic sheets within the model correspond to the ultramafic parts of the Betts Cove Ophiolite (Units 1 to 4 of De Grace *et al.*, 1976) whereas the similarly dense but non-magnetic structural units correspond to the non-ultramafic parts of the ophiolite together with the conformably overlying volcanic and sedimentary rocks of the Snooks Arm Group.

Section GH

Section GH (Fig. 6) also follows a ship's track orthogonal to the geophysical trends and lies between Section IJ and Cape St. John. The gravity profile is more noisy than profile IJ because of poor marine conditions when its data were collected.

The constraints on the model that evolved were almost identical to that of Section IJ. In order to fit the observed anomalies, it is necessary to have:

(a) an overall east-dipping, high density body containing high and low magnetic zones;

(b) a gently dipping, shallow western limb;

(c) a steep contact (at location 'a') between that shallow western limb and a central zone that extends to a depth of over 4 km;

(d) an abrupt eastern termination of the surface proximity of the high density block (at location 'b');

(e) at least two magnetic sheets ('a' and 'b') within that zone; and

(f) other sheets outside that central zone, two of which (at locations 'c' and 'd') may outcrop.

The biggest problem faced during the iterative approach to the final model was in reconciling the bathymetric and magnetic expression of the ultramafic sheets with their lack of significant gravitational expression. Even with the thin ultramafic sheet included in Section GH at location 'd', a gravity anomaly of a few milligals would be expected, but none was detected at that location above the noise level of the measured profile.

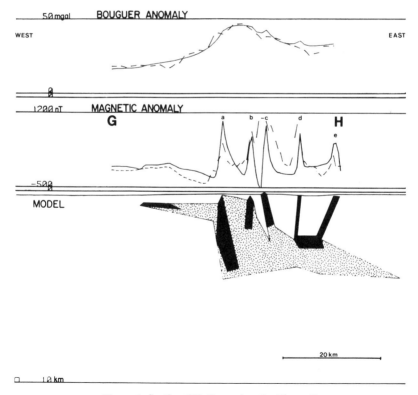

Figure 6. Section GH. Legend as for Figure 5.

Summary of Subsurface Structures, Northwest Notre Dame Bay

Each of the profiles modelled agrees at its western extremity with the geological situation in the Nippers Harbour-Betts Cove area of Burlington Peninsula where flat-lying ophiolitic rocks are faulted against steeply dipping ophiolitic rocks. However, each of the modelled profiles also indicates that east of that fault contact, several sheets of magnetic rock either outcrop or reach close to the surface. We envisage that each such sheet represents the basal ultramafic section of an ophiolite suite which was broken into fragments, each of which was thrust over its neighbour to the west to form an imbricate thrust sequence. Only one slice of this imbricate thrust sequence is seen on Burlington Peninsula, but the width of the sequence exposed there (2 to 5 km) is approximately the same as the horizontal distance between successive thrust faults (3 to 6 km between 'a' and 'b') in models IJ and GH. The steeply dipping imbricate sequence has a width of approximately 7 km and its continuity between profiles and thence to the mainland on the basis of correlation between individual anomalies (Haworth et al., 1976) is shown in Figure 1. Correlation between the location of the western, gently dipping sequence (west of locations 'a' on the profiles) is also shown on Figure 1, parallel to the trend of the imbricate block and trending towards Cape St. John. We suggest that the Cape St. John Group overlies this unit on Cape St. John as it does southwest of Nippers Harbour (De Grace et al., 1976). The flat lying ophiolites are therefore interpreted as being truncated against the imbricate thrust block southeast of the letters CSJ in Figure 1 just as they are in the vicinity of Nippers Harbour.

East of the imbricate thrust block, the individual magnetic slivers interpreted in Sections IJ and GH (locations 'c', 'd', and 'e', Figs. 5, and 6) may be correlated on a trend parallel to that block. They lie above the high density ophiolitic slab which falls between 2 and 5 km in depth, and we interpret them to be additional thrust slices of ophiolite within the sedimentary rocks that once overlay them. The extent of thrusting and degree of imbrication is probably less in that location, which is farther away from the Grenvillian land mass against which all these units were being thrust, as was envisaged by De Grace et al. (1976).

Overall, the models are extremely similar to the section across Burlington Peninsula (Fig. 7) as interpreted by Williams (1977). The Nippers Harbour ophiolite presumably never escaped from the source area as did the Bay of Islands and Coney Head complexes. In addition, there are several more imbricate thrust sheets of ophiolite east of the units shown in Williams' Middle Ordovician model (Fig. 7).

SOUTHEAST NOTRE DAME BAY

Southwest of Twillingate, on the islands north of the Chanceport-Lobster Cove fault, are exposed the rocks attributed to a Pre-Caradocian volcanic island sequence correlative with those of the Snooks Arm Group in northwestern Notre Dame Bay (Fig. 1; Dean, 1978). The Chanceport-Lobster Cove fault is probably the major thrust fault in Notre Dame Bay (Dean and Strong, 1977) and its northeastward extension offshore from Twillingate can be clearly seen as a bathymetric scarp (Fig. 2). The Reach Fault (Kay, 1976) has been interpreted to be the suture line marking the closure of Iapetus at the end of the early Devonian (McKerrow and Cocks, 1977).

Post-Acadian movement on that Fault is shown by its truncation of the Loon Bay intrusion (Fig. 1; Dean and Strong, 1977) one of the several intrusions lying southeast of the Chanceport-Lobster Cove Fault. The Gander River ultra-basic belt (GRUB line) is a discontinuous sequence of outcropping units having major ultramafic component that trends roughly northeastward across central and northeast Newfoundland (Kennedy, 1975; Blackwood, 1978; Dean, 1978; Currie et al., 1979; Strong, 1979). In northeastern Newfoundland the GRUB line has been interpreted as separating geological units having a stable shelf lithology from those of a deep water turbidite volcanogenic facies (Blackwood, 1978). Currie et al. (1979) interpret the ultramafic rocks to have been thrust from the west as a result of obduction. An extensive area of granites (Strong et al., 1974b; Jayasinghe and Berger, 1976; Bell et al., 1977; Strong and Dickson, 1978) lies southeast of the GRUB line within the Gander Zone (Williams et al., 1974; Kennedy, 1975; Blackwood, 1978) which has been interpreted as representing a shelf environment in Early Ordovician or earlier time. With the complications of the intrusions in southeastern Notre Dame Bay, it is difficult to construct 'typical' geological profiles. It is also impossible with a single straight geophysical profile to cut all the important geological features perpendicularly, and there are few ship's tracks or roads perpendicular to any of them. Therefore, the compromise cross-sections to be discussed in the following pages have to be considered collectively to determine the overall structural environment of southeast Notre Dame Bay.

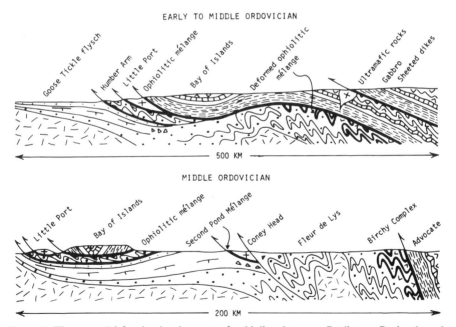

Figure 7. Thrust model for the development of ophiolite sheets on Burlington Peninsula and their transported equivalents at Bay of Islands (from Williams, 1977). The high angle thrust sheets at Betts Cove represent an equivalent sequence farther east.

Section CD

Section CD (Fig. 8) crosses the Long Island Batholith (Fig. 1) with its associated gravity low (Fig. 3) and magnetic high (Fig. 4), and continues north-northwest across the gravity and magnetic high which partially coincides with the island-arc volcanic rocks of the Moretons Harbour Group (Dean, 1978). The section has been extended to C' north of Cape St. John to provide the composite section of the entire Bay discussed later (Fig. 11). The geophysical profiles were digitized at a 1 km interval from contoured data (Figs. 3 and 4). Several features of the interpreted model (Fig. 8) need explanation:

(a) The prominent gravity high in the centre of the section is markedly asymmetric. By analogy with the asymmetric gravity high on Section IJ, which crosses cor-relative volcanic rocks, a northwestward dipping high density block was initially modelled in order to reproduce the anomaly. Using the same density contrasts as on the northwest side of the Bay, the high density unit was deduced to have a thickness similar to that on Sections IJ and GH. That high density unit is ap-parently truncated at its southeastern end approximately coincident with the Chanceport-Lobster Cove Fault. If the high density unit crosses the fault as

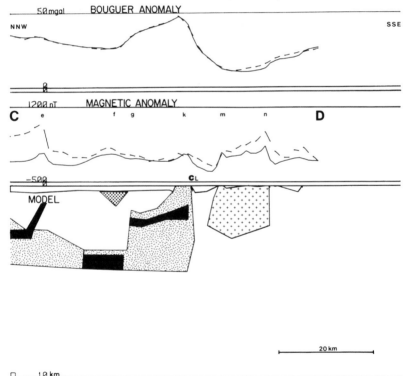

Figure 8. Section CD. Legend as for Figure 5. Chanceport-Lobster Cove Fault = CL. Dot pattern, near surface at f indicates low density sedimentary rocks lying approximately on the axis of the Notre Dame Bay synclinorium.

suggested by Miller and Deutsch (1976), then it must do so at a depth of greater than 5 km because of the abruptness of the gravity gradient on the southeast side of the high.

(b) On the northwest side of the high (location 'g') the abrupt change in gravity level indicates that the high density unit is faulted down to a depth of 4 km or more. The gravity gradient at 'g' is so steep that it is also necessary to include a low density, near-surface body at location 'f' in the model. We ascribe such a low density body to the presence of Carboniferous sediments in a fault bounded basin as shown by seismic profiles in the Bay (Haworth *et al.*, 1976; Jacobi and Kristoffersen, 1976).

(c) In the vicinity of location 'k' where the high density block is close to or at the surface, the magnetic anomalies are lower than those encountered on the north-western side of the Bay. The main magnetic high lying parallel to and north of the Chanceport Fault (Fig. 4), and crossed by profile CD at 'k', can be modelled by a magnetic zone which is truncated together with the northwest-dipping high density slab at the Chanceport-Lobster Cove Fault.

(d) Another magnetic high that runs east-west (at approximately 49°36'N) and heads onshore near Twillingate intersects profile CD 2 km northwest of the gravity low interpreted in (b) above as a sedimentary 'low'. If that magnetic high is due to ultramafic rocks at depth within the high density slab, their magnetization must be approximately double that modelled elsewhere.

(e) Southeast of Chanceport Fault is a gravity low which coincides with outcrop of the Long Island granite. The aeromagnetic data clearly show a positive magnetic anomaly associated with its southeastern margin at 'n' and a marked negative northwest of its northwest edge at 'm'. This requires that the granite be positively magnetized with respect to its surroundings. The location of the associated magnetic anomalies defines the area of outcrop of the granite. Since the assumed density for the granite is a minimum (see section on Data) the interpreted depth of 2.5 km for the Long Island granite is also a minimum. The magnetization ($10 \cdot 10^{-3}$ A m^{-1}) used in fitting the magnetic anomaly over this 'minimum depth' body, will not be significantly reduced if the depth is increased. This may seem to be a high magnetization, but it is the same as that deduced for the granite body northeast of Fogo Island, and well within the range of other Canadian Appalachian granites (McGrath *et al.*, 1973).

Section A'B

Gravity data collected along the road between Moreton's Harbour and Gander Bay South were projected onto the section AB joining those locations. All other data for Section A'B (Fig. 9) were digitized at 1 km intervals from the contour maps (Figs. 3 and 4).

Many of the interpreted features of the section are similar to those of Section CD:

(a) The asymmetry of the gravity profile indicates that a subsurface high density unit dips to the northwest.

(b) The location of the steep southeast face of the asymmetric gravity profile indicates that the subsurface high density unit is truncated at the Chanceport-Lobster Cove Fault.

(c) The 'notch' in the peak of the gravity profile 3 km northwest of location 'k' is caused by the proximity to the outcrop of the low density Twillingate granodiorite (Fig. 1).

(d) Between locations 'k' and 'h' the wavelength of the magnetic anomalies appears to decrease, and their amplitudes increase, indicating that the ultramafic units are closer to the surface offshore. The lack of gravity anomalies associated with them implies that these units are either of lower density or are quite thin. To be consistent with the interpretation elsewhere they are shown in Figure 9 as thin, high density units. In support of this interpretation, note that where the zone of anomalies crossed at 'h' and 'i' reaches Twillingate Island, the outcrop is of the (dominantly pillow lava) Sleepy Cove Formation, which has been correlated with parts of the ophiolite sequences on Burlington Peninsula (Strong and Payne, 1973; Dean, 1978).

(e) There is no local geophysical expression of the Reach Fault unlike the suggestion by Miller and Deutsch (1976). A minor gravity anomaly 4 km southeast of the location of the Reach Fault is caused by the end effect of the Loon Bay granite projected obliquely across the Fault.

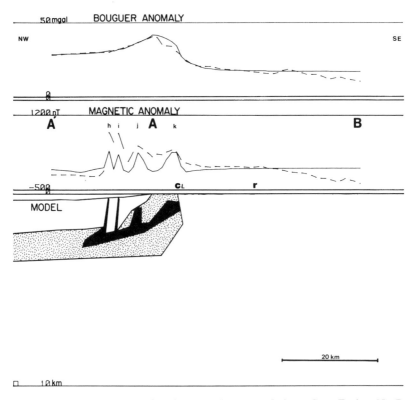

Figure 9. Section A'B. Legend as for Figure 5. Chanceport-Lobster Cove Fault – CL, Reach Fault – r.

(f) The general decrease in the level of the gravity and magnetic field at the southeastern end of the profile is typical of the Botwood and Gander zones (Jacobi and Kristoffersen, 1976; Miller, 1977; Haworth, 1980).

Section EF

Section EF (Fig. 10), perpendicular to the geophysical trends of eastern Notre Dame Bay, uses data from a continuous ship's track, but a 1 km digitizing interval has been used to be consistent with the other profiles from which structures are compared.

The features of this profile are in some ways different from those of other southeastern sections:

(a) The gravity high is more symmetric than on profiles CD and AB, but nevertheless is still fitted best by a northwestward dipping slab.

(b) The shortest wavelength, highest amplitude magnetic anomalies are coincident with the gravity high indicating that the ultramafic part of the main body is closest to the surface at location 'k'. The anomalies crossed at 'k' are part of a northeast trending zone of short wavelength anomalies of amplitude up to 800 nT that coin-

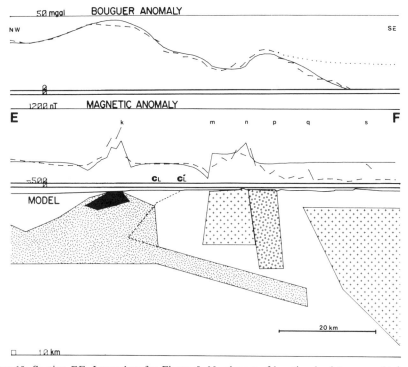

Figure 10. Section EF. Legend as for Figure 5. Northwest of location k, data were obtained from ship's track oblique to EF following dashed line in Figures 1 to 4. Chanceport-Lobster Cove fault if at the seafloor, lies at CL′, or as far west as CL if subsurface. The dotted Bouguer anomaly profile is that calculated for the model without the granite subsurface southeast of 'q'.

cide with a distinct bathymetric scarp on which the ultramafic unit probably outcrops (short dashed line, Fig. 1). Sampling attempts with a 4 m drill have as yet been unsuccessful because, although the ultramafics get closer to the bedrock surface with increasing distance northeastwards (shown by the anomalies becoming shorter in wavelength), the surficial sedimentary cover over bedrock also increases (Dale and Haworth, 1979).

(c) If that high density slab is truncated abruptly at a near-vertical fault as on CD and AB, then the edge of the slab has to be covered by nearly 1 km of lower density rocks at a location of the fault indicated by 'CL'. In order to satisfy the gravity data, the slab could only continue in outcrop southeast of 'k' if it had a thrust sheet cross-section as indicated by the short dashed line in Figure 10, in which case the Chanceport-Lobster Cover Fault would be in location CL' or location 'CL''.

(d) In either case the level of the gravity field southeast of the Chanceport-Lobster Cove Fault is higher than was the case in southern Notre Dame Bay and necessitates the inclusion of higher density rock in model EF. The choice to include it as a thin sheet of the high density material northwest of the fault is an arbitrary one. It might be due to a general increase in diversity of the matrix rocks southeast of 'CL' but there is no evidence between profiles A'B and EF for such a change.

(e) The outcrop of a granite body between locations 'm' and 'n', northeast of Fogo Island (Fig. 1), was defined by the magnetic anomalies at its margins, and the areal extent of the gravity low associated with it (Fig. 3). The magnetic anomalies indicate that its outcrop is not continuous with that of the Fogo Island granite, although the continuity of low gravity values suggest they may be connected at depth.

(f) The gravity high between 'n' and 'p' on the southeastern side of the granite body cannot readily be explained without the existence of a localized high density body marginal to the granite. The density used (2.79 g cm^{-3}) is the minimum that is necessary to produce the observed anomaly, higher densities for a shallower body give an equally good fit. It is possible that the body may be dioritic, just as diorite is adjacent to the granite along strike on Fogo Island (Baird, 1958).

(g) The gravity field falls over 35 mgal at the southeastern end of the profile. To demonstrate the significance of the crustal change when crossing into the Gander zone, the decrease has been fitted by a body of granite having the same minimum density value as used elsewhere in the models. Without the presence of such a granite, the calculated anomaly would follow the dotted gravity profile in Figure 10. This model has, however, assumed no change in structure at a depth greater than 10 km, an assumption that may not be warranted. Not all granites within the Gander zone are positively magnetized. The northwest margin of the megacrystic Deadman's Bay pluton can, however, be extrapolated offshore (Fig. 1) on the basis of the magnetic high associated with it. This is crossed by section EF at location 's'.

(h) A linear magnetic anomaly close to the margin of Deadman's Bay pluton has a trend similar to that of the GRUB line. This anomaly interesects the eastern margin of Figure 1 at 49°43'N, 53°30'W and is associated with a bathymetric high from which an ultramafic sample was drilled. It presumably represents a northeastward extension of the GRUB line which has been intersected by the younger pluton.

CONCLUSION

'Factual' Summary

On the western side of Notre Dame Bay, the structural models IJ and GH were consistent in showing:

(a) A high density, east-to-southeast-dipping body, consisting of
  (i) a gently dipping western limb at shallow depth,
  (ii) a deep central section having steep contacts at least at its western edge, and
  (iii) a gently dipping eastern section with shallower angle of dip at a depth of at least 1 or 2 km.

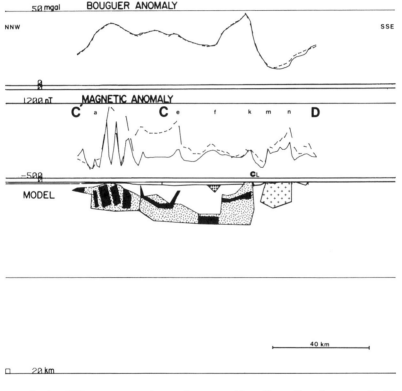

Figure 11. Section C′D, a representative section across Notre Dame Bay. Legend as for Figure 5 except model scales are 0 to 20 km depth, and 162 km width, maintaining the vertical exaggeration of 3.8 as in all other sections. The axis of the Notre Dame Bay synclinorium is in the vicintiy of 'f'. The ophiolites are truncated on the northeast side of the synclinorium at 'a' correlative with the western faulted margin of the Betts Cove ophiolites (Fig. 1). Southeast of 'f', the ophiolitic units are truncated at the Chanceport-Lobster Cove fault (CL). A zone of shallow granites, here represented between 'm' and 'n' by the Long Island granite, lies southeast of the Chanceport-Lobster Cove Fault, succeeded farther to the southeast by a major crustal change possibly characterized by the thick granites shown in Figure 10.

(b) The high density unit contains thin sheet-like, high magnetic zones.

(c) Thin magnetic zones also occur within the lower density rocks overlying the high density body in the centre of Notre Dame Bay.

From the models CD, A'B, and EF across the southeastern margin of Notre Dame Bay, it is seen that:

(d) The high density body dips consistently northwestward. Combined with result (a) above, this implies that Notre Dame Bay has an overall synclinal structure as shown in Figure 11.

(e) The high density body is closest to the surface adjacent to, and is truncated by, the Chanceport-Lobster Cove Fault.

(f) The magnetic zones within the high density unit get closer to the surface with increasing distance northeastwards, but never outcrop at the sea floor.

(g) There is no local geophysical expression of the Reach Fault, but

(h) the decrease in gravity and magnetic field southeast of the Gander River Ultra-basic Belt indicates an overall crustal change in this vicinity, possibly because of the presence of granites reaching a depth of over 10 km.

(i) A zone of granites, striking northeast from Fogo Island, is bordered to the southeast by a zone of rocks having a higher density than any of the other cover rocks.

'Interpretative' Summary

In the previous section we attempted to concentrate on the conclusive aspects of the modelling. In this section we briefly suggest the possible geological implications of that work. This is therefore purely interpretative and not definite on the basis of our work. Other geological interpretations may be possible but they must stay within the geophysical constraints.

The ophiolite suite at Betts Cove continues offshore as part of a north-northeasterly trending zone that is the western limb of a major synclinorium (Notre Dame Bay synclinorium, after Williams, 1978) whose northeast trending axis lies beneath the centre of Notre Dame Bay. That western limb consists of an imbricate thrust stack of complete ophiolite suites. The imbricate stack contrasts with the flat-lying ophiolites immediately to the west and suggests that the imbrication resulted from compression against a substantial structural barrier, with the flat-lying ophiolites having been thrust (obducted) out of the imbricated zone. This model (Fig. 7) is very similar to that proposed by Williams (1977) but we do not know whether the Baie Verte line or the Betts Cove line represents the eastern margin of the craton against which the ophiolites were thrust. If the Baie Verte line does represent that margin (Williams, 1977), it is difficult to envisage what structure to the east (i.e., east of the continental margin) would be a sufficient barrier against which imbricate thrusts could develop.

Imbricate thrusting of the ophiolite suite has apparently also occurred on the southeastern margin of the Notre Dame Bay synclinorium, where the Chanceport-Lobster Cove Fault is the eastern equivalent of the steep thrust fault between Nippers Harbour and Betts Cove. The lack of any geophysical expression directly associated with the Reach Fault is difficult to reconcile with its interpretation as the suture *line* between opposite sides of Iapetus as proposed by McKerrow and Cocks (1977). The only geophysically significant boundaries that have a surface expression close to the Reach Fault are the Chanceport-Lobster Cove Fault (as a thrust boundary) and the northwestern limit of the thick granites in the Gander zone. The location of the Gander River Ultrabasic Belt thrust (Blackwood, 1979) relative to the Chanceport-Lobster Cove Fault, and the location of the Baie Verte Line relative to the western thrust margin of the Betts Cove ophiolites, are symmetrical about the centre of the Notre Dame Bay synclinorium. The ophiolites in these thrust sheets are interpreted to be the remnants of the oceanic crust protected in the wide zone between the Newfoundland re-entrant on the Grenville margin and the Hermitage re-entrant on the Avalon margin (Williams, 1979). However, the symmetry in their near-surface structure may preclude a definitive interpretation of the causative direction of subduction. On the basis of more regional geophysical considerations (Haworth *et al.*, 1978) we believe that the Notre Dame Bay synclinorium is a localized feature within an overall southeastward dipping zone of ultramafic rocks, which indicates that Iapetus was subducted to the southeast.

REFERENCES

Baird, D.M., 1958, Fogo Island map area, Newfoundland (2E/9, mainly): Geol. Survey Canada Memoir 301, 43 p.

Bell, K., Blenkinsop, J. and Strong, D.F., 1977, The geochronology of some granite bodies from eastern Newfoundland and its bearing on Appalachian evolution: Canadian Jour. Earth Sci., v. 14, p. 456-476.

Bird, J.M. and Dewey, J.F., 1970, Lithosphere plate-continental margin tectonics and the evolution of the Appalachian orogen: Geol. Soc. America Bull., v. 81, p. 1031-1060.

Blackwood, R.F., 1978, Northeastern Gander Zone, Newfoundland: in Gibbons, R.V., ed., Report of Activities for 1977: Newfoundland Dept. Mines and Energy, Mineral Development Division Report 78-1, p. 72-79.

_____, 1979, Geology of the Gander River area (2E/2), Newfoundland: in Gibbons, R.V., ed., Report of Activities for 1978: Newfoundland Dept. Mines and Energy, Mineral Development Division Report 79-1, p. 38-42.

Church, W.R. and Stevens, R.K., 1971, Early Paleozoic ophiolite complexes of the Newfoundland Appalachians as mantle-oceanic crust sequences: Jour. Geophys. Research, v. 76, p. 1460-1466.

Currie, K.L., Pajari, G.E., Jr. and Pickerill, R.K., 1979, Tectonostratigraphic problems in the Carmanville area, northeastern Newfoundland: in Current Research Part A, Geol. Survey Canada Paper 79-1A, p. 71-76.

Dale, C.T. and Haworth, R.T., 1979, High resolution reflection seismology studies on the late Quaternary sediments of the northeast Newfoundland Continental Shelf: in Current Research Part B, Geol. Survey Canada Paper 79-1B, p. 357-364.

Dean, P.L., 1978, The volcanic stratigraphy and metallogeny of Notre Dame Bay, Newfoundland: Memorial University of Newfoundland, Geology Report 7, 204 p.

Dean, P.L. and Strong, D.F., 1977, Folded thrust faults in Notre Dame Bay, Central Newfoundland: American Jour. Sci., v. 277, p. 97-108.

DeGrace, J.R., Kean, B.F., Hsu, E. and Green, T., 1976, Geology of the Nippers Harbour map area (2E/13), Newfoundland: Newfoundland Department of Mines and Energy, Mineral Development Division, Report 76-3, 73 p.

Dewey, J.F. and Bird, J.M., 1971, Origin and emplacement of the ophiolite suite: Appalachian ophiolites in Newfoundland: Jour. Geophysical Research, v. 76, p. 3174-3266.

Folinsbee, R.A., Haworth, R.T. and MacIntyre, J.B., 1978, Marine gravity and magnetic maps, northeast of Newfoundland: Geol. Survey Canada Open File 525.

Geodetic Reference System, 1967, Special Publication No. 3: International Association of Geodesy, Paris, France.

Harland, W.B. and Gayer, R.A., 1972, The Arctic Caledonides and earlier oceans: Geol. Magazine, v. 109, p. 289-314.

Haworth, R.T., 1980, Appalachian structural trends northeast of Newfoundland and their trans-Atlantic correlation: Tectonophysics, v. 64, p. 111-130.

Haworth, R.T., LeFort, J.P. and Miller, H.G., 1978, Geophysical evidence for an east-dipping Appalachian subduction zone beneath Newfoundland: Geology, v. 6, p. 522-526.

Haworth, R.T. and Wells, I., 1980, Interactive computer graphics method for the combined interpretation of gravity and magnetic data: Marine Geophys. Research, v. 4, p. 277-290.

Haworth, R.T., Poole, W.H., Grant, A.C. and Sanford, B.V., 1976, Marine geoscience survey northeast of Newfoundland: Geol. Survey Canada Paper 76-1A, p. 7-15.

Hood, P.J. and Reveler, D.A., 1977, Magnetic anomaly maps of the Atlantic Provinces: Geol. Survey Canada Open File 496.

IAGA, 1969, International Geomagnetic Reference Field 1965: Jour. Geophys. Research, v. 74, p. 4407-4408.

Jacobi, R. and Kristoffersen, Y., 1976, Geophysical and geological trends on the continental shelf off northeastern Newfoundland: Canadian Jour. Earth Sci., v. 13, p. 1039-1051.

Jayasinghe, N.R. and Berger, A.R., 1976, On the plutonic evolution of the Wesleyville area, Bonavista Bay, Newfoundland: Canadian Jour. Earth Sci., v. 13, p. 1560-1570.

Kay, M., 1976, Dunnage Melange and subduction of the Protacadic Ocean, Northeast Newfoundland: Geol. Soc. America Spec. Paper 175, 49 p.

Kean, B.F. and Strong, D.F., 1975, Geochemical evolution of an Ordovician island arc of the central Newfoundland Appalachians: American Jour. Sci., v. 275, p. 97-118.

Kennedy, M.J., 1975, Repetitive orogeny in the northeastern Appalachians – new plate models based upon Newfoundland examples: Tectonophysics, v. 28, p. 39-87.

Kidd, W.S.F., 1977, The Baie Verte Lineament, Newfoundland: Ophiolite complex floor and mafic volcanic fill of a small Ordovician marginal basin: in Talwani, M., and Pittman, W.C., eds., Island arcs, deep sea trenches and back arc basins: American Geophys. Union, Maurice Ewing Series 1, p. 407-418.

McGrath, P.H., Hood, P.J. and Cameron, G.W., 1973, Magnetic surveys of the Gulf of St. Lawrence and the Scotian Shelf: in Hood, P.J., ed., Earth Science Symposium on Offshore Eastern Canada: Geol. Survey Canada Paper 71-23, p. 339-358.

McKerrow, W.S. and Cocks, L.R.M., 1977, The location of the Iapetus Ocean suture in Newfoundland: Canadian Jour. Earth Sci., v. 14, p. 488-495.

Miller, H.G., 1977, Gravity zoning in Newfoundland: Tectonophysics, v. 38, p. 317-326.

Miller, H.G. and Deutsch, E.R., 1973, A gravity survey of eastern Notre Dame Bay, Newfoundland: in Hood, P.J., ed., Earth Science Symposium on Offshore Eastern Canada: Geol. Survey Canada Paper 71-23, p. 389-406.

Miller, H.G. and Deutsch, E.R., 1976, New gravitational evidence for the subsurface extent of oceanic crust in north-central Newfoundland: Canadian Jour. Earth Sciences, v. 13, p. 459-469.

————————————, 1978, The Bouguer anomaly field of the Notre Dame Bay area, Newfoundland, with map No. 163 – Notre Dame Bay: Earth Physics Branch, Ottawa, Gravity Map Series No. 163.

Mitchell, A.H. and Reading, H.G., 1971, Evolution of island arcs: Jour. Geol., v. 79, p. 253-284.

Morelli, C., Gantar, C., Honkasalo, T., McConnell, R.K., Szabo, B., Tanner, J.G., Uotila, U. and Whalen, C.T., 1971, The International Gravity Standardization Net 1971 (IGSN71) Special Publication No. 4: International Assoc. Geodesy, Paris, France.

Strong, D.F., 1973, Lushs Bight and Roberts Arm Groups of central Newfoundland: Possible juxtaposed oceanic and island arc volcanic suites: Geol. Society America Bull., v. 84, p. 3917-3928.

_____, 1979, The enigmatic ultramafic and associated rocks of eastern Newfoundland (Abst.): Geol. Assoc. Canada Program with Abstracts, v. 4, p. 81.

Strong, D.F. and Dickson, W.L., 1978, Geochemistry of Paleozoic granitoid plutons from contrasting tectonic zones of northeast Newfoundland: Canadian Jour. Earth Sci., v. 14, p. 145-156.

Strong, D.F. and Payne, J.G., 1973, Early Paleozoic volcanism and metamorphism of the Moretons Harbour-Twillingate area, Newfoundland: Canadian Jour. Earth Sci., v. 10, p. 1363-1379.

Strong, D.F., Dickson, W.L., O'Driscoll, C.F., Kean, B.F. and Stevens, R.K., 1974a, Geochemical evidence for an east-dipping Appalachian subduction zone in Newfoundland: Nature, v. 248, p. 37-39.

Strong, D.F., Dickson, W.L., O'Driscoll, C.F. and Kean, B.F., 1974b, Geochemistry of Eastern Newfoundland granitoid rocks: Newfoundland Dept. Mines and Energy, Mineral Development Division Report 74-3, 140 p.

Talwani, M. and Heirtzler, J.R., 1964, Computation of magnetic anomalies caused by two dimensional structures of arbitrary shape: in Parks, G.A., ed., Computers in the Mineral Industries: published by School of Earth Sciences, Stanford University.

Talwani, M., Worzel, J.L. and Landisman, M., 1959, Rapid gravity computations for two dimensional bodies with application to the Mendocino Submarine Fracture Zone: Jour. Geophys. Research, v. 64, p. 49-59.

Upadhyay, H.D., Dewey, J.F. and Neale, E.R.W., 1971, The Betts Cove ophiolite complex, Newfoundland: Appalachian oceanic crust and mantle: Geol. Assoc. Canada Proceedings, v. 24, p. 27-34.

Weaver, D.F., 1968, Preliminary results of the gravity survey of the island of Newfoundland: Dominion Observatory, Ottawa, Gravity Map Series No. 53-57.

Weir, H.C., 1970, A gravity profile across Newfoundland: M.Sc. Thesis, Memorial University, St. John's, Newfoundland.

Wells, I., 1979, MAGRAV User's Guide: A computer program to create two-dimensional gravity and/or magnetic models: Geol. Survey Canada Open File 597.

Williams, H., 1975, Structural succession, nomenclature, and interpretation of transported rocks in western Newfoundland: Canadian Jour. Earth Sci., v. 12, p. 1874-1894.

_____, 1977, Ophiolitic mélange and its significance in the Fleur de Lys Supergroup, northern Appalachians: Canadian Jour. Earth Sci., v. 14, p. 987-1003.

_____, 1978, Tectonic lithofacies map of the Appalachian Orogen: Memorial University of Newfoundland, Map No. 1.

_____, 1979, Appalachian orogen in Canada: Canadian Jour. Earth Sci., v. 16, p. 792-807.

Williams, H., Kennedy, M.J. and Neale, E.R.W., 1974, The northeastward termination of the Appalachian orogen: in Nairn, A.E.M., and Stehli, F.G., eds., The Ocean Basin and Margins, V. 2: New York, Plenum Publishing, p. 79-123.

Manuscript Received October 3, 1979
Revised Manuscript Received November 20, 1980

ZONE BOUNDARIES AND FAULTS

Major Structural Zones and Faults of the Northern Appalachians, edited by
P. St-Julien and J. Béland, Geological Association of Canada Special Paper 24, 1982

THE BAIE VERTE-BROMPTON LINE: EARLY PALEOZOIC CONTINENT-OCEAN INTERFACE IN THE CANADIAN APPALACHIANS

Harold Williams
Department of Geology, Memorial University of Newfoundland, St. John's, Newfoundland A1B 3X5

Pierre St-Julien
Department of Geology, Laval University, Québec, Québec G1K 7P4

ABSTRACT

A narrow zone of steeply dipping, east facing ophiolitic complexes is traceable along the west flank of the Canadian Appalachians from Northeast Newfoundland to the Québec Eastern Townships, a distance of 1500 km. The surface trace of the ophiolitic belt is known as the Baie Verte-Brompton Line. The ophiolitic complexes are in tectonic contact with intensely deformed and metamorphosed rocks toward the west, and they are overlain by shaly megaconglomerates and olistostromal melanges toward the east.

Throughout its length, the Baie Verte-Brompton Line maintains a constant position with respect to Cambrian-Ordovician facies belts and Ordovician structural features. In contrast, the line bears no obvious relationship to facies patterns or structures in Silurian and younger rocks.

The Baie Verte-Brompton Line represents a tectonic zone of Ordovician ophiolite emplacement. Subsequently, it has been affected by mid-Paleozoic orogenesis that has modified its course and complicated its geometry.

Cambrian and Ordovician rocks and structures to the west of the line record the evolution and destruction of the Early Paleozoic continental margin of eastern North America. Ophiolitic complexes along the line, and ophiolitic complexes and volcanic rocks farther east record the generation of oceanic crust and thick volcanic arc sequences. Accordingly, the Baie Verte-Brompton Line is interpreted as the surface trace of a structural junction between deformed rocks of an ancient continental margin and bordering ocean.

IGCP Project 27   *Canadian*
IUGS UNESCO *Caledonide*   *Contribution*
Orogen   *No. 27*

RÉSUMÉ

Une étroite zone de complexes ophiolitiques fortement inclinés et faisant face à l'est longe le flanc ouest des Appalaches Canadiennes depuis le nord-est de Terre-Neuve jusqu'aux Cantons-de-l'Est du Québec, soit une distance de 1500 km. La trace en surface de cette zone est la Ligne Baie Verte-Brompton. Du côté ouest, les complexes ophiolitiques sont en contact tectonique avec des roches très déformées et métamorphisées; du côté est, elles sont recouvertes par des mégaconglomérats argileux et des mélanges olistostromes.

Sur toute sa longueur, la Ligne de Baie Verte-Brompton occupe la même position par rapport aux facies cambro-ordoviciens et les structures ordoviciennes; par contre elle ne montre aucune relation systématique avec les facies ou structures siluriennes ou plus récentes.

La Ligne de Baie Verte-Brompton représente une zone de mise en place tectonique d'ophiolites ordoviciennes subséquemment affectée par une orogénèse du Paléozoïque moyen et cette orogénèse a compliqué la position géométrique originelle de la zone.

Les roches cambro-ordoviciennes et leurs structures du côté ouest de la ligne témoignent de l'évolution et de la destruction de la marge continentale de l'Amérique du Nord orientale au cours du Paléozoïque inférieur. Les complexes ophiolitiques le long de la ligne et ceux plus à l'est avec leurs cortèges de roches volcaniques sont les preuves de la génération d'une croûte océanique et de séquences volcaniques d'îles en arc. Conséquemment la Ligne de Baie Verte-Brompton est interprétée comme le pointement superficiel d'une jonction structurale entre les roches déformées de l'ancienne marge continentale et le domaine océanique adjacent.

INTRODUCTION

A narrow steep structural zone, marked by discontinuous ophiolitic complexes and bounded by similar lithic units along its length, can be recognized throughout the Canadian Appalachians. The surface trace of the steep ophiolitic zone is called the Baie Verte-Brompton Line (Fig. 1), colloquially the B-B Line, after the geographic localities of its extremities in Canada; Baie Verte, Newfoundland (northeast) and Brompton Lake, Quebec (southwest). Ophiolitic complexes that define the line are bordered to the west by mainly metaclastic rocks with superimposed deformations. These rocks overlie a crystalline Grenvillian basement. To the east, the ophiolitic rocks are bordered by coarse conglomerates and olistostromal melanges (hereafter referred to as olistostromes) that overlie an ophiolitic basement.

The Baie Verte-Brompton Line is a feature that resulted from the Ordovician emplacement of ophiolite complexes. Subsequent Devonian (Acadian) and Carboniferous (Alleghanian) deformations accentuate the feature and are responsible for steep attitudes along its length.

Like other geological features that are referred to as 'lines' on geologic maps of the Appalachian Orogen, for example Logan's Line, Hollins Line, Martic Line and Stonewall Line (Williams, 1978a), the Baie Verte-Brompton Line is merely the surface trace of an initial planar or tabular zone, subsequently folded and faulted. Consequently, its present course may be quite intricate on detailed geologic maps.

The Baie Verte-Brompton Line is easily recognized in places where it is marked by mafic-ultramafic plutons that separate metaclastic rocks on the west from olistostromes on the east. Where ophiolitic rocks are absent, the line separates multideformed metaclastic rocks to the west and less deformed olistostromes or volcanic suites to the east. Where the stratigraphic record is completely obliterated, as in southwestern Newfoundland, the line is drawn at the eastern limit of Grenvillian basement. The Baie Verte-Brompton Line is concealed by Silurian-Devonian cover rocks throughout the Gaspé Peninsula and it is hidden locally by Carboniferous cover rocks in Newfoundland.

The Baie Verte-Brompton Line has an aeromagnetic signature that results mainly from the presence of its ophiolitic rocks. This can be seen on 1:50,000 and 1:250,000 scale maps published by the Geological Survey of Canada, and it can be recognized even on 1:1,000,000 scale maps (Zietz *et al.*, 1980). It also lies close to a pronounced gradient in the Bouguer gravity anomaly field, between negative values in the west and positive values in the east (Miller and Deutsch, 1976; Haworth *et al.*, 1976, 1980).

In addition to its tectonic significance, the Baie Verte-Brompton Line is of special economic importance because the ultramafic rocks that occur along it are host to the asbestos deposits that up to recent years made Canada the world's leading producer of asbestos fibre (Vagt, 1976). Even today, asbestos production and reserves in this single restricted zone make the Baie Verte-Brompton Line the world's richest asbestos belt.

Structural and stratigraphic relationships among the rock groups along the Baie Verte-Brompton Line are analysed in this paper in an attempt to establish correlations along it and to assess its tectonic significance. The analysis is based mainly upon our own field work, but in some areas we draw heavily upon the work of others. Several reciprocal visits between Newfoundland and Quebec over the past few years have served to identify and help correlate rocks and structures at the extremities of the line (Williams and St-Julien, 1978). Of course the local relationships are not without controversy, so that for some areas a review of previous ideas and current interpretations is included.

LOCATION AND EXTENT

Along the Baie Verte Peninsula, the Baie Verte-Brompton Line is marked by a zone of nearly continuous mafic-ultramafic plutons that can be traced from Baie Verte approximately 100 km southward to Sandy Lake (Fig. 1). Farther south in the vicinity of Howley, it is concealed by overlying Carboniferous rocks. A narrow poorly exposed belt of ultramafic and volcanic rocks immediately east of Deer Lake (Don Burbeck, pers. commun., 1979; John O'Loughlin, pers. commun., 1980) may be its southward continuation, but it is offset toward the west. The line reappears at Glover Island in Grand Lake where it is again marked by a narrow zone of ophiolitic rocks in a structural setting similar to that on the Baie Verte Peninsula (Williams and St-Julien, 1978; Knapp *et al.*, 1979). At the south end of Grand Lake, the line is displaced sharply eastward. It may be coincident with the morphologic depression of the Lloyds River valley, which is bordered to the east by the Annieopsquotch ophiolitic complex (Herd and Dunning, 1979; Dunning and Herd, 1980). Southward from there, the diagnostic rocks that define the line in Newfoundland are absent, but it is inferred to follow the Cape Ray Suture (Brown, 1973; Williams, 1978a).

The Baie Verte-Brompton Line reappears in the mainland Appalachians along the North Port-Daniel River in Gaspé where it is marked by local ultramafic occurrences and olistostromes. It is hidden for the next 350 km westward by an unconformable cover of Upper Ordovician and Silurian rocks of the Aroostook-Matapédia

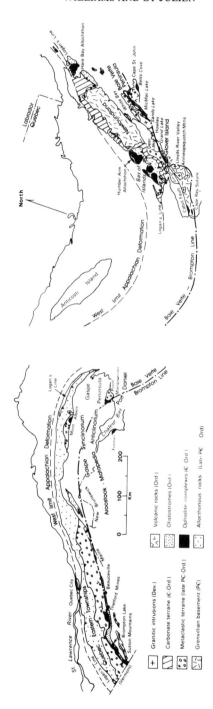

Figure 1. Distribution of rocks and structures along the west flank of the Canadian Appalachians and the location of the Baie Verte-Brompton Line.

Anticlinorium and Silurian and Devonian strata of the Gaspé Synclinorium. On the western side of the synclinorium and between the Quebec-Maine border and Thetford Mines, ultramafic bodies are few and widely separated, though characteristic olistostromes are common. From Thetford Mines southwestward, the line is marked by a wide zone of ophiolitic rocks, commonly referred to as the Quebec Serpentinite Belt. These are virtually continuous from Thetford Mines 150 km southwestward to Brompton Lake and the Canada-United States border.

Although it is exposed only in widely separated segments, the Baie Verte-Brompton Line is characterized by similar and distinctive geologic relationships. Furthermore, the segments that are exposed occur in a consistent position with respect to major facies belts and structural zones within the orogen. For these reasons the Baie Verte-Brompton Line is interpreted as continuous where it is concealed by a cover of younger rocks. Some of the gaps in its continuity are the result of truncation by younger structures.

LOCAL GEOLOGIC RELATIONSHIPS AT THE BAIE VERTE-BROMPTON LINE

Descriptions of rocks and relationships at the best exposed and best studied segments of the Baie Verte-Brompton Line are presented as a basis for structural comparisons and correlations of rock units along its length. Four areas are selected, which from north to south are the Baie Verte Peninsula and Glover Island in Newfoundland and the North Port-Daniel River and Eastern Townships of Quebec.

Baie Verte Peninsula

Since the early studies of Fuller (1941), Watson (1947), Baird (1951), and Neale and Nash (1963), the Baie Verte Peninsula has been an area of intense geological research activity, e.g. Neale and Kennedy (1967), Church (1969), Church and Stevens (1971), Dewey and Bird (1971), Kidd (1974, 1977), Kennedy (1975), Bursnall and De Wit (1975), DeGrace *et al.* (1976), Williams (1977), Williams *et al.* (1977), Kidd *et al.* (1978) and Hibbard (1978, 1981). The recent work is of special significance to our analysis because interpretations regarding the distribution of lithic units, proposed correlations across the Baie Verte-Brompton Line and the timing of structural events have led to a variety of conflicting conclusions.

The geology of the Baie Verte Peninsula as interpreted by Williams *et al.*(1977) is summarized in Figure 2. The Baie Verte-Brompton Line in this part of Newfoundland is defined by the presence of mafic-ultramafic rocks of the Advocate Complex (Kennedy, 1975; Bursnall and De Wit, 1975; Williams *et al.* 1977). The Advocate Complex is bounded to the west by metaclastic rocks of the Fleur de Lys Supergroup (Fuller, 1941; Church, 1969) and greenschists and mafic-ultramafic rocks of the Birchy Complex (Fuller, 1941; Williams, 1977). It is bounded to the east by olistostromes that form the base of the Flatwater Group of Williams (1977) and Williams *et al.* (1977) and were included in the Baie Verte Group of Watson (1947) and Kidd (1974).

There are two opposing models for the origin and tectonic setting of the ophiolitic rocks at Baie Verte. Dewey and Bird (1971), Kennedy (1975), Kidd (1977) and Kidd *et al.* (1978) have interpreted the ophiolitic rocks of the Advocate and Point Rousse Complexes (Fig. 2) and their overlying rocks as the oceanic floor and vol-

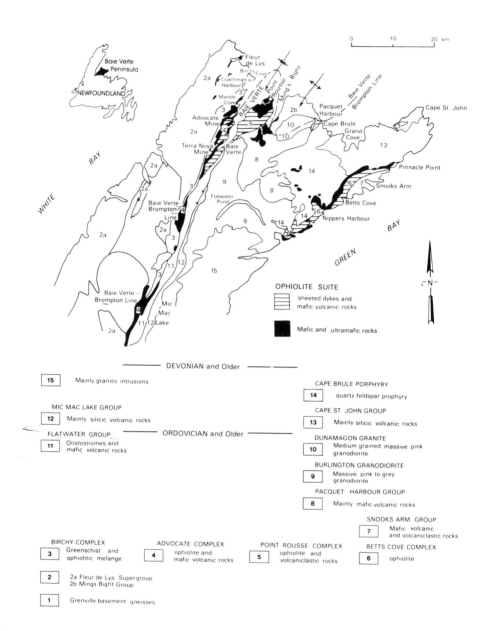

Figure 2. The Baie Verte-Brompton Line and relationships among rock groups at the Baie Verte Peninsula, Newfoundland.

canic fill of a marginal ocean basin that was generated within and bounded by already-deformed and metamorphosed rocks of the Fleur de Lys Supergroup. A corollary of this interpretation is that each ophiolite belt of the Baie Verte Peninsula originated in a separate ocean basin, and furthermore that the Bay of Islands ophiolite complex in western Newfoundland originated to the west of the Baie Verte-Brompton Line at the present site of White Bay (Dewey and Bird, 1971). Another interpretation (Stevens, 1970; Church and Stevens, 1971), favoured by us, is that all of the Baie Verte ophiolitic complexes, including the Bay of Islands Complex, are correlatives and that they all originated in a single ocean whose margins were undeformed before ophiolite transport. A corollary of this interpretation is that the Bay of Islands Complex originated at or near the Baie Verte-Brompton Line, and that deformation at the ancient continental margin to the west accompanied and followed ophiolite obduction (Bursnall and De Wit, 1975; Williams, 1977).

Local Relationships along the Baie Verte-Brompton Line. Structural complexity, intensity of penetrative deformation and metamorphic grade all increase westward across the series of mafic-ultramafic complexes that occur along the Baie Verte-Brompton Line and adjacent to it (Betts Cove, Point Rousse, Advocate and Birchy Complexes). Thus toward the east, the Betts Cove Complex (Upadhyay *et al.*, 1971) and the Point Rousse Complex (Williams *et al.*, 1977; Kidd, 1977; Kidd *et al.*, 1978) are only mildly deformed and metamorphosed and exhibit all essential components of the ophiolite suite. Farther west at the line, the Advocate Complex is imbricated and in places intensely foliated and consists of separate structural slices that contain only one or two components of the ophiolite sequence. Still farther west, mafic-ultramafic rocks and ophiolitic melange (Williams, 1977) of the Birchy Complex are metamorphosed to greenschist facies and affected by the same multiple deformations recognized in metaclastic rocks of the Fleur de Lys Supergroup.

Where the order of stratigraphic units in the steeply dipping ophiolitic suites is clear, e.g. Betts Cove, parts of the Point Rousse and parts of the Advocate Complex, the units face toward the east or southeast. Thus the overall structural arrangement involves westward increase in intensity of deformation and metamorphism across westward verging imbricate slices of eastward facing ophiolite sequences. Later Acadian deformation, involving steep upright folds and local eastward thrusting, has complicated this simple picture (Kidd, 1974; Bursnall, 1975; Hibbard, 1981).

Mafic-ultramafic rocks and ophiolitic melanges of the Birchy Complex are separated from the Advocate Complex on structural style and because greenschists of the Birchy Complex are interlayered with Fleur de Lys metaclastic rocks that are absent in the Advocate Complex. Yet the distinction between the Birchy and Advocate Complexes is not everywhere obvious, and this has been a problem for interpretations involving a significant age difference and contrasting tectonic histories for the juxtaposed complexes (see also Bursnall and De Wit, 1975). The Advocate Complex is geographically distinct from the Point Rousse Complex, which is less complex structurally and contains a more complete ophiolite sequence. Moreover, the cover sequence above the Advocate Complex contains basal olistostromes, whereas volcanic rocks and finer grained sedimentary rocks are conformable and gradational above the Point Rousse Complex.

The Advocate Complex, which defines the Baie Verte-Brompton Line, has a steep northeast-trending foliation in most places and it is cut by numerous steep northeast-trending shear zones across which the rock units are repeated (Bursnall, 1975; Bursnall and De Wit, 1975). A large ultramafic body at the northwest margin, near the coast at Baie Verte, consists of foliated and intensely fractured serpentinite that is host to the Advocate asbestos deposit. Gabbros of the Advocate Complex vary from massive to intensely foliated and some have been altered to distinctive white calc-silicate rocks with green fuchsite on foliation planes, e.g., those at Marble Cove. Sheeted dykes and pillow lavas are recognizable locally.

An occurrence of garnetiferous amphibolite in the western wall of the advocate open pit (Bursnall, 1975) at a tectonic contact between serpentinite (east) and greenschist (west) suggests local development of a Bay of Islands-type metamorphic aureole (Williams and Smyth, 1973) beneath the Advocate Complex. This implies that the rocks at the Baie Verte-Brompton Line were emplaced by tectonic transport rather than originating by in situ generation.

Black slates occur at steep tectonic contacts within the Advocate Complex, structurally comingled with the segmented ophiolite, e.g., occurrences near Marble Cove. Other black shaly rocks, unseparated in Figure 2, are clearly sedimentary breccias and olistostromes like those of the Flatwater Group. One example of the latter, which occurs near the entrance gate to the Advocate Mine, has volcanic fragments and large gabbro blocks in a black shaly matrix. Another example in a stream bed at the abandoned Terra Nova Mine of Baie Verte contains large brecciated ultramafic blocks and a local block of massive sulphides (Williams and St-Julien, 1978).

Local Relationships to the west of the Baie Verte-Brompton Line. The Fleur de Lys Supergroup to the west of the Baie Verte-Brompton Line consists mainly of psammitic to semipelitic schists that are metamorphosed to greenschist and amphibolite facies and affected by polyphase deformation. Thin bedded marbles and limestone breccias are common in western exposures and these rock types can be equated with similar rocks of Cambrian to Middle Ordovician age in the Humber Arm Allochthon of western Newfoundland (Bursnall and De Wit, 1975; Williams, 1975). The recognition of reworked basement gneisses deformed with the Fleur de Lys cover, a basal conglomerate to the cover sequence, and metamorphosed mafic intrusions in thick basal parts of the group (De Wit, 1972; De Wit and Strong, 1975) all support the contention that the Fleur de Lys represents a clastic sequence built up at the ancient margin of a rifted continent (Bird and Dewey, 1970; Williams and Stevens, 1974). Easterly parts of the Fleur de Lys Supergroup, near the Baie Verte-Brompton Line, consist mainly of greenschists with thin psammitic and black pelitic units, comprising the Birchy Complex (Williams, 1977). These metamorphic rocks include large bodies of schistose gabbro, brecciated serpentinized peridotite and ophiolitic melange. All workers concur that the Birchy Complex has been involved in the same deformations that affected the Fleur de Lys metaclastic rocks to the west.

Several phases of intense deformation are recognized in the Fleur de Lys Supergroup. The most conspicuous foliation is a second phase schistosity related to major west-facing recumbent folds. No major first phase folds are recognized, but the first deformation produced major tectonic dislocations that locally predate the earliest schistosity (Kennedy, 1975). Steep third and fourth phase structures increase in intensity southeastward toward the Baie Verte-Brompton Line (Kennedy, 1975; Bursnall and De Wit, 1975).

The ophiolitic Coachmans Melange (Williams, 1977) of the Birchy Complex has a schistose black pelitic matrix with conspicuous deformed and recrystallized ultramafic blocks, now represented by bright green fuchsite-actinolite schist. Sedimentary blocks with ill-defined outlines are common, and in some places large brecciated and serpentinized ultramafic blocks, foliated gabbro blocks, and marble are also present. The black pelites and the discrete blocks contained within them show all of the deformational structures of the Birchy Complex, thus formation of the melange must have preceded deformation and metamorphism of the Fleur de Lys terrane.

Worldwide, occurrences of ophiolitic melanges beneath ophiolites have been related to transport of the ophiolites (Gansser, 1974). Melanges beneath the ophiolites at the Baie Verte-Brompton Line and their similarity to melanges of the Humber Arm and Hare Bay Allochthons in western Newfoundland (Williams, 1975) imply ophiolite transport across the Fleur de Lys terrane. The formation of the Coachmans Melange is probably related to the westward transport of ophiolite complexes such as the Bay of Islands Complex, from a root zone at or near the Baie Verte-Brompton Line.

Metamorphosed and deformed small ultramafic and gabbroic bodies in the central psammitic part of the Fleur de Lys terrane are also thought to be ophiolitic remnants localized along early structural surfaces (Williams and Talkington, 1977; Bursnall, 1979) rather than representing high level small intrusions (Kennedy and Philips, 1971). Their presence implies further disruption and imbrication across the Fleur de Lys terrane.

Local Relationships to the east of the Baie Verte-Brompton Line. South of Baie Verte, the Advocate Complex is bordered on its eastern side by olistostromes that occur at the base of the Flatwater Group. Detailed mapping between Flatwater Pond and Mic Mac Lake (Kidd, 1974) has shown that the Flatwater Group is a steeply-dipping, easterly-facing sequence that is faulted against the Advocate Complex to the west. Upper formations of the Flatwater Group include mafic volcanic rocks and volcanic breccias, with silicic volcaniclastic rocks near the top. It is overlain with local angular unconformity by bright red volcanic rocks and clastic sedimentary rocks of the Silurian-Devonian Mic Mac Lake Group (Kidd, 1974). Farther east, on the east side of a faulted syncline, westfacing Mic Mac Lake rocks overlie an older intrusion, the Burlington Granodiorite (Baird, 1951; Neale and Kennedy, 1967) with marked nonconformity. The distinctive types of clasts in the basal Flatwater olistostromes, which include granitic rocks like the Burlington Granodiorite, serpentinite, a variety of foliated gabbroic rocks, and virginite (iron-magnesite-quartz-fuchsite alteration of ultramafic rocks), suggest that the Flatwater Group is younger than the nearby Advocate Complex to the west and the Burlington Granodiorite to the east. The conglomerates also contain tough highly indurated boulders of quartzose greywacke, clasts of vein quartz, limestone, and a large block of pelitic schist at Mic Mac Lake that is equated with the deformed Fleur de Lys Supergroup (Kidd, 1974). The structural gradation between the Fleur de Lys (Birchy Complex) and Advocate Complex (Bursnall, 1975), coupled with this sedimentological evidence, implies that the Flatwater Group was deposited after tectonic emplacement of the Advocate Complex. We therefore interpret the Flatwater olistostromes as chaotic rocks deposited upon disturbed oceanic crust, or deposited upon orogenically deformed oceanic crust; rather than the olistostromal and volcanic fill of an undisturbed small ocean basin (Kidd, 1977).

The Fleur de Lys Supergroup as defined by Church (1969) and used by Dewey and Bird (1971), Kennedy (1975), Kidd (1977) and Kidd *et al.* (1978) included a variety of rocks east of Baie Verte (the eastern Fleur de Lys Supergroup) that resemble the western Fleur de Lys Supergroup in structural style and metamorphic grade. Some of these are psammitic schists with local black pelitic zones containing bright green actinolite lenses (Mings Bight Group), which bear a remarkable resemblance to the Fleur de Lys psammites and mélange at Coachmans Harbour. Most others are volcanic rocks and intrusive porphyry that are unlike rocks at Fleur de Lys, i.e. Pacquet Harbour and Grand Cove Groups (Church, 1969), Cape St. John Group and Cape Brulé Porphyry (Baird, 1951). Correlations of rocks and structures to the east of Baie Verte with the Fleur de Lys Supergroup to the west of Baie Verte led to the concept of similar terranes, deformed together, on both sides of the Baie Verte-Brompton Line. Furthermore, the early deformations were interpreted to predate the generation of the Advocate and Point Rousse ophiolite complexes (Church, 1969; Dewey and Bird, 1971; Kidd, 1977).

Deformation and metamorphism in rocks at Mings Bight and Pacquet Harbour decrease abruptly southeastward. At Nippers Harbour on the shores of Green Bay, the Cape St. John Group (eastern Fleur de Lys) clearly overlies ophiolitic rocks with nonconformity (Schroeter, 1973); and at Pinnacle Point it overlies the fossiliferous Lower Ordovician Snooks Arm Group with angular unconformity (Neale, 1957; Neale *et al.* 1975; DeGrace *et al.* 1976; Williams *et al.* 1977). These well established unconformities are critical, for they indicate that some of the rocks that previously were assigned to the eastern Fleur de Lys terrane are younger, rather than older, than nearby ophiolite suites. On the other hand, if the Mings Bight Group of the eastern Fleur de Lys is indeed the same rock stratigraphic unit as the western Fleur de Lys-continental margin clastics, it should lie to the west, rather than to the east, of the Baie Verte ophiolitic belt, i.e., the Baie Verte-Brompton Line. Furthermore, a structural break is implied between the Mings Bight clastics and nearby volcanic groups.

This enigma has plagued most syntheses of the geology of the Baie Verte Peninsula. However, it can be avoided, following the interpretation of Hibbard (1981), who has proposed a sharp flexure that involves the ophiolites along the Baie Verte-Brompton Line, changing its course from a straight line extending northward from Baie Verte to a 'Z' shaped line projecting seaward and northeastward near Pacquet Harbour. Thus the Fleur de Lys psammites at Coachmans Harbour and the Mings Bight psammites at Mings Bight are equated and interpreted to form the opposing limbs of a tight, steeply plunging flexure, which wraps around the north end of Point Rousse as outlined in Figure 2. Furthermore, a structural break is proposed between the Mings Bight Group and nearby ophiolitic and volcanic rocks near Pacquet Harbour (Hibbard, 1981). Intuitively, this break will be no more obvious at Pacquet Harbour than the equivalent contact between the Advocate and Birchy Complexes near Coachmans Harbour. The situation thus emphasizes the futility in attempts to separate similar lithic assemblages on structural style.

Glover Island

A narrow zone of ophiolitic rocks occurs on the west side of Glover Island, Grand Lake where it separates gneisses and schists to the west from volcanic and volcaniclastic rocks to the east (Fig. 3). This spatial distribution of rock groups is interpreted to define the Baie Verte-Brompton Line in this area (Williams and St-Julien, 1978; Knapp et al., 1979). Studies are in progress and the descriptions that follow are taken partly from Kanpp et al. (1979).

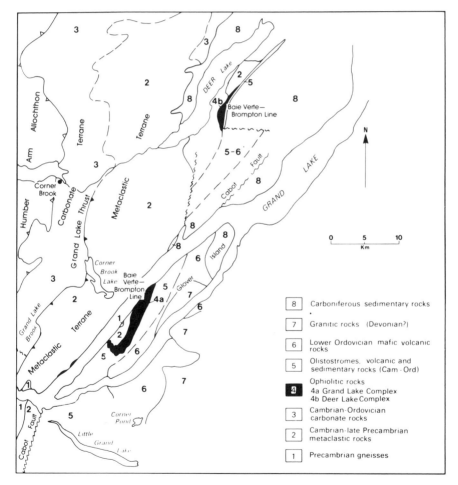

Figure 3. Relationships among rock groups at the Baie Verte-Brompton Line, Glover Island and Deer Lake, Newfoundland (partly after Knapp et al., 1979).

The ophiolitic rocks are referred to the Grand Lake Complex (Knapp, pers. commun., 1981). Toward the north, it consists of 150 m of serpentinized peridotite overlain by 800 m of layered cumulate gabbro and massive gabbro. This part of the complex is steeply-dipping, east-facing and exposed on the eastern limb of a south-plunging anticline. The ophiolitic units are truncated to the west by the Cabot Fault, a major fault affecting Carboniferous rocks and traceable across the island of Newfoundland.

Rocks at Glover Island and to the west of the Grand Lake Complex consist of quartzo-feldspathic gneisses overlain by metasedimentary rocks that are affected by polyphase deformation. Mafic dykes, now amphibolites, cut the gneisses. The overlying rocks consist of a basal conglomeratic unit overlain by marble and graphitic pelites, in turn overlain structurally by the Grand Lake Complex. A zone of high strain occurs at the base of the ophiolitic rocks and it is marked by greenschists and a variety of highly sheared, strongly retrograded mafic and ultramafic rocks.

West of Grand Lake, a wide belt of unnamed psammitic schists in amphibolite facies are of higher metamorphic grade than equivalent rocks at Glover Island. The psammitic schists exhibit early isoclinal folds and later tight upright folds that are also recognized in deformed carbonate rocks farther west, i.e., Grand Lake Brook Group (Walthier, 1949; Kennedy, 1980). Since the Grand Lake Brook Group is part of the west Newfoundland carbonate bank sequence, the relationship suggests continuity of deformation between the metamorphic terrane here and the Cambrian-Ordovician carbonate sequence farther west.

The Grand Lake Complex is overlain to the east by a sequence of dominantly mafic volcanic rocks with a shaly olistostromal unit at the base. Green and black shales with gabbro pebbles and cobbles overlie gabbro of the Grand Lake Complex toward the north. A breccia, 100 m thick, with a green shale matrix overlies pillow lavas and sheeted dykes at the south tip of Glover Island. The greenish breccia contains mostly volcanic clasts, bright red jasper and purplish shale clasts. Most clasts are small, averaging 5 cm diameter, but locally they are greater than 1 m diameter. East of Glover Island at Corner Pond, a thick volcanic sequence with black shale units is dated by graptolites as Arenigian (Dean, 1976; Williams and St-Julien, 1978).

Lithic similarities, sequence of rock units and faunal correlation of rock groups at Glover Island with those of the Baie Verte Peninsula are obvious; viz, basal quartzo-feldspathic gneisses are comparable to Grenvillian inliers of the Fleur de Lys terrane (De Wit, 1972), the overlying multideformed metasedimentary sequence correlates with the Fleur de Lys Supergroup, the Grand Lake Complex occupies the same structural position as the Advocate Complex, shaly olistostromes and volcanic rocks above the Grand Lake Complex correlate with the Flatwater Group, and the Arenigian volcanic sequence east of Grand Lake is of precisely the same age and lithology as the Snooks Arm Group.

The Glover Island ophiolite occurrence is significant because it shows that the ophiolitic rocks along the Baie Verte-Brompton Line may be involved in later folding or they may be cut out by later faulting.

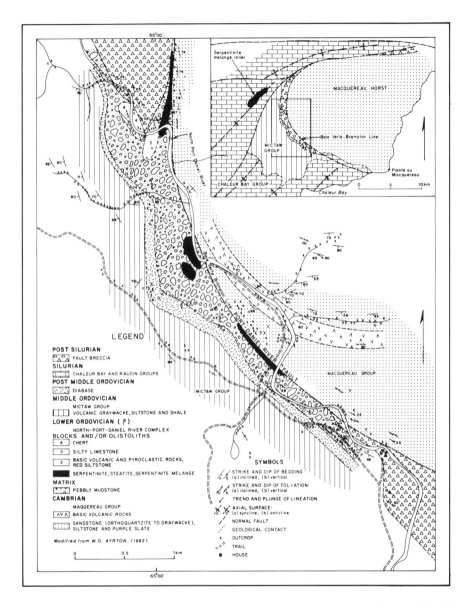

Figure 4. Relationships among rock groups at the Baie Verte-Brompton Line, Port Daniel, Gaspé Peninsula, Quebec.

North Port-Daniel River

An olistostromal melange with ophiolitic blocks up to 500 m long marks the general location of the Baie Verte-Brompton Line in the North Port-Daniel River area. The melange, or North Port-Daniel-River Complex (Ayrton, 1967) separates a horst of metasedimentary rocks, Maquereau Group, to the east from a Middle Ordovician flysch sequence, Mictaw Group, to the west. All of these rocks occur in an inlier that is surrounded by Silurian rocks of the Chaleur Bay Group (Fig. 4).

The North Port-Daniel-River Complex has a matrix of pebbly mudstone with blocks of serpentinite, volcanic rocks, chert, greywacke, silty limestone and red siltstone. Rocks of the Maquereau Group to the east are metamorphosed psammitic to pelitic schists with multideformation, thus contrasting in structural style with nearby groups. Regional correlations suggest they are Cambrian, and older than the Middle Ordovician Mictaw Group and North Port-Daniel River Complex. Locally, a fault breccia marks the contact between the Maquereau Group and North Port-Daniel-River Complex. The breccia contains angular schistose fragments of the Maquereau Group and white quartz fragments. These average 5 to 20 cm diameter, and some exceed 50 cm diameter. The fault breccia is related to the uplift of the Maquereau horst, which is a relatively young feature compared to the time of metamorphism and multideformation of the Maquereau Group.

A small inlier of serpentinite melange is surrounded by Silurian rocks immediately west of the Mictaw-Maquereau inlier. It is 3.5 x 0.6 km and contains blocks up to 1 km long. These consist of serpentinized peridotite, amphibolite, quartzite and granite; all set in a matrix of sheared greasy serpentinite. The periphery of the inlier is marked by fault breccias and the serpentinite melange is interpreted as a diapir, protruding from an underlying thrust sheet of ultramafic rocks.

The Maquereau Group is equated with similar groups that occur to the west (or north) of the Baie Verte-Brompton Line. The North Port-Daniel-River Complex is thought to be equivalent to olistostromes that border ophiolite complexes, where present, along the course of the Baie Verte-Brompton Line. Furthermore, the lithic units and structural relations within the Mictaw-Maquereau inlier are identical to those of the Thetford Mines area. Therefore the Baie Verte-Brompton Line is interpreted to lie at the western (and southern) end of the Maquereau structural dome.

Québec Eastern Townships

The Québec Eastern Townships has been an area of intense geological activity for more than 40 years (Cooke, 1937, 1950; Fortier, 1946; Ambrose, 1942, 1943, 1957; Riordon, 1954; De Romer, 1960; St-Julien, 1963; Osberg, 1965; and Rickard, 1965). Results of more recent studies, which involve the mafic and ultramafic complexes that

Figure 5. The Baie Verte-Brompton Line and major rock units of the Quebec Eastern Townships: 1) Grenville basement (P∈); 2) St. Lawrence platform (∈-0); 3) Foreland thrust belt (0); 4) Allochthones of the external domain (∈-0); 5) Allochthones of the internal domain (mainly Sutton-Bennett schists) (∈-0); 6) Caldwell Gr. and Mansonville Fm. (∈); 7) ophiolites (∈); 8) St. Daniel and Brompton Fms. (Mélange) (L.0); 9) St. Victor synclinorium (Magog Gr.) (M.0); 10) Ascot-Weedon Fms (L 0-M.0); 11) Connecticut Valley-Gaspé synclinorium (S-D); 12) Frontenac Fm. (0?); 13) Chain Lakes Massif (Helikian); 14) Ordovician granites; 15) Devonian granites; 16) Mesozoïc alkaline intrusive rocks.

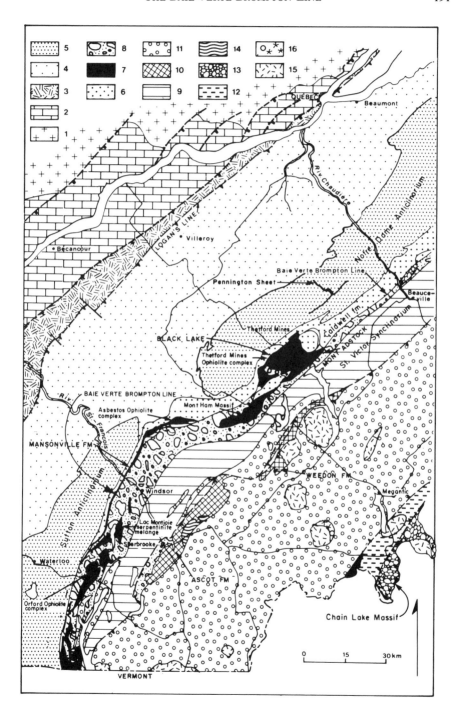

define the Baie Verte-Brompton Line in this area, are contained in Lamarche (1972, 1973), Hébert (1974), Laurent (1975a, 1975b, 1977), Blackburn (1975), St-Julien and Hubert (1975), Laurent and Hébert (1977), and Hébert (1979). Most of these authors interpret the mafic-ultramafic complexes as ophiolitic, comprising slivers or fault slices of ocean floor and upper mantle rocks. Nearby chaotic rocks to the south are interpreted as olistostromal cover to the ophiolite sequence.

The geology of the Eastern Townships is summarized in Figure 5. The Baie Verte-Brompton Line strikes northeasterly and it is marked by several mafic-ultramafic complexes, e.g., Thetford Mines, Asbestos and Orford. Polydeformed metamorphic rocks occur to the northwest of the line, i.e., Oak Hill, Rosaire, Ottauquechee, Caldwell and Mansonville Groups; and their more metamorphosed equivalents of the Sutton-Bennett Schists. These consist of phyllite and quartzite, and feldspathic sandstone, slates and volcanic rocks. Small mafic-ultramafic bodies, mostly serpentinite sheets, occur along major faults within the metaclastic terrane. To the southeast, the Baie Verte-Brompton Line is bounded by olistostromes of the St-Daniel and Brompton Formations (St-Julien, 1963; Lamothe, 1978), or the Brompton Rocks (Fortier, 1946). Farther southeast, a Middle Ordovician turbidite sequence of the Magog Group occupies the St-Victor Synclinorium (St-Julien, 1967). This is bordered by calc-alkaline volcanic rocks of the Ascot-Weedon Formation farther southeastward.

Local Relationships along the Baie Verte-Brompton Line. Ophiolite suites along this section of the Baie Verte-Brompton Line vary from complete to dismembered and imbricated. The various sizes of these bodies and the presence of intervening shale zones suggest that much of the ophiolitic terrane is little more than a huge steep zone of ophiolitic melange (Gansser, 1974). The backbone of this zone is formed by the large ophiolite complexes at Thetford Mines, Asbestos and Orford.

The Thetford Mines ophiolite complex is composed of the Black Lake, Mont Adstock and Mont Ham massifs (Figs. 5 and 6). Rock units strike northeast, dip vertical to steeply northwest, and are repeated in a series of slices bounded by steeply dipping faults. The ophiolitic sequences face southeast, suggesting imbrication by northwestward thrusting.

The Black Lake ophiolite massif extends 25 km in a northeasterly direction from Breeches Lake, southwest of Coleraine, to Thetford Mines (Fig. 6). From northwest to southeast, the ascending sequence of rock units is harzburgite, dunite, pyroxenite, gabbro, and pillow basalts interbedded with pyroclastic rocks. In the massif, the rocks strike northeast and dip steeply to the northwest beneath southeast-facing beds of the Caldwell Group.

The Mont Adstock ophiolite massif extends 20 km in a northeasterly direction from Disraeli to Mont Adstock, located north of the town of St-Daniel (Fig. 6). The Mont Ham ophiolite massif can be traced in a northeasterly direction for 25 km from Mont Ham (Fig. 5). Both massifs are composed of a broken and strongly serpentinized cumulate unit, which is overlain by basaltic volcanic rocks (lower volcanics) and andesitic pyroclastic rocks and pillowed basalts (upper volcanics). The cumulate unit consists of schistose serpentinite that contains blocks of serpentinized dunite at its base, blocks of brecciated and serpentinized pyroxenite in its middle, and blocks of gabbro and diabase at its top. A horizon of red and green cherty argillite separates

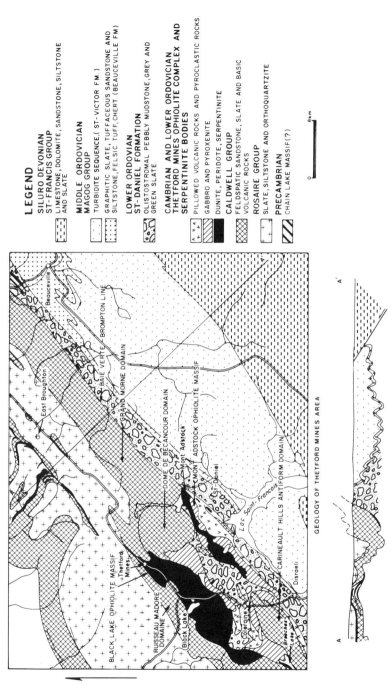

Figure 6. Relationships between rock groups and structures at the Baie Verte-Brompton Line, Thetford Mines, Quebec.

the tholeiitic basalts of the lower volcanic unit from the upper volcanic unit (Laurent and Hébert, 1977). Blocks of schistose and metamorphosed Caldwell feldspathic sandstone occur in the red and green cherty argillite. This implies that the main regional deformation and metamorphism of the psammitic and pelitic schists to the northwest of the Baie Verte-Brompton Line preceded deposition of the cherty argillite unit and overlying volcanic rocks.

An occurrence of garnetiferous amphibolite at the tectonic contact between harzburgite of the Black Lake ophiolite massif and Caldwell feldspathic sandstone to the west (Feininger, pers. commun., 1979) suggests the local development of a Bay of Islands-type metamorphic aureole (Laurent, 1975a), implying transport of the ophiolitic rocks to their present position.

The Asbestos ophiolite complex lacks most of the upper sequence of metagabbro, diabase and volcanics, but has well developed harzburgite and dunite units (Laliberté et al., 1979; Laurent, 1977; Lamarche, 1973). The rocks dip southward and face in the same direction. The ophiolites are thrust over the Caldwell feldspathic sandstones and phyllites that lie to the north. The contact between the Asbestos ophiolite complex and the St-Daniel Formation to the south has been interpreted both as a thrust fault (Laliberté et al., 1979) and a normal stratigraphic contact (Lamarche, 1973). We favour the latter interpretation and suggest that an initial stratigraphic contact is fault modified.

From Asbestos southwestward to the Vermont border, the Baie Verte-Brompton Line is marked by a vertical sheet of serpentinite associated with local mafic and ultramafic rocks that face east. These rocks are followed eastward by the Brompton Formation, which consists of an olistostromal pebbly mudstone at its base and a megaconglomerate or coarse olistostromal melange at its top. The chaotic rocks, exposed over a width of 10 km in the Orford area, contain the Lac Montjoie serpentinite melange (Lamothe, 1978) and the Orford ophiolite complex (Fig. 5), which face southeast and east, respectively. These local relationships at Orford suggest that the ophiolites of the Baie Verte-Brompton Line are repeated by thrusting and in each slice the ophiolitic rocks are steeply dipping and east facing (De Römer, 1960; St-Julien, 1963; Lamothe, 1978; Rodrique, 1979).

The Lac Montjoie serpentinite melange is interpreted as a diapir of serpentinite, probably derived from an underlying slice of ultramafic rocks (Lamothe, 1978). Its largest blocks are clinopyroxenite, harzburgite, amphibolite, gabbro, volcanic breccia, greenstone, and pebbly mudstone; all set in a sheared serpentinite matrix. In the southwest prolongation of the serpentinite terrane, multideformed and metamorphosed Mansonville feldspathic sandstone occurs in a horst-like structure.

The Orford ophiolite complex is thrust onto the chaotic melange of the Brompton Formation to the west. It strikes north-northeast and dips vertical. The complex is a cumulate suite of dunite, pyroxenite and gabbro, with sheeted diabase and pillowed volcanics (Rodrique, 1979). A thrust sheet of diabase and lower volcanics occurs farther south, followed by another thrust slice of upper volcanics (Rodrique, 1979).

Ophiolite complexes of the Eastern Townships are only partly recrystallized in the greenschist facies of regional metamorphism. This suggests that they were emplaced directly onto a continental margin by obduction (Laurent, 1977).

Local Relationships west of the Baie Verte-Brompton Line. The Oak Hill, Rosaire, Ottauquechee, Caldwell and Mansonville Groups, and their higher grade metamorphic equivalents, the Sutton-Bennett Schists, are mainly psammitic to pelitic schists, which are multideformed and locally metamorphosed to upper greenschist facies. In places, these rocks have associated mafic volcanics, ie., Tibbit Hill volcanics of the Oak Hill Group and mafic volcanics of the Caldwell Group. Grenvillian inliers in correlatives of the Sutton Schists in Vermont, and the presence of a sliver of Grenvillian gneisses at the sole of a thrust fault in front of the Bennett Schists southeast of Quebec City (Vallières *et al.*, 1978) support the contention that the psammitic and pelitic rocks northwest of the Baie Verte-Brompton Line represent the eastern part of a clastic wedge that accumulated on a rifted Grenvillian basement (Bird and Dewey, 1970; St-Julien and Hubert, 1975).

Locally, sodic amphiboles, magnesium riebeckite and crossite occur in mafic volcanic rocks associated with the pre-Ordovician psammitic and pelitic schists west of the Baie Verte-Brompton Line (Doolan *et al.,* 1973; Trzcienski, 1976). The presence of these minerals suggests crystallization at high pressure in a regime possibly representing a transition from greenschist to blueschist metamorphic facies (Laurent, 1977). This metamorphic assemblage may result from special conditions related to ophiolite emplacement (Laurent, 1977).

Except for local complications at Thetford Mines, prominent schistosities are inclined toward the Baie Verte-Brompton Line. They dip southeast on the northwest side of the line, and northwest on the southeast side of the line. The facing direction of folds related to these prominent fabrics is away from the line (St-Julien, 1967).

Throughout its course in Québec, the Baie Verte-Brompton Line maintains a constant position to the southeast of a series of anticlinoria, i.e., Maquereau horst, Notre Dame Mountains, Sutton Mountains and Green Mountains. A seismic profile across the Bennett Schists north of Thetford Mines suggests that the Notre Dame Anticlinorium is made up of a pile of thrust sheets that root at or near the Baie Verte-Brompton Line. Metamorphosed and deformed serpentinite sheets and small mafic-ultramafic bodies in the Sutton-Bennett Schists are thought to be localized along thrust faults (St-Julien *et al.,* 1972; Laurent, 1975) rather than representing small intrusions (Cooke, 1937; Riordon, 1953).

At Thetford Mines, the Caldwell Group can be separated into four distinct structural domains, namely the Grand Morne, Madore Brook, Bécancour dome and Carineault Hills antiform domains (Fig. 6). The Grand Morne and Madore Brook domains occur northeast and northwest of the Thetford Mines ophiolite complex, respectively. The Bécancour dome and Carineault Hills antiform occur southeast of Thetford Mines and south of Coleraine, respectively.

Outcrop patterns and top determinations in the Grand Morne domain show that the Caldwell beds, although gently to steeply dipping to the northeast and northwest are overturned and constitute a series of synformal anticlines and antiformal synclines that plunge either northeast or southwest. The Caldwell sandstones of the Madore Brook domain are more deformed and metamorphosed but they can be traced uninterruptedly from the Grand Morne domain. Local top determinations in the Grand Morne domain also show that the rocks are overturned. The structural relationships indicate that in the Grand Morne and Madore Brook domains the Caldwell beds form the inverted limb of a major recumbent nappe.

Caldwell rocks of the Bécancour dome and Carineault Hills antiform were emplaced by uplift from beneath the Thetford Mines ophiolite complexes. The feldspathic sandstones of these domains are more metamorphosed compared to those at Grand Morne and they exhibit two subparallel schistosities that are subparallel to bedding. In the Bécancour area these planar elements are bent into an open dome.

The Bécancour dome and Carineault Hills antiform are surrounded by curvilinear normal faults that juxtapose the uplifted Caldwell rocks with the Thetford Mines ophiolite complexes, with olistostromes of the St-Daniel Formation, and with other Caldwell rocks of the nearby Grand Morne domain to the northeast. The bounding normal faults are marked by zones of coherent tectonic breccia up to several metres wide. The breccias contain angular fragments of schistose metasandstones and vein quartz in a matrix of finer fragmental material. Their time of formation and the emplacement of the domes followed the deformation of the Caldwell Group, the emplacement of the ophiolite complexes, and the deposition of the St-Daniel Formation.

The structure and stratigraphy in the vicinity of the Bécancour dome are identical to those in the North Port-Daniel River area of Gaspé. In both areas, circular horsts of polydeformed and metamorphosed feldspathic sandstones are surrounded by a characteristic fault breccia and by olistostromal melanges with ophiolite components. The melanges, in turn, are overlain by Middle Ordovician flysch sequences; the Magog Group in the Thetford Mines area, and the Mictaw Group in southern Gaspé (St-Julien and Hubert, 1979).

Local Relationships to the southeast of the Baie Verte-Brompton Line. Olistostromes of the St-Daniel and Brompton Formations, which occur on the southeast side of the Thetford Mines, Asbestos and Orford ophiolite complexes are steeply-dipping, east-facing and stratigraphically above the upper volcanic rocks of the ophiolite complexes (St-Julien, 1963, 1965, 1970; Lamarche, 1973). Toward the south, these formations are overlain with angular unconformity by sedimentary rocks of the Middle Ordovician Magog Group (St-Julien, 1967; St-Julien et al., 1972). Farther east, calc-alkaline volcanic rocks of the Ascot-Weedon Formation (St-Julien, 1967; St-Julien and Hubert, 1975) are thrust above the Magog Group with olistostromal pebbly shale at the thrust contact.

The olistostromal rocks of the St-Daniel and Brompton Formations contain fragments of various type and size. These include pebbles and cobbles of green and dark grey shale and argillite, chert, siltstone and dolomitic siltstone, quartz arenite and quartzite, gabbro, and fragments of volcanic rocks and serpentinite with dimensions from a few metres to more than one kilometre. The fragments are angular to subrounded, unsorted, and set in a matrix of dark grey and green shale. The presence of ophiolitic blocks tens of metres across shows that the olistostromes are younger than the nearby Thetford Mines, Asbestos and Orford ophiolite complexes. As mentioned previously, cherty argillites between the lower and upper volcanics of the ophiolite complexes contain blocks of polydeformed and metamorphosed psammitic schist. This indicates that the main regional deformation and metamorphism that affected the metaclastic terrane to the northwest of the Baie Verte-Brompton Line preceded deposition of the olistostromal rocks (Laurent et al., 1979).

A 10 km long sliver of tough fragmental metamorphic rocks occurs within the olistostromes of the St-Daniel Formation 6 km north of Beauceville. Metamorphic rocks of the sliver are similar to rocks of the nearby Chain Lakes complex in the Boundary Mountain Anticlinorium of Maine. The latter is dated isotopically at 1500 Ma (Naylor *et al.*, 1973), and it forms the basement to nearby Ordovician and Silurian rocks. The occurrence of similar metamorphic rocks in the St-Daniel mélange suggests proximity to this easterly basement terrane. The relationship further implies that the Early Paleozoic ocean represented by the ophiolitic rocks of the Baie Verte-Brompton Line was nearly closed during deposition of the St-Daniel Formation in this locality.

SUMMARY OF SALIENT FEATURES AND REGIONAL CORRELATIONS

The foregoing analyses illustrate several consistent features that are recognizable along the full length of the Baie Verte-Brompton Line. Some of these are apparent from a cursory study of regional geologic maps, i.e., Tectonic Lithofacies Map of the Appalachian Orogen (Williams, 1978a); others are indicated by more detailed stratigraphic, structural and sedimentologic comparisons.

The *mafic-ultramafic complexes that define the Baie Verte-Brompton Line* are all lithologically similar, and where the complexes are sufficiently well preserved, it is obvious that their sequences of lithic units are typical of the ophiolite succession. Local stratigraphic relationships and lithic correlations indicate that the complexes are Early Ordovician or latest Cambrian.

Deformation in the ophiolitic complexes varies from moderate to intense, from east to west, and some westerly examples exhibit the full sequence of local deformations recognized in nearby metasedimentary rocks. Almost everywhere the mafic-ultramafic complexes are steeply dipping, and in some places they are overturned; but their sequences of rock units face consistently toward the east and southeast. Locally the ophiolitic sequence of rock units is repeated by thrusting.

Some of the ophiolitic complexes have high grade metamorphic rocks at their bases, mainly amphibolites, that seem out of context compared to low grade rocks nearby. The presence and structural setting of the high grade rocks indicate the local development of dynamothermal aureoles, similar to those developed beneath highly allochthonous ophiolite suites such as the Bay of Islands Complex (Williams and Smyth, 1973; Malpas, 1979).

These considerations, combined with a consistent structural and stratigraphic position of the ophiolitic rocks with respect to nearby groups, indicate that all of the mafic-ultramafic complexes along the Baie Verte-Brompton Line are correlatives. This conclusion is supported also by their petrologic and chemical characteristics (Norman and Strong, 1975; Church, 1977; Laurent *et al.*, 1979). It is also supported by the fact that the serpentinized ultramafic parts of the complexes are host to major asbestos deposits, that are rare or absent in similar ophiolitic suites elsewhere in the Appalachian Orogen.

West of the Baie Verte-Brompton Line, Grenvillian gneisses are overlain by thick clastic sequences. Mafic dykes that cut the gneisses and basal parts of the clastic cover are feeders for nearby volcanic rocks, e.g. Tibbit Hill volcanics of Québec. These relationships are clear in westerly undeformed parts of the Appalachians (Strong and Williams, 1972). Near the Baie Verte-Brompton Line, both basement and cover rocks are affected by multiphase deformation and metamorphism.

The clastic sequences to the west of the line are mainly of late Precambrian to Early Ordovician age and they can be traced the full length of the Appalachian Orogen (Williams and Stevens, 1974). The rocks are lithologically similar and everywhere exhibit a complex structural style of multiple deformation and metamorphism, e.g., Fleur de Lys Supergroup, Maquereau Group, Sutton-Bennett Schists. The earliest folds in the clastic cover sequences are isoclinal, commonly recumbent and westward facing. The locus of peak metamorphism in the clastic cover rocks is parallel to the Baie Verte-Brompton Line, but lies up to 10 km westward. Several lines of indirect evidence indicate that these structures and accompanying metamorphism are of Ordovician age (Taconian). This conclusion is most obvious between the Québec Eastern Townships and Gaspé Peninsula, where complex structures and metamorphism characteristic of Ordovician and earlier rocks do not affect overlying Silurian and younger rocks of the Chaleur Bay and Connecticut Valley-Gaspé Synclinoria. As well, metamorphic rock fragments occur in olistostromes above the ophiolite complexes along the Baie Verte-Brompton Line in Quebec and Newfoundland.

The small mafic to ultramafic plutons that occur within the multideformed psammitic sequences to the west of the Baie Verte-Brompton Line are metamorphosed and deformed with the surrounding rocks. Some of these are serpentinite screens along structural surfaces, e.g., Pennington sheets of Thetford Mines area. They are all probably dismembered ophiolitic rocks, emplaced tectonically, rather than high level early intrusions. Similar small mafic to ultramafic plutons, which are associated with similar polydeformed and metamorphosed clastic rocks, can be traced along the western flank of the entire Appalachian Orogen (Williams and Talkington, 1977).

East of the Baie Verte-Brompton Line, olistostromes that occur above the ophiolite suites are a diagnostic feature. These rocks contain ophiolitic clasts, a variety of sedimentary clasts and sparse metamorphic rock fragments. Their ubiquitous presence at the line and their absence above ophiolite suites nearby to the east, e.g., above the Point Rousse and Betts Cove Complexes in Newfoundland, suggest that the olistostromal rocks bear some important spatial relationship to rocks at the line. Local stratigraphic relationships indicate that these rocks are Early or Early Middle Ordovician and they are correlated throughout the length of the Baie Verte-Brompton Line.

The consistent pattern of Ordovician and earlier rocks with respect to the Baie Verte-Brompton Line, combined with their local continuity and correlations proposed here, all indicate that the line is an Ordovician feature, although modified by younger deformations. Its course is parallel to the sinuous pattern of Ordovician and earlier facies belts and tectonic elements along the western flank of the orogen. In contrast, the Baie Verte-Brompton Line has no expression in the facies distribution of Silurian and younger rocks; and Silurian facies belts cross the line at acute angles in the Gaspé Synclinorium.

Throughout its sinuous course, the Baie Verte-Brompton Line is parallel to and equidistant from Logan's Line. The latter marks the leading edge of Taconian allochthons and is roughly coincident with the west limit of Taconian and Acadian deformations in the Appalachian Orogen. The Baie Verte-Brompton Line also defines the west limit of Ordovician plutonism and important volcanism in the Canadian Appalachians. Almost all of the Silurian-Devonian major granitic intrusions and volcanic belts occur to the east of the line, except at Gaspé Peninsula and western Newfoundland.

The nature of deep crustal structure as determined seismically bears a geometric relationship to the position of the Baie Verte-Brompton Line. In northeast Newfoundland, slow seismic refraction velocities indicate a light thin (30 km) crust to the west of the line and faster velocities indicate a denser thicker (45 km) crust to the east (Sheridan and Drake, 1968; Dainty et al., 1966). This variation in crustal density and thickness is also supported in northeast Newfoundland by local gravity measurements (Miller and Deutsch, 1976).

More regionally throughout its full length, the line is parallel and lies close to a pronounced gradient in the Bouguer gravity anomaly field (Haworth et al., 1980). This gradient between negative values to the west and positive values to the east is thought to reflect a transition between ancient continental and oceanic crust, just as the edges of modern continental margins have marked gravity gradients.

SIGNIFICANCE AND INTERPRETATION

The most important feature of the Baie Verte-Brompton Line is its continuity for like other continuous stratigraphic and structural features in the Appalachian Orogen, it provides a link for distant correlation and implies a similar overall tectonic development for the western flank of the orogen. Thus, an awareness of the length of such features as the Baie Verte-Brompton Line provides some measure of scale when formulating geologic models based upon local relationships.

All recent workers agree that the ophiolitic occurrences along the Baie Verte-Brompton Line represent a sampling of oceanic crust and mantle. Whether or not they represent the sampling of a major wide ocean or a small marginal ocean remains debatable. However, the length of the feature and attempts at palinspastic restoration imply an ocean of several thousand kilometres in length and 1000 km or more wide (Williams, 1980a).

Grenvillian gneisses to the west of the line represent the rifted basement at the Early Paleozoic continental margin. Mafic intrusions, which cut the gneisses, and coeval volcanic rocks and associated clastic cover sequences relate to the initial rifting episode (Strong and Williams, 1972). The position of thick clastic sequences to the west of the line, between a Cambrian-Ordovician carbonate bank sequence to the west and ophiolitic complexes to the east, suggests that these rocks represent an original slope-rise assemblage built up at the rifted ancient continental margin of eastern North America (Williams and Stevens, 1974). Marbles and limestone breccias associated with the clastic rocks in westerly exposures may represent distal facies of the carbonate bank sequence to the west (Rodgers, 1968; Williams and Stevens, 1974). The sinuous course of the Baie Verte-Brompton Line is interpreted to reflect an original zig zag or orthogonal continental margin bounded by rifts and transform faults (Thomas, 1977), comparable to that of the modern Atlantic margin at the Grand Banks of Newfoundland (Williams and Doolan, 1979).

Ophiolitic melange in eastern parts of the metaclastic terrane, affected by the full

range of local deformations, indicates ophiolite transport across an initially undeformed continental margin. This event is interpreted to represent the passage of
ophiolitic suites such as the Bay of Islands and Mount Albert complexes from the
oceanic tract to their present westerly positions above the carbonate bank sequence
(Williams, 1977). The Baie Verte-Brompton Line is viewed therefore as the most
westerly possible root zone for highly allochthonous ophiolitic complexes farther
west. Ophiolitic complexes at the Baie Verte-Brompton Line are also allochthonous,
judging by the local occurrences of dynamothermal aureoles of Bay of Islands-type.

Olistostromes above the ophiolite suites at the Baie Verte-Brompton Line contain abundant ophiolitic blocks and local schistose metasedimentary rock fragments.
This suggests deposition following deformation at the continental margin. The
coarseness and composition of the olistostromal rocks, and their absence above
ophiolite suites farther east, implies local derivation from the disturbed and over-
steepened oceanic crust at the deformed continental edge. Local volcanic suites
above the shaly chaotic rocks, which resemble volcanic arc sequences above nearby
easterly ophiolite suites, indicates continuing arc volcanism during this early phase of
destruction of the continental margin.

Thus the stratigraphic and structural features of rocks at or near the Baie
Verte-Brompton Line can be interpreted in terms of an evolving continental margin
that was destroyed in the Ordovician by obduction of oceanic lithosphere. One
possible model for the evolution of the Baie Verte-Brompton Line, as represented in
Newfoundland, is developed schematically in Figure 7. This model emphasizes the
Early Paleozoic development of the west flank of the Appalachian Orogen and the
destruction of the ancient continental margin of eastern North America by ophiolite
obduction during Ordovician Taconian Orogeny. According to this model, the Baie
Verte-Brompton Line represents the surface trace of a steep ophiolitic zone at a
continent-ocean interface within the deformed rocks of the Appalachian Orogen.

Following ophiolite emplacement and destruction of the Ordovician continental
margin, Silurian and later rocks of the Appalachian Orogen bear no obvious relationship to earlier facies belts. This is interpreted to reflect an almost complete destruction of the oceanic tract and its Early Paleozoic continental margins during the
Taconian Orogeny.

Conceptually, the Baie Verte-Brompton Line should be traceable along the entire length of the Appalachian-Caledonides System, just as Early Paleozoic facies
belts can be traced throughout the entire length of the system (Williams, 1978a,
1978b). Large ophiolitic complexes at Baltimore, Maryland (Morgan, 1977) and Staten Island, New York occur in a structural setting somewhat analogous to ophiolite
occurrences along the Baie Verte-Brompton Line in Canada. Similarly in the British
Caledonides, small mafic-ultramafic plutons along the Highland Boundary Fault and
in the northeast Shetland Islands of Unst and Fetlar may define this important structural boundary.

An absence of characteristic ophiolites and olistostromes of the Baie Verte-
Brompton Line in the New England Appalachians may reflect complete suturing and
deep erosion, as in southwestern Newfoundland. In the southern Appalachians,
similar rocks may be hidden by crystalline nappes of the Inner Piedmont.

Recent seismic reflection studies in the Southern Appalachians indicate a sub-

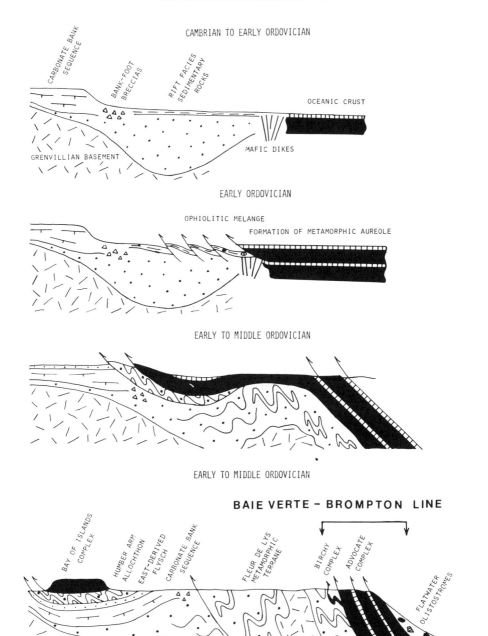

Figure 7. Model for the development of the Baie Verte-Brompton Line in Newfoundland.

horizontal reflector that is traceable in subsurface beneath interior parts of the exposed orogen (Cook *et al.*, 1979), and possibly extends much farther eastward (Harris and Bayer, 1979). The reflector is interpreted as a structural décollement (Cook *et al.*, 1979; Harris and Bayer, 1979), implying that westerly parts, and possibly the entire orogen, are allochthonous. This prospect of a completely allochthonous Appalachian Orogen raises the possibility that the rocks that define the Baie Verte-Brompton Line occur well to the west of the subsurface position of the edge of the ancient continental margin of eastern North America.

Several lines of evidence indicate that this thin-skinned model for the Southern Appalachians is not directly applicable to the Canadian Appalachians (Williams, 1980b), and as follow: (a) parallelism and proximity of the Baie Verte-Brompton Line and the gravity gradient throughout the Canadian Appalachians, (b) parallelism of the Baie Verte-Brompton Line with the Taconian and Acadian structural fronts in the Canadian Appalachians, (c) parallelism of the Baie Verte-Brompton Line and Early Paleozoic facies belts that are at least partly autochthonous in the Canadian Appalachians, (d) a seismic reflector that implies a décollement surface beneath westerly parts of the Quebec Appalachians appears to terminate near the Baie Verte-Brompton Line (Granger *et al.*, 1980), (e) coincidence of the Baie Verte-Brompton Line with a marked change in crustal thickness and density as determined by seismic refraction studies in Northeast Newfoundland, and (f) the absence of an important Alleghanian foreland thrust zone in the Northern Appalachians and its presence in the Southern Appalachians.

The Canadian Appalachians are viewed therefore as an orogen involving profound structural telescoping, but essentially rooted. Thus the Baie Verte-Brompton Line and adjacent facies belts in the Canadian Appalachians are not predicted to be displaced by several hundred kilometres, with respect to the subsurface edge of Grenvillian basement rocks, as proposed for the Southern Appalachians.

ACKNOWLEDGEMENTS

We wish to thank J.P. Hibbard for previews of his unpublished maps and papers and for detailed criticism and useful discussion. We also wish to thank D. Knapp, R.A. Price, P.H. Osberg and R.K. Stevens for reviewing the manuscript and helpful remarks. We are grateful for support of our field work in the Canadian Appalachians by the Natural Sciences and Engineering Research Council of Canada, the Department of Energy, Mines and Resources and the Killam Program of the Canada Council.

REFERENCES

Ambrose, J.W., 1942, Preliminary Map of Mansonville Map-area, Quebec: Geol. Survey Canada, Paper 42-1

———————, 1943, Preliminary Map of Stanstead Map-area, Quebec: Geol. Survey Canada Paper 43-12.

———————, 1957, The age of the Bolton Lavas, Memphremagog District, Quebec: Naturaliste Canadien, v. LXXXIV, p. 161-170.

Ayrton, W.G., 1967, Chandler-Port Daniel Area, Bonaventure and Gaspe South Counties: Ministère des Richesses Naturelles, Quebec, Geological Report 120, 91 p.

Baird, D.M., 1951, The geology of Burlington Peninsula, Newfoundland: Geol. Survey Canada Paper 51-21.

Bird, J.M. and Dewey, J.F., 1970, Lithosphere plate-continental margin tectonics and the evolution of the Appalachian Orogen: Geol. Soc. America Bull., v. 81, p. 1031-1060.

Blackburn, M., 1975, Analyse structurale des assises du Groupe de Caldwell à l'Est du complexe ophiolitique de Thetford Mines: Thèse de M.Sc., Université Laval, 61 p.

Brown, P.A., 1973, Possible cryptic suture in southwest Newfoundland: Nature, Physical Science, v. 245, no. 140, p. 9-10.

Bursnall, J.T., 1975, Stratigraphy, structure and metamorphism west of Baie Verte, Burlington Peninsula, Newfoundland: Ph.D. Thesis, Cambridge University, 337 p.

_____, 1979, Geology of part of the Fleur de Lys Map-area (121/1E^1/2), Newfoundland: in Gibbons, R.V., ed., Newfoundland Dept. Mines and Energy, Report of Activities for 1978, p. 68-74.

Bursnall, J.T. and De Wit, M.J., 1975, Timing and development of the orthotectonic zone in the Appalachian Orogen of Northwest Newfoundland: Canadian Jour. Earth Sci., v. 12, p. 1712-1722.

Church, W.R., 1969, Metamorphic rocks of Burlington Peninsula and adjoining areas of Newfoundland, and their bearing on continental drift in North Atlantic: in Kay, M., ed., North Atlantic-Geology and Continental Drift: American Assoc. Petroleum Geol. Memoir 12, p. 212-233.

_____, 1977, The ophiolites of Southern Quebec: oceanic crust of Betts Cove type: Canadian Jour. Earth Sci., v. 14, p. 1668-1673.

Church, W.R. and Stevens, R.K., 1971, Early Paleozoic ophiolite complexes of the Newfoundland Appalachians as mantle-oceanic crust sequences: Jour. Geophys. Research, v. 76, no. 5, p. 1460-1466.

Cook, F.A., Albaugh, D.S., Brown, L.D., Kaufman, S. and Oliver, J.E., 1979, Thin-skinned tectonics in the crystalline southern Appalachians: COCORP seismic-reflection profiling of the Blue Ridge and Piedmont: Geology, v. 7, p. 563-567.

Cooke, H.C., 1937, Thetford, Disraéli, and eastern half of Warwick Map-areas, Quebec: Geol. Survey Canada Memoir 211, 176 p.

_____, 1950, Geology of the southwestern part of the Eastern Townships of Quebec: Geol. Survey Canada Memoir 257, 142 p.

Dainty, A.M., Keen, C.E., Keen, M.J. and Blanchard, J.E., 1966, Review of geophysical evidence on crust and upper mantle structure on the eastern seaboard of Canada: American Geophys. Union Monography 10, p. 349-369.

Dean, W.T., 1976, Some aspects of Ordovician correlation and trilobite distribution in the Canadian Appalachians: in Bassett, M.G., ed., The Ordovician System: Cardiff University of Wales Press and National Museum of Wales, p. 227-250.

DeGrace, J.R., Kean, B.F., Hsü, E. and Green, T., 1976, Geology of the Nippers Harbour Map-area (2E/13), Newfoundland: Newfoundland Dept. Mines and Energy, Rept. 76-3, 73 p.

DeRömer, H.S., 1960, Geology of the Eastman-Orford Lake Area, Eastern Townships, Quebec: Ph.D. Thesis, McGill University, Montreal, 397 p.

Dewey, J.F. and Bird, J.M., 1971, Origin and emplacement of the ophiolite suite: Appalachian ophiolites in Newfoundland: Jour. Geophys. Research, v. 76, p. 3179-3206.

De Wit, M.J., 1972, The geology around Bear Cove, Eastern White Bay, Newfoundland: Ph.D. Thesis, Cambridge University, England, 232 p.

De Wit, M.J. and Strong, D.F., 1975, Eclogite-bearing amphibolites from the Appalachian Mobile Belt, Northwest Newfoundland: Dry versus wet metamorphism: Jour. Geol., v. 83, p. 609-627.

Doolan, B.L., Drake, J.C. and Crocker, D., 1973, Actinolite and subcalcic hornblende from a greenstone of the Hazens Notch Formation, Lincoln Mountain Quadrangle, Vermont: Geol. Soc. America, Abstracts with Programs, v. 5, p. 157.

Dunning, G.R. and Herd, R.K., 1980, The Annieopsquotch ophiolite complex, southwest Newfoundland, and its regional relationships: in Current Research, Part A, Geol. Survey Canada Paper 80-1A, p. 227-234.

Fortier, Y.O., 1946, Geology of the Orford Area, Quebec: Ph.D. Thesis, Stanford University, 248 p.

Fuller, J.O., 1941, Geology and mineral deposits of the Fleur de Lys Area: Newfoundland Geol. Survey Bull. 15, 41 p.

Gansser, A., 1974, The ophiolitic melange, a world-wide problem on Tethyan examples: Eclogae Geologicae Helvetiae, v. 67, p. 479-507.

Granger, B., St-Julien, P. and Slivitsky, A., 1980, A seismic profile across the southwestern part of the Quebec Appalachians: Geol. Soc. America, Abstracts with Programs, v. 12, no. 7, p. 435.

Harris, L.D. and Bayer, K.C., 1979, Sequential development of the Appalachian Orogen above a master decollement – A hypothesis: Geology, v. 7, p. 568-572.

Haworth, R.T., Poole, W.H., Grant, A.C. and Sanford, B.V., 1976, Marine geoscience survey northeast of Newfoundland: in Report of Activities, Geol. Survey Canada Report 76-1A, p. 7-15.

Haworth, R.T., Daniels, D.L., Williams, H. and Zietz, I., 1980, Bouguer Gravity Anomaly Map of the Appalachian Orogen: Memorial University Newfoundland, Map No. 3.

Hébert, R., 1979, Etude pétrologique des roches ophiolitiques d'Asbestos et du Mont Ham (Ham Sud), Québec: Thèse de M.Sc., Université Laval.

Hébert, Y., 1974, Etude pétrographique et chimique de la coupe du lac de l'Est dans le complexe ophiolitique de Thetford Mines, Québec: Thèse de M.Sc., Université Laval, 109 p.

Herd, R.K. and Dunning, G.R., 1979, Geology of Puddle Pond Map-area, southwestern Newfoundland: in Current Research, Geol. Survey Canada Paper 79-1A, p. 305-310.

Hibbard, J.P., 1978, Geology east of the Baie Verte Lineament: in Gibbons, R.V., ed., Newfoundland Dept. Mines and Energy, Report of Activities for 1977, Report 78-1, p. 103-109.

_____, 1981, Significance of the Baie Verte Flexure, Newfoundland: Geol. Soc. America Bull., (in press).

Kennedy, D.P., 1980, Geology of the Corner Brook Lake area, Western Newfoundland: in Current Research, Part A, Geol. Survey Canada Paper 80-1A, p. 235-240.

Kennedy, M.J., 1975, Repetitive orogeny in the Northeastern Appalachians – new plate models based upon Newfoundland examples: Tectonophysics, v. 28, p. 39-87.

Kennedy, M.J. and Philips, W.E., 1971, Ultramafic rocks of Burlington Peninsula, Newfoundland: Geol. Assoc. Canada Proceedings, A Newfoundland Decade, v. 24, p. 35-46.

Kidd, W.S.F., 1974, The evolution of the Baie Verte Lineament, Burlington Peninsula, Newfoundland: Ph.D. Thesis, University of Cambridge, England, 294 p.

_____, 1977, The Baie Verte Lineament, Newfoundland: ophiolite complex floor and mafic volcanic fill of a small Ordovician marginal basin: in Talwani, M. and Pitman, W.C., eds., Island Arcs, Deep Sea Trenches and Back-Arc Basins, American Geophys. Union, Maurice Ewing Series, v. 1, p. 407-418.

Kidd, W.S.F., Dewey, J.F. and Bird, J.M., 1978, The Mings Bight ophiolite complex Newfoundland: Appalachian oceanic crust and mantle: Canadian Jour. Earth Sci., v. 15, p. 781-804.

Knapp, D., Kennedy, D. and Martineau, Y., 1979, Stratigraphy, structure and regional correlation of rocks at Grand Lake, Western Newfoundland: in Current Research, Geol. Survey Canada Paper 79-1A, p. 317-325.

Laliberté, R., Spertini, F. and Hébert, R., 1979, The Jeffrey Asbestos Mine and the ophiolitic complexes at Asbestos, Quebec: Geol. Assoc. Canada, Field Trip B-3 and A-17, Université Laval, 23 p.

Lamarche, R.Y., 1972, Ophiolites of Southern Quebec: Department of Energy, Mines and Resources, Ottawa, Earth Physics Branch Publication 42, no. 3, p. 65-69.

_____, 1973, Géologie du complexe ophiolitique d'Asbestos, Cantons de l'Est: Ministère des Richesses Naturelles, Québec, Document Public GM 28558.

Lamothe, D., 1978, Analyse structurale du mélange ophiolitique du lac Montjoie: Thèse de M.Sc., Université Laval, 81 p.

Laurent, R., 1975a, Occurrences and Origin of the ophiolites of Southern Quebec, Northern Appalachians: Canadian Jour. Earth Sci., v. 12, p. 443-455.

_____, 1975b, Petrology of the Alpine-type serpentinites of Asbestos and Thetford Mines, Quebec: Schweizerische Mineralogische Petrographische Mitteilungen, v. 55, no. 3, p. 431-455.

_____, 1977, Ophiolites from the Northern Appalachians of Quebec: in Coleman, R.G. and Irwin, W.P. eds., North American Ophiolites: Oregon Dept. Geology and Mineral Industries, Bulletin 95, p. 25-40.

Laurent, R. and Hébert, Y., 1977, Features of submarine volcanism in ophiolites from the Quebec Appalachians: in Baragar, W.R.A., Coleman, L.C. and Hall, J.M. eds., Volcanic Regimes in Canada: Geol. Assoc. Canada Spec. Paper 16, p. 91-109.

Laurent, R., Hébert, R. and Hébert, Y., 1979, Tectonic setting and petrological features of the Quebec Appalachian ophiolites: in Malpas, J. and Talkington, R.W., eds., Ophiolites of the Canadian Appalachians and Soviet Urals: Memorial University of Newfoundland, Geology Dept. Rept. 8, p. 53-77.

Malpas, J., 1979, The dynamothermal aureole of the Bay of Islands ophiolite suite: Canadian Jour. Earth Sciences, v. 16, p. 2086-2101.

Miller, H.G. and Deutch, E.R., 1976, New gravitational evidence for the surface extent of oceanic crust in North-Central Newfoundland: Canadian Jour. Earth Sci., v. 13, p. 459-469.

Morgan, B.A., 1977, The Baltimore Complex, Maryland, Pennsylvania and Virginia: in Coleman, R.G. and Irwin, W.P., eds., North American Ophiolites: Oregon Dept. Geology and Mineral Industries, Bulletin 95, p. 41-49.

Naylor, R.S., Boone, G.M., Boudette, E.L., Ashenden, D.D. and Robinson, P., 1973, Pre-Ordovician rocks in the Bronson Hill and Boundary Mountain Anticlinoria, New England, U.S.A.: Trans. American Geophys. Union, Abstracts, v. 54, No. 4, p. 495.

Neale, E.R.W., 1957, Ambiguous intrusive relationships of the Betts Cove-Tilt Cove Serpentinite Belt, Newfoundland: Geol. Assoc. Canada Proceedings, v. 9, p. 95-107.

Neale, E.R.W. and Kennedy, M.J., 1967, Relationships of the Fleur de Lys Group to younger groups of the Burlington Peninsula, Newfoundland: in Neale, E.R.W. and Williams, H. eds., Geology of the Atlantic Region, Hugh Lilly Memorial Volume: Geol. Assoc. Canada Spec. Paper 4, p. 139-169.

Neale, E.R.W. and Nash, W.A., 1963, Sandy Lake (east half), Newfoundland: Geol. Survey Canada Paper 62-28, 40 p.

Neale, E.R.W., Kean, B.F. and Upadhyay, H.D., 1975, Post-ophiolite unconformity, Tilt Cove-Betts Cove area, Newfoundland: Canadian Jour. Earth Sci., v. 12, p. 880-886.

Norman, R.E. and Strong, D.F., 1975, The geology and geochemistry of ophiolitic rocks exposed at Mings Bight, Newfoundland: Canadian Jour. Earth Sci., v. 12, p. 777-797.

Osberg, P.H., 1965, Structural geology of the Knowlton-Richmond area, Quebec: Geol. Soc. America Bull., v. 76, p. 223-250.

Rickard, M.J., 1965, Taconic Orogeny in the Western Appalachians, experimental application of microtextural studies to isotopic dating: Geol. Soc. America Bull., v. 76, p. 523-536.

Riordon, P.H., 1953, Geology of Thetford Mines-Black Lake Area, with particular reference to the asbestos deposits: Ph.D. Thesis, McGill University, Montréal, 236 p.

————————, 1954, Thetford Mines-Black Lake Area: Department of Natural Resources, Quebec, Preliminary, Report No295.

Rodgers, J., 1968, The eastern edge of the North American Continent during the Cambrian and Early Ordovician: in Zen, E-an White, W.S., Hadley, J.B. and Thompson, J.B., Jr., eds. Studies of Appalachian Geology: Northern and Maritime, John Wiley and Sons, New York, p. 141-150.

Rodrigue, G., 1979, Etude pétrologique des roches ophiolitiques du Mont Orford: Thèse de M.Sc., Université Laval, Québec, 169 p.

St-Julien, P., 1963, Géologie de la région d'Orford-Sherbrooke, Québec: Thèse de Ph.D., Université Laval, Québec, 369 p.

————————, 1965, Orford-Sherbrooke Area: Dept. Natural Resources, Québec, Map 1619.

————————, 1967, Tectonics of part of the Appalachian Region of Southeastern Quebec (Southwest of the Chaudière River): Royal Soc. Canada, Special Publ. 10, p. 41-47.

————————, 1970, Géologie de la région de Disraéli (moitié est): Ministère des Richesses Naturelles, Québec Rapport Préliminaire No587, 23 p.

St-Julien, P. and Hubert, C., 1975, Evolution of the Taconian Orogen in the Quebec Appalachians: in Tectonics and Mountain Ranges, John Rodgers Volume: American Jour. Science, v. 275-A, p. 337-362.

————————, 1979, Structural setting of the Thetford Mines ophiolite complexes: Geol. Assoc. Canada, Field Trip B.10, Université Laval, 27 p.

St-Julien, P., Hubert, C., Skidmore, B. and Béland, J., 1972, Appalachian structure and stratigraphy, Québec: 24th Internatl. Geol. Congress, Montréal, 1972, Guidebook, AC-56, 99 p.

Schroeter, T.M., 1973, Geology of the Nippers Harbour Area, Newfoundland: M.Sc. Thesis, University of Western Ontario, 88 p.

Sheridan, R.E. and Drake, C.L., 1968, Seaward extension of the Canadian Appalachians: Canadian Jour. Earth Sci., v. 5, p. 337-373.

Stevens, R.K., 1970, Cambro-Ordovician flysch sedimentation and tectonics in west Newfoundland and their possible bearing on a Proto-Atlantic Ocean: in Lajoie, J. ed., Flysch Sedimentology in North America: Geol. Assoc. Canada Spec. Paper 7, p. 165-177.

Strong, D.F. and Williams, H., 1972, Early Paleozoic flood basalts of northwest Newfoundland: their petrology and tectonic significance: Geol. Assoc. Canada Proceedings, v. 24, no. 2, p. 43-54.

Thomas, W.A., 1977, Evolution of Appalachian-Ouachita salients and recesses from reentrants and promontories in the continental margin: American Jour. Sci., v. 277, p. 1233-1278.

Trzcienski, W.E., 1976, Crossitic amphibole and its possible tectonic significance in the Richmond Area, Southeastern Quebec: Canadian Jour. Earth Sci., v. 13, p. 711-714.

Upadhyay, H.D., Dewey, J.F. and Neale, E.R.W., 1971, The Betts Cove ophiolite complex, Newfoundland: Appalachian oceanic crust and mantle: Geol. Assoc. Canada Proceedings, A Newfoundland Decade, v. 24, p. 27-34.

Vagt, G.O., 1976, Asbestos: Canadian Minerals Yearbook, Publishing Centre, Dept. Supply Services, Ottawa.

Vallières, A., Hubert, C. and Brooks, C., 1978, A slice of basement in the western margin of the Appalachian Orogen, Saint-Malachie, Quebec: Canadian Jour. Earth Sci., v. 15, p. 1242-1249.

Walthier, T.N., 1949, Geology and mineral deposits of the area between Corner Brook and Stephenville, Western Newfoundland: Newfoundland Geol. Survey Bull. 35, Part I, p. 1-62.

Watson, K. dep., 1947, The geology and mineral deposits of the Baie Verte-Mings Bight Area: Newfoundland Geol. Survey Bulletin 21, 48 p.

Williams, H., 1975, Structural succession, nomenclature and interpretation of transported rocks in Western Newfoundland: Canadian Jour. Earth Sci., v. 12, p. 1874-1894.

——————————, 1977, Ophiolitic melange and its significance in the Fleur de Lys Supergroup, Northern Appalachians: Canadian Jour. Earth Sci., v. 14, p. 987-1003.

——————————, 1978a, Tectonic Lithofacies Map of the Appalachian Orogen: Memorial University of Newfoundland, Map No. 1.

——————————, 1978b, Geological development of the Northern Appalachians: its bearing on the evolution of the British Isles: in Bowes, D.R. and Leake, B.E., eds., Crustal Evolution in Northwestern Britain and Adjacent Regions, Geol. Jour. Spec. Issue 10, Seal House Press, Liverpool, England, p. 1-22.

——————————, 1980a, Structural telescoping across the Appalachian Orogen and the minimum width of the Iapetus Ocean: in Stranway, D.W., ed., The Continental Crust and Its Mineral Deposits: Geol. Assoc. Canada Spec. Paper 20, p. 421-440.

——————————, 1980b, Thin-Skinned tectonics in the crystalline southern Appalachians: COCORP seismic-reflection profiling of the Blue Ridge and Piedmont; and Sequential development of the Appalachian Orogen above a master decollement – A hypothesis; Discussion: Geology, v. 8, no. 5, p. 211-212.

Williams, H. and Doolan, B.L., 1979, Evolution of Appalachian-Ouachita salients and recesses from reentrants and promontories in the continental margin, Discussion: American Jour. Sci., v. 279, p. 92-95.

Williams, H. and St-Julien, P., 1978, The Baie Verte-Brompton Line in Newfoundland and regional correlations in the Canadian Appalachians: in Current Research, Geol. Survey Canada Paper 78-1A, p. 225-229.

Williams, H. and Smyth, W.R., 1973, Metamorphic aureoles beneath ophiolite suites and Alpine peridotites: tectonic implications with West Newfoundland and examples: American Jour. Sci., v. 273, p. 564-621.

Williams, H. and Stevens, R.K., 1974, The ancient continental margin of Eastern North America: in Burk, C.A. and Drake, C.L., eds., The Geology of Continental Margins: New York, Springer-Verlag, p. 781-796.

Williams, H. and Talkington, R.W., 1977, Distribution and tectonic setting of ophiolites and ophiolitic melanges in the Appalachian Orogen: in Coleman, R.G. and Irwin, W.P., eds., North American Ophiolites: Oregon Dept. Geology and Mineral Industries Bull. 95, p. 1-11.

Williams, H., Hibbard, J.P. and Bursnall, J.T., 1977, Geologic setting of asbestos-bearing ultramafic rocks along the Baie Verte Lineament, Newfoundland: in Report of Activities, Geol. Survey Canada Paper 77-1A, p. 351-360.

Zietz, I., Haworth, R.T., Williams, H. and Daniels, D.L., 1980, Magnetic Anomaly Map of the Appalachian Orogen: Memorial University of Newfoundland, Map No. 2.

Manuscript Received May 2, 1980
Revised Manuscript Received February 16, 1981

Major Structural Zones and Faults of the Northern Appalachians, edited by
P. St-Julien and J. Béland, Geological Association of Canada Special Paper 24, 1982.

OPHIOLITE ASSEMBLAGE OF EARLY PALEOZOIC AGE
IN CENTRAL WESTERN MAINE

Eugene L. Boudette
United States Geological Survey, Denver, Colorado 80225

ABSTRACT

Ophiolite of Cambrian(?) to Early Ordovician(?) age is exposed near the Canadian border in central western Maine. The ophiolite rests on the south side of the Precambrian Chain Lakes massif, and together they constitute the lower two of four major tectonostratigraphic divisions that span Proterozoic X or Y (Helikian, 1.6 Ga) to Early Devonian time. The Chain Lakes massif is composed of inclusion-rich metasedimentary granofels, metamorphosed to high rank in Precambrian time, which acted as a buttress throughout the Phanerozoic.

The ophiolite sequence is nearly complete except for a sheeted dike complex. It is also rather thin (3 km or thinner). At the base of the sequence are: serpentinite (locally with distinctive relict layering) and pyroxenite, which are overlain by epidiorite, gabbro, autobreccia, clinopyroxenite, and plagiogranite all of the Boil Mountain Complex (new name), in turn grading upward to greenstone volcanics, commonly pillowed. The ophiolite sequence is contiguous with metadacite and cherty iron formation, which are conformably overlain by lower Paleozoic (pre-Middle Ordovician) quartzwacke and flysch.

The ophiolite sequence is welded to the underlying Chain Lakes massif on what is probably an obduction surface. As suggested by evidence of subtle metamorphism of the Chain Lakes and by ductile faulting at the base of the sequence, the ophiolite was thrust northwestward upon the Chain Lakes.

Nearby, but outside the principal ophiolitic body, serpentinite associated with soapstone and chromiferous muscovite-carbonate rock has been emplaced diapirically along faults. Soapstone and serpentinite bodies exposed on-strike in western New Hampshire and central Maine may be tectonically smeared remnants of ophiolite, and possibly an ophiolitic mélange zone.

RÉSUMÉ

Une ophiolite du Cambrien (?) Ordovicien inférieur (?) affleure dans le centre-ouest du Maine, près de la frontière canadienne. L'ophiolite repose sur la partie sud du massif de Chain Lakes d'âge précambrien et avec ce dernier constituent les deux divisions inférieures des quatre divisions tectonostratigraphiques majeures s'étendant du Protérozoïque X ou Y (Hélikien, 1.6 Ga) au Dévonien inférieur. Le massif de Chain Lakes est formé de roches métasédimentaires granoblastiques, riches en inclusions et qui ont été fortement métamorphisées durant le Précambrien. Il a servi de butoir durant le Phanérozoïque.

La séquence ophiolitique est quasiment complète; seul manque le complexe de dykes. Elle est toutefois plutôt mince (3 km ou moins). A la base de la séquence s'observe le complexe de "Boil Mountain" (nouveau nom) comprenant: une serpentinite (localement montrant un vestige de rubanement) et de la pyroxénite recouvertes par de l'épidiorite, du gabbro, de l'autobrèche, de la clinopyroxénite et du plagiogranite. Au sommet, affleurent des roches métavolcaniques vertes coussinées. La séquence ophiolitique est contiguë à de la métadacite et à des formations de fer cherteuses lesquelles sont recouvertes en concordance par des wackes quartzeux et des flyschs du Paléozoïque inférieur (pre-Ordovicien moyen).

La séquence ophiolitique est soudée au massif de Chain Lakes et la surface de contact est interprétée comme une surface d'obduction. Un faible métamorphisme rétrograde du massif et des failles ductiles à la base de la séquence suggèrent que l'ophiolite a chevauché vers le nord-ouest sur le massif.

Au voisinage, mais à l'extérieur du massif principal d'ophiolite, de la serpentinite associée à de la stéatite et à une roche chromifère à carbonate et muscovite résulte d'une montée diapirique le long de failles.

Dans l'ouest du New Hampshire et au centre du Maine, des amas de stéatite et de serpentinite laminées situés dans le prolongement de la séquence ophiolitique sont peut-être des restants d'ophiolite et, possiblement une zone de mélange ophiolitique.

INTRODUCTION

Serpentinite in central western Maine has received intermittent attention for more than 25 years as a source for asbestos fiber (see Wing and Dawson, 1949; Wing, 1951a, 1951b; Hurley and Thompson, 1950), but its tectonic significance was not evaluated until more recent times when it was proposed as part of an ophiolite sequence (see Boudette, 1970, 1978; Boudette and Boone, 1976). Ophiolite in the orogenic strike belt represented by the rocks of central western Maine opens penetrating questions relating not only to the circumstances of its own emplacement, but to the tectono-stratigraphic relationships of the rocks of the Connecticut River valley to the southwest, and central Maine to the northeast, which are contained in the same strike belt. Lyons *et al.* (this volume) present evidence of an ultramafic trend extending from the ophiolite in this report to southern New England, based on the appearance of a line of soapstone and serpentinite bodies on strike. A similar serpentinite line may extend to the northeast through central Maine to the New Brunswick border. Northeast and southwest of central western Maine no other complete ophiolite sequence is known; the serpentinite bodies are either cold diapirs or remnants of the plutonic components (Boil Mountain Complex, new name) of the ophiolite described in this report. Volcanic components of the ophiolite and related iron-formation (Jim Pond Formation, new name) do, however, continue northeast and southwest in well-defined strike belts. The purpose of this paper is to present the anatomy of a nearly complete ophiolite sequence and speculate on its implications in the tectonic and stratigraphic history of the Caledonian-Hercynian orogen.

REGIONAL GEOLOGY

The rocks of central western Maine are within the Acadia geosyncline of Poole (1967, p. 11, 17-18). The pre-Silurian sequences belong to the Gander zone and the Silurian and Devonian sequences to the "successor basin" tectonic-lithofacies categories of Williams (1978).

The rocks of central western Maine are for the most part complexly folded, generally trend northeast, and are faulted. Major faults striking parallel to the regional tectonic grain appear to have originated as thrusts, which were subsequently rotated; some, such as the Squirtgun (Fig. 1), then passed into right-lateral transcurrent motion. These major strike faults appear to correlate with the fault system of the Connecticut River valley to the southwest.

Two important features are present in these rocks (Boudette and Boone, 1976; Boudette, 1978): 1) the Chain Lakes massif, a segment of Helikian (Proterozoic Y or X) continental crust at least 1.6 Ga in age (Naylor, 1973) that has ophiolite of Cambrian(?) or Ordovician(?) age welded to its southeastern margin; and 2) a tectonic hinge zone separating interlayered Ordovician and Silurian rocks on the southeast from Silurian and Devonian rocks deposited on an angular unconformity on the northwest. The proximity of the two features in central western Maine is probably not fortuitous, but suggests a major tectonic role played by the Precambrian massif during development of the Taconian unconformity.

Metamorphic rocks younger than the Chain Lakes massif in central western Maine pass gradually from the supracrustal regime (regional greenschist facies with locally overprinted hornfels) to their infracrustal equivalents (mostly amphibolite facies) from northwest to southeast across the region. Metamorphism in the Chain Lakes massif varies from the upper amphibolite to granulite facies and is probably a Precambrian event (Boudette, 1978, p. 209-290). The Chain Lakes massif was probably detached from either Europe or Africa after this prograde metamorphism. The massif apparently behaved as a buttress throughout Paleozoic time after its arrival in North America. Because of its age, lithology, and tectonic history, the massif is not to be confused with rocks of Middle Ordovician age in the Oliverian domes of central New England.

As shown in Table I, rock successions of the region are assigned to four tectono-stratigraphic divisions (Boudette and Boone, 1976): (I) The Chain Lakes massif basement, which is composed principally of diamictite, aquagene metavolcanic rocks, and metasedimentary rocks; (II) a Cambrian(?) and Ordovician(?) ophiolite (principally pillowed greenstone) and volcanic-flysch succession; (III) Middle Ordovician to Upper(?) Ordovician flysch; and (IV) Silurian and Devonian molasse-like deposits grading up into turbidite. Each of these divisions can, in turn, be broken into formations or stratigraphic units as shown on Figure 2.

The layered metamorphic rocks contain four major irruptive suites: 1) the plutonic (or stratigraphically lowermost) components of the ophiolite; 2) calc-alkalic rocks of the Attean pluton that have an age of approximately 445 Ma (John B. Lyons, personal communication, 1981); 3) quartz porphyry with a probable age of about 445 Ma (F.C. Canney, unpublished data); and 4) an Early Devonian age calc-alkalic gabbro-diorite-quartz monzonite suite. Small bodies of metadiorite on the eastern side of the Chain Lakes massif and near the Quebec boundary along the Maine-New Hampshire state line (Fig. 2) are apparently equivalent to the ophiolite in age and thus are correlated with it. These irruptive suites are subdivided for mapping purposes into discrete sill or dike sequences, plutons, or intrusive complexes (see Boudette, 1978). This report is specifically concerned with the nearly complete ophiolite sequence which is comprised of the oldest intrusive suite and the pillowed greenstone (lowermost unit of Division II).

The subdivision of the ophiolite into plutonic and tectono-stratigraphic components for purposes of mapping is convenient, but is obviously cumbersome. The principal occurrence of ophiolite is mapped as a layer sequence. Grading occurs from the uppermost plutonic portion of the ophiolite into aquagene mafic metavolcanic rocks which are typified by the pillowed greenstone.

The boundary between the Chain Lakes massif and ophiolite is interpreted as a tectonic surface and displays evidence of mild thermal metamorphism and cataclasis. The nature of the boundary between Divisions II and III is presently an equivocal matter, but it is interpreted by Boudette and Boone (1976) and by Boudette (1978) to be either a tectonic break, such as the base of a stacking slice, or an unconformity. The boundary between Divisions III and IV, as indicated above, is either gradational or an unconformity, depending upon its relative position with reference to the tectonic hinge zone. The principal emphasis herein is on the lower part of Division II, as outlined above. An important point to keep in mind, however, is that all of the rocks of Division II form a seemingly orderly layered succession essentially comparable to those of modern island-arc–lower-rise–abyssal-plain depositional environments. The observed stratigraphic sequence determined in the field from depositional structures is consistent with most sedimentologic models for trench and lower oceanic regimes. The sequence in Division II in ascending order is (1) pillowed greenstone with interlayered metadacite near its top, (2) cherty iron-formation, (3) metaquartzwacke and flysch, (4) euxinic flysch with exotic rafts and argile-à-blocs facies, and (5) quartzose and calcareous flysch. Units were probably deposited mainly in the island-arc environment, in a trench, and on a abyssal plain, respectively.

The ophiolite is profoundly dislocated by the Squirtgun fault. The Squirtgun is an uninterrupted fault that extends more than 90 km on strike from the vicinity of the Quebec-Maine-New Hampshire common corner northeast to Moosehead Lake. Observations along the Squirtgun have always presented a paradox. On one hand, the regional lithostratigraphic pattern south of the Chain Lakes massif (Fig. 2) indicates that the southeastern limb of the Moose River synclinorium is clipped off by the Squirtgun fault. On the other hand, observations of sections of the fault north of Flagstaff Lake commonly show vertical cataclastic (slip) foliation in which the drag-sense mechanics of deformed foliation indicate right-lateral slippage and, therefore, last fault motion. The search for the missing segment of the synclinorium, along the strike of the fault and its projections, has been unsuccessful; therefore, initial thrusting (southeast over northwest) is implied.

Sections of the Squirtgun south of the ophiolite show fault-zone segments con-

Figure 1. Major tectonic features and plutonic suites of the Rangeley Lakes-Dead River basin region, central western Maine. Geology modified from Boudette and Boone (1976) and Boone *et al.* (1970).

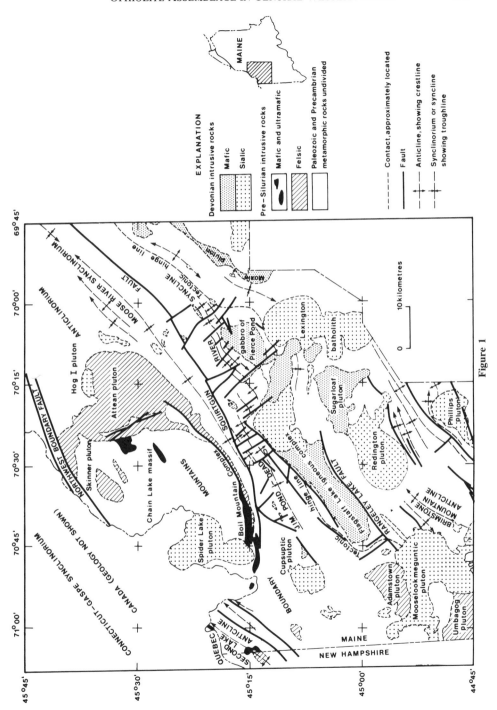

EXPLANATION

Devonian intrusive rocks

Mafic

Sialic

Pre-Silurian intrusive rocks

Mafic and ultramafic

Felsic

Paleozoic and Precambrian metamorphic rocks undivided

- - - - - Contact, approximately located

————— Fault

Anticline, showing crestline

Synclinorium or syncline showing troughline

Figure 1

TABLE I

MAJOR TECTONO-STRATIGRAPHIC DIVISIONS AND PLUTONIC SUITES IN THE RANGELEY LAKES-DEAD RIVER BASIN REGION, CENTRAL WESTERN MAINE

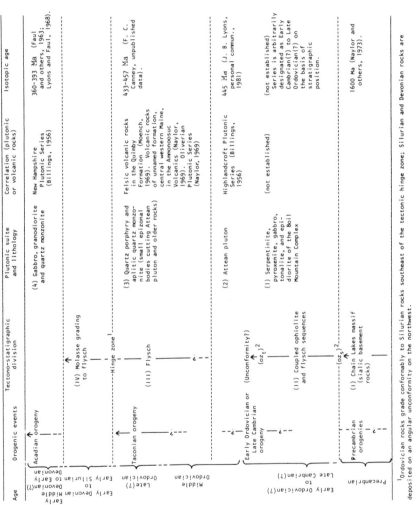

Age	Orogenic events	Tectono-statigraphic division	Plutonic suite and lithology	Correlation (plutonic or volcanic rocks)	Isotopic age
Early Devonian to Middle Devonian(?)	Acadian orogeny	(IV) Molasse grading to flysch	(4) Gabbro, granodiorite and quartz monzonite	New Hampshire Plutonic Series (Billings, 1956)	360-393 Ma (Faul and others, 1963; Lyons and Faul, 1968).
Early Silurian to Early Devonian(?)		— Hinge zone[1]			
Ordovician Late(?)	Taconian orogeny	(III) Flysch	(3) Quartz porphyry and aplitic quartz monzonite (small epizonal bodies cutting Attean pluton and older rocks)	Felsic volcanic rocks in the Quimby Formation (Moench, 1969). Volcanic rocks of unnamed formation, central western Maine, in the Ammonoosuc Volcanics (Naylor, 1969). Oliverian Plutonic Series (Naylor, 1969)	433-457 Ma (F. C. Canney, unpublished data).
Middle Ordovician	Early Ordovician or Late Cambrian orogeny	(Unconformity?) (oz_L)[2]	(2) Attean pluton	Highlandcroft Plutonic Series (Billings, 1956)	445 Ma (J. B. Lyons, personal commun., 1981)
Late Cambrian(?) to Early Ordovician(?)		(II) Coupled ophiolite and flysch sequences	(1) Serpentinite, pyroxenite, gabbro, tonalite, and epidiorite of the Boil Mountain Complex	(not established)	(not established) Series is arbitrarily designated as Early Cambrian(?) to Late Ordovician(?) on the basis of stratigraphic position.
Precambrian	Precambrian orogenies	(oz_S)[2] — (I) Chain Lakes massif (sialic basement rocks)			1600 Ma (Naylor and others, 1973).

[1] Ordovician rocks grade conformably to Silurian rocks southeast of the tectonic hinge zone; Silurian and Devonian rocks are deposited on an angular unconformity on the northwest.

[2] (oz_L) time and (oz_S) stratigraphic position of obduction zone of Early Ordovician(?) age.

taining discrete fault planes dipping between 40° and 55° southeast, decorated with slickensides and mullion structure, and having minor structure indicating southeast over northwest motion. A few outcrops observed near the fault trace in other places show a combination of vertical flow foliation, slip cleavage, and fold bands comparable to those in the section north of Flagstaff Lake. Thus, the Squirtgun is interpreted to be a complex fault developed in three stages: thrust, rotation and strike-slip motion. It is possible, however, that great variations of motion occurred simultaneously in each stage.

Plutonic rocks of the ophiolite sequence persist on regional strike for only about 16 km. To the west these plutonic rocks become more segmented. To the east, dislocation also occurs, but the plutonic part of the ophiolite appears to be mainly covered by Silurian and Devonian rocks of the Moose River synclinorium (Fig. 1) and is divided at its western end into two principal segments. This segmentation is believed to be the result of fault repetition either during emplacement or later during activity on the Squirtgun fault system, as is shown on Figure 3. Thus, any distinction between ophiolite of the principal segments and other ophiolites in their strike belt becomes arbitrary.

The volcanic units of the ophiolite are, by contrast to the plutonic component, persistent in their regional strike belt grain throughout and beyond central western Maine. The Silurian and Devonian rocks of the Moose River synclinorium also cover most of the volcanic and iron-formation components in the northeast.

The complete ophiolite sequence has about 3 km maximum thickness where mapped. This is relatively thin compared to most other ophiolites. Greenstone dikes arbitrarily correlated with the volcanic ophiolite are found within the plutonic rocks, but no sheeted dike complex is known.

The exact age of the ophiolite is not known, but it is younger than the Helikian-age Chain Lakes massif, to which it is welded, and older than the Attean pluton, of probable Middle to Late Ordovician age, which intrudes it.

BOIL MOUNTAIN COMPLEX

Plutonic rocks of the ophiolite sequence are herein formally designated the Boil Mountain Complex named for Boil Mountain (shown on Figure 3) located 14 km north of the outlet of Kennebago Lake in northern Franklin County (Fig. 2). Lithologic units include serpentinite, pyroxenite, gabbro, tonalite, and epidiorite (Table I). The type section is composite and is described in detail by Boudette (1978); it coincides with the two east-west elongate contiguous parts of the complex about 15 km north of Kennebago Lake in northern Franklin County (Fig. 1).

Rocks of the Boil Mountain Complex are extensively altered, but relatively pristine examples may be found in the northeast (Fig. 3). Deformation and dislocation of the rocks of the complex increase toward the southwest along with alteration, especially in the serpentinite. This effect becomes especially notable where the Squirtgun fault system becomes tangential to the ophiolite. Extreme tectonism and metamorphism is believed to have remobilized segments of the serpentinite of the complex to produce the diapiric variety. The serpentinite of the Boil Mountain Complex contains antigorite and is moderately hydrated. Diapiric serpentinite is more hydrated and is composed of clinochrysotile and magnesian chlorite. The diapiric serpentinite has apparently detached itself from the parent complex and has migrated more than 7 km from apparent source areas. Antigorite and clinochrysotile do not coexist in these rocks. Soapstone and a distinctive marble-like carbonate rock (arnoldite), composed mostly of dolomite-magnesite with subordinate chromiferous muscovite, chromite, and magnetite, are usually associated with the clinochrysotile serpentinite.

Representative analyses of calcic gabbro, sodic (feldspathic) gabbro, and clinochrysotile serpentinite are given in Table II. The "anomalous" high alumina content for the serpentinite is superficially troublesome, but indicates major modal magnesian chlorite. Molecular quantities of the phases present are approximately: Clinochrysotile 68 per cent, Mg-rich chlorite 31 per cent, epidote 0.5 per cent, and Fe-Ti oxides 0.5 per cent.   G. M. Boone (oral commun., 1979) finds that equivalent rocks in central western Maine contain sufficient magnesian chlorite to yield such an analysis, and magnesian chlorite is optically almost indistinguishable from clino-

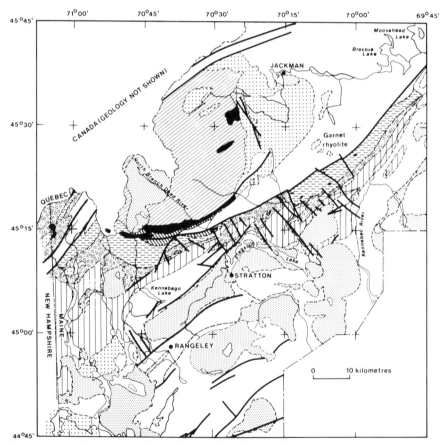

Figure 2. Generalized regional bedrock map of the Rangely Lakes-Dead River region, central western Maine, showing distribution of the principal lithostratigraphic sequences and plutonic suites. Geology modified from Boudette and Boone (1976) and Boone et al. (1970).

chrysotile. Comparative analyses of antigorite serpentinite are not presently available, but modal data indicate it is not apt to be anomalously aluminous. It is therefore inferred that alumina enrichment, possibly accompanied by lime depletion, has occurred in the diapiric serpentinite if its source is assumed to be in the Boil Mountain Complex. FeO, MgO, Cr, and Ni abundances are relatively higher in the gabbro of the Boil Mountain Complex than in the Devonian-age gabbro of calc-alkalic affinity of the Flagstaff Lake Igneous Complex (Boudette, 1978). These chemical contrasts support the assignment of the Boil Mountain rocks to an ophiolite affiliation.

The distribution of rock types within the Boil Mountain Complex is relatively uneven and, to some degree, they are mixed. The two-fold subdivision of the ultramafic and mafic components shown on Figure 3 is generalized and reflects the dominant rock types in each. The northeastern part of the complex has a stratigraphy wherein the rocks are part of an upright sequence as much as 1.6 km thick facing southeast and are relatively enriched in Mg at their base along the northwestern

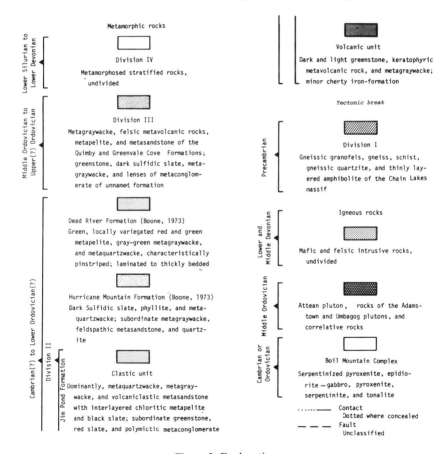

Figure 2. Explanation

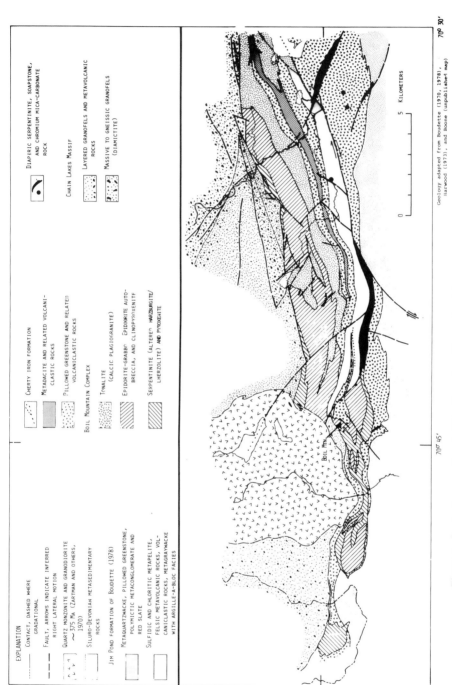

EXPLANATION

········· CONTACT, DASHED WHERE
GRADATIONAL

─ ─ ─ FAULT, ARROWS INDICATE INFERRED
RIGHT LATERAL MOTION

QUARTZ MONZONITE AND GRANODIORITE
~375 MA (ZARTMAN AND OTHERS,
1970)

SILURO-DEVONIAN METASEDIMENTARY
ROCKS

JIM POND FORMATION OF BOUDETTE (1978)

METAQUARTZWACKE, PILLOWED GREENSTONE,
POLYMICTIC METACONGLOMERATE AND
RED SLATE

SULFIDIC AND CHLORITIC METAPELITE,
FELSIC METAVOLCANIC ROCKS, VOL-
CANICLASTIC ROCKS, METAGRAYWACKE
WITH ARGILLE-A-BLOC FACIES

CHERTY IRON FORMATION

METADACITE AND RELATED VOLCANI-
CLASTIC ROCKS

PILLOWED GREENSTONE AND RELATED
VOLCANICLASTIC ROCKS

BOIL MOUNTAIN COMPLEX

TONALITE
(CALCIC PLAGIOGRANITE)

EPIDIORITE-GABBRO, EPIDIORITE AUTO-
BRECCIA, AND CLINOPYROXENITF

SERPENTINITE (ALTERED HARZBURGITE/
LHERZOLITE) AND PYROXENITE

DIAPIRIC SERPENTINITE, SOAPSTONE,
AND CHROMIUM MICA-CARBONATE
ROCK

CHAIN LAKES MASSIF

LAYERED GRANOFELS AND METAVOLCANIC
ROCKS

MASSIVE TO GNEISSIC GRANOFELS
(DIAMICTITE)

70° 45'　　　　　　　　　　　　　　　　　　　　70° 30'

0 _____ 5

KILOMETERS

Geology adapted from Boudette (1970, 1978),
Harwood (1973), and Boone (unpublished map)

Figure 3. Geologic map of part of the Boundary Mountains Highland, central western Maine, showing the distribution of ophiolite of Early Paleozoic age and subjacent rocks.

TABLE II

MAJOR OXIDES, NORMATIVE COMPOSITION, AND ELEMENTAL COMPOSITION OF SELECTED OPHIOLITIC ROCKS, CENTRAL WESTERN MAINE

[Gabbro samples are from the southeastern flank of Boil Mountain in the type section of the complex. The serpentinite sample is from a diapiric body along a fault about 4 km southwest of Jim Pond (fig. 3). Major oxide analysis by rapid methods--Hezekiah Smith, Lowell Artis, and Floyd Brown, analysts. Trace element analysis by semi-quantitative emission spectrographic method--Joseph L. Harris, analyst]

Oxide	Calcic gabbro 1	Sodic gabbro 2	Chrysotile serpentinite
	Data in percent		
SiO_2	43.8	50.7	34.0
Al_2O_3	15.9	15.7	14.8
Fe_2O_3	3.6	1.5	2.3
FeO	7.3	5.1	3.3
MgO	10.4	10.1	32.5
CaO	15.6	11.7	.80
Na_2O	.18	2.2	.04
K_2O^+	.14	.46	.02
H_2O^+	2.1	2.5	12.1
H_2O^-	.02	.06	.35
TiO_2	.21	.36	.24
P_2O_5	.01	.04	.01
MnO	.12	.10	.02
CO_2	.02	-.04	.02
Total less H_2O^-	99.38	100.5	100.15

Normative molecule	Calcic gabbro 1	Sodic gabbro 2	Chrysotile serpentinite
	Data in percent		
Q	0	0	0
C	0	0	15.8
or	.85	2.78	.13
ab	1.57	19.02	.39
an	43.50	32.30	4.30
cp[1]	29.62	21.39	--
op[1]	2.02	14.62	22.15
ol	21.95	8.99	57.26
il	.41	.70	.52
ap	.02	.10	.03
cc	.05	.09	.05

Element	Calcic gabbro 1	Sodic gabbro 2	Chrysotile serpentinite
	Data in parts per million		
	(-- = below detectability)		
Ag	<1	--	--
B	10	--	13
Ba	3	35	--
Co	39	21	26
Cr	121	67	263
Cu	71	19	6
Dy	--	5	--
Mn	1410	1370	623
Mo	3	2	1
Ni	34	53	127
Pr	--	--	3
Sc	46	31	3
Sr	152	89	1
V	337	116	65
Y	5	1	--
Yb	1	47	<1
Zn	60	47	26
Zr	--	13	--

[1]cp = clinopyroxene, op = orthopyroxene.

margin. The lower zone is represented by the principal occurrence of the antigorite serpentinite (altered harzburgite and dunite) and pyroxenite. The Mg-rich rocks are generally overlain by massive epidiorite which in turn grades into gabbro, epidiorite autobreccia (epidiorite and subordinate calcic plagiogranite), calcic plagiogranite, and minor clinopyroxenite. Greenstone septa (or possibly dikes) are common in the plagiogranite facies. Repetition of lithologies is common near the base, where distinctive igneous layering is also seen.

The base of the Boil Mountain Complex is exposed in the Jim Pond 7-1/2-minute quadrangle (Fig. 4). Here epidiorite, gabbro, and serpentinite of the ophiolite are in sharp, generally concordant, contact with rocks of the Chain Lakes massif. The latter appear as discrete septa within the ophiolitic rocks and show subtle effects of thermal recrystallization. Thus the basal contact is difficult to characterize at the scale of outcrops, but some cataclasis of the ophiolitic rocks suggests that ductile faulting accompanied their emplacement and is the pre-eminent relationship. Elsewhere along the base of the complex, the contrast in mechanical competence between the rocks of the Chain Lakes massif and those of the Boil Mountain Complex has resulted in the localization of post-intrusion fault dislocations characterized by brittle deformation, and relationships are usually obscure.

The southeastern contact of the Boil Mountain Complex, probably the stratigraphic top, is mostly dominated by the relatively thick layer of tonalite which may or may not be properly part of the ophiolite. This tonalite thins and cuts discordantly downward into the Boil Mountain Complex in the west. This is the only known place where the rocks of the complex and the greenstone of the Jim Pond Formation are in undisturbed contact, but this contact is not well exposed. One series of outcrops here shows transition from epidiorite into greenstone, possibly with some repetition of units, in a zone as much as 150 m across. The greenstone and interlayered metagreywacke here show development of amphibole and reaction hornfels texture probably related to a thermal prograde effect, but it is not clear if this effect resulted from the Spider Lake pluton or the Boil Mountain Complex. Contacts between the rocks of the complex and greenstone septa are relatively sharp with only minor reaction effects and marginal recrystallization of the greenstone. As indicated above, some (or all) of these septa could be dikes.

JIM POND FORMATION

The Jim Pond Formation consisting of the volcanic and metasedimentary components of the ophiolite sequence is herein named for Jim Pond located about 8 km above the confluence of the North Branch of the Dead River with Flagstaff Lake, northern Franklin County (Figs. 2 and 3). The type section is composite along the North Branch (Boudette, 1978).

The basal part (Volcanic Unit, Fig. 2) is composed essentially of chlorite-albite-epidote-actinolite greenstone with minor amounts of mafic-rich metagraywacke, metamorphosed dacite, maroon phyllite, and hematitic chert (jasper). The greenstone member is discontinuous along strike in belts northwest and southeast of the Squirtgun fault. Northwest of this fault, the greenstone member, 150 to 500 m thick is divided into lower and upper strike units by a metadacite member in the east and a

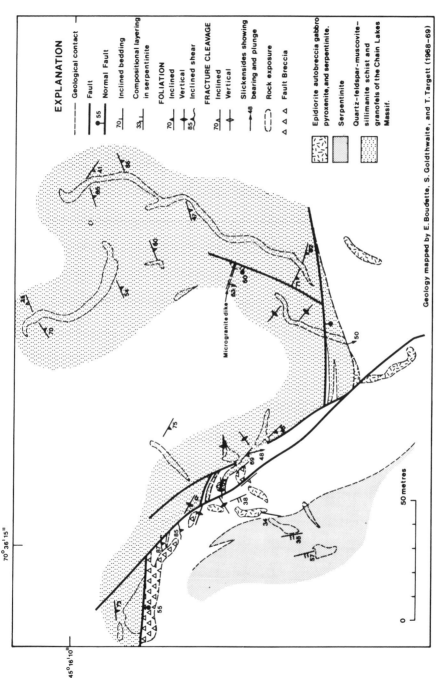

Figure 4. Detailed geologic map of part of the Jim Pond 7-1/2-minute quadrangle, Maine, showing boundary relationships between ophiolite and rocks of the Chain Lakes Massif.

metagreywacke member in the west. Southeast of the fault, the greenstone member constitutes all but about 150 m of the Jim Pond section in the east, and gives way by short-ranged facies change to a metaquartzwacke member toward the west. Greenstone is present only in relatively small lenses within the metaquartzwacke member. Local patches of altered amphibolite of the Jim Pond Formation occur in contact with the southeastern belt of tonalite. The greenstone is thickly layered with lenses throughout characterized by uniformly and well developed pillows. The thick units are probably individual flows that are 15 m or more thick. Mafic lapillite, in layers 1-20 cm thick, and volcanic breccia compose less than 10 per cent of the greenstone member and are found throughout interlayered with pillowed and massive flows.

Metamorphosed dacite and jasper compose less than two per cent of the greenstone member, and other members of the Jim Pond Formation, too small to map separately, also occur within the greenstone member. A mafic metagreywacke is more common in the belt northwest of the Squirtgun fault, where it makes up less than two per cent of the Jim Pond Formation and appears to be an epiclastic lithology gradational between the lapilli greenstone and the metaquartzwacke and metagreywacke members. A maroon, iron-rich phyllite is found only southeast of the Squirtgun fault within the mafic metavolcanic member, and in lenses as much as 60 m thick near its top. This phyllite could, in part, be the equivalent of the iron formation member in the belt northwest of the Squirtgun fault.

In addition to metamorphosed dacite, the metadacite member of the Jim Pond Formation contains sodic quartz-latite flow rock and related ash-flow rock, breccia, and epiclastic rock. Some interlayered beds of the quartzwacke member and metachert beds of the iron-formation member are included in the upper part of the metadacite member. A lense of the metadacite member underlies rocks of the greenstone member in at least one area northwest of the Squirtgun fault. Metadacite also makes at least one appearance within the metagreywacke member to the southeast. It generally thickens northeastward.

The thickness of the metadacite member varies from 0 to more than 500 m. The metadacite occurs in layers that are about 15 m thick or more. Ash-flow deposits are finely laminated in beds 1 to 10 mm thick. The boundary between the greenstone of the mafic member and the metadacite is sharp. Repetitive sequences of the two volcanic rocks occur with individual flows averaging about 30 m in thickness. The metadacite and iron-formation members are almost everywhere closely associated. In the northeast on regional strike, the main belt of metadacite is succeeded (toward the southeast) by iron formation with interlayered metaquartzwacke and metagreywacke. In the type section of the Jim Pond Formation (Boudette, 1978) the top of the metadacite member is marked by the appearance of the iron formation member interlayered with representatives of the metagreywacke and quartzwacke members.

Iron-formation is also found in lenses associated with the greenstone member and within the metagreywacke member. These lenses, for the most part, appear to be subjacent to metadacite lenses. The iron formation ranges up to about 20 m in thickness, and its contacts with other rocks are sharp. The member is relatively erodable and, therefore, does not crop out well; thus its extent may be greater than suspected.

PETROGENESIS

The combination within the Boil Mountain Complex of the following features: stratiform geometry, layering along gradational contacts, absence of chaotic flow structure, magnesium-enriched base, and layering suggests that the rocks of the complex were derived from a parent magma that was differentiated by crystal settling. Unequivocal intrusive superposition is observed in the ophiolite where tonalite (calcic-plagiogranite) crosscuts gabbro and epidiorite. Greenstone crosscuts epidiorite and tonalite, and occurs as septa within them. The differentiation suite is interpreted to be peridotite-gabbro-epidiorite-tonalite, and this sequence mostly coincides with the observed layering sequence. Gabbro and epidiorite intergrade well, and progressive clinopyroxene transformation to hornblende from gabbro to epidiorite opens the question of metasomatic alteration of gabbro to epidiorite.

The autobreccia is a distinctive facies formed when open space in brecciated epidiorite or gabbro became filled with tonalite. The autobreccia is distributed throughout the gabbro and epidiorite, but tends to be more abundant along the southeastern margin of the Boil Mountain Complex which is believed to be the stratigraphic top. Three questions arise concerning the autobreccia: 1) Was it formed by a repetitive process, as its distribution suggests, or in a single event affecting the entire complex? 2) Is the process of autobrecciation tectonic or internal to the complex? That is, is it related to ophiolite emplacement or to a process of explosive release such as would occur owing to unroofing or initiation of tectonic transport at the ophiolite source environment of magmatic crystallization in either isothermal or polythermal decreasing P_{load} conditions? 3) If a process of explosive release is possible, could the responsible rest liquids be petrologically implicated in a process of transformation of gabbro to epidiorite? The answer to (3) is affirmative if early paragenetic hydrous minerals formed in the epidiorite as is believed. Postemplacement, pervasive saussuritization and propylitization of the complex was probably associated with events in the Acadian orogeny, but earlier alteration, such as that associated with the Taconian or earlier orogenies (Table I), could also have left a cryptic imprint.

Upadhyay and others (1971), Kennedy and Phillips (1971), Church (1972), Lamarche (1972), Laurent (1975, 1977), and Laurent and Hébert (1977) have described other possibly coeval northern Appalachian ophiolites, which may have been emplaced by mechanisms similar to those responsible for the ophiolite of central western Maine. The ductile faulting relationships between the complex and the rocks of the Chain Lakes massif and the subtle metamorphic effects suggest that the complex came to rest at some advanced stage of its crystallization, perhaps as a hot slab (Harwood, 1973, p. 53, proposed a "crystal mush" model), and did not represent a very effective thermal reservoir. If the complex was transported tectonically any appreciable distance (see for example Williams, 1971), it probably was as a unit. The ophiolite of central western Maine is not indicated by gravity signature to extend to depth, and may well be rootless (W. A. Bothner, oral commun., 1981). The ophiolite of central western Maine is separated from the ophiolite of southern Quebec by 120 km across regional strike. The two ophiolite sequences share many common features, however, including: 1) lack of sheeted dike units, 2) tectonite along structurally lower boundaries, and 3) apparent lack of roots. Lamarche (1972) interprets the ophiolite of southern Quebec to be autochthonous sequences emplaced on a eugeosynclinal ocean floor in an extensional tectonic environment. Laurent (1975, 1977) and Laurent and Hébert (1977), on the other hand, advocate a thrust-slice model of emplacement wherein fragments of oceanic crust on or near a rapidly spreading ridge were obducted upon a continental margin contemporaneously with the development of a subduction zone. This model has been referred to by Laurent (1975) as the "Vourinos-type", and it is comparable in most aspects to Malpas' (1979, p. 2100) model "a" for the Bay of Islands complex in western Newfoundland. Malpas interprets that the Bay of Islands complex was hot at the time of emplacement, but Laurent (1975) believes that the southern Quebec ophiolites were emplaced cold. The ophiolite of central western Maine, like the Bay of Islands complex, is believed to have been emplaced hot; but otherwise conforms well to the Vourinos model which is preferred herein. It is possible that much of the harzburgite (depleted mantle) layer (Laurent, 1975, Fig. 2; Malpas, 1979, Fig. 6) is missing in the Boil Mountain Complex which accounts for its abnormally thin appearance among Appalachian ophiolites and the lack of an amphibolite layer at its structural base. Alternatively, the ophiolite of central western Maine could represent a segment of autochthonous oceanic crust, such as Lamarche (1972) proposes for the southern Quebec ophiolite, formed in an ensialic environment provided by the Chain Lakes massif during continental rifting events such as those described by Leonardos and Fyfe (1974) for the genesis of the continental margin of eastern Brazil at a much later time. In a rifting model for the ophiolite of central western Maine according to G. M. Boone (written commun., 1981), the Jim Pond and overlying formations (oceanic layers 1 and 2) would have decoupled, and these were obducted upon the plutonic ophiolite (oceanic layer 3) during later events of collision such as those proposed by Dewey (1976). Rifting emplacement could account for the lack of a sheeted dike unit and is consistent with the presence of tectonite at the structural base of the ophiolite. The surface of decoupling and obduction is required to be cryptic in the rifting model, but could be the zone now occupied by plagiogranite sheets.

The minor element signature and the contents of FeO, Fe_2O_3, and Al_2O_3 (Table 2) all show the allochthonous-type serpentinite (clinochrysotile-bearing) to be class A, or formed from ultramafic plutonic rocks (Faust and Fahey, 1962, p. 86). Shearing, relict dunite texture, preference for fault control of these bodies, the occurrence of the more hydrated form of serpentine (clinochrysotile) in place of antigorite (Page and Coleman, 1967, p. B107), steatitization, and the development of chromiferous muscovite-carbonate rock (arnoldite) suggest either direct emplacement of parent ultramafic plutons into active fault zones or remobilization of serpentinite from the autochthonous counterparts within the Boil Mountain Complex. The latter hypothesis is preferred here, because mapped spatial relations and apparent tectonic communication with the Squirtgun fault both suggest that the sheared serpentinite was volumetrically supplied at the expense of the Boil Mountain Complex, and steatitization and fractionation of the arnoldite was a cogenetic process which occurred during diapiric transport of the serpentinite along active fault zones.

The arnoldite is superficially similar to carbonatite and rodingite. W. B. Thompson (unpublished report, 1968) investigated these similarities and found that rodingite mineral assemblages are quite different from those in arnoldite; they are characterized by the minerals grossularite and zoisite, with or without prehnite. Most occurrences reported in the literature and investigated by Thompson are interpreted to be hydrothermally altered gabbroic dikes. None contain the dolomite-magnesite-talc-chromiferous muscovite-chromite assemblage of the arnoldite. Carbonatite is generally more alkali-rich and has crystallized at temperatures in excess of the stability range of the clinochrysotile associated with the arnoldite. The spatial occurrence and comparative minor element profiles of the arnoldite suggest either that it was totally fractionated from the precursors of the clinochrysotile-bearing serpentinite or that it was produced by some unknown, partially allochemical, process during the serpentinization-steatitization process and crystallized in proximity. This crystallization probably occurred at a time close to or coincident with the termination of migration of the sheared serpentinite bodies.

The timing of the emplacement of the sheared serpentinite bodies is uncertain; it could have occurred at the time of intrusion of the Boil Mountain Complex or much later. The apparent involvement of these sheared bodies with the tectonics of the Squirtgun fault system suggests movement at least as late as Devonian time, because the Squirtgun is known to have dislocated rock at least that young. It is possible that the system was active after Devonian time and that these bodies were mobile well into Mesozoic time during documented tectonic dislocations in the Appalachians.

The tonalitic rocks invite comparison to plagiogranite (Coleman and Peterman, 1975) and trondhjemite, especially that in the Bay of Islands Complex of western Newfoundland as described by Malpas (1978). It is possible that calcium has been introduced into the tonalite from alteration of enclosing rocks; indeed, the protolith of the tonalite rock could have originally qualified as a soda granite. If the tonalite originated as a melt produced by fractional crystalization, as believed, it conforms closely to the genesis described by Malpas (1978) for the Bay of Islands trondhjemite. The tonalite presents a close comparison to aspects of the Rockabema Quartz Diorite of northern Maine (Ekren and Frischknecht, 1967, p. 9-10).

Pillow structures in greenstone, the bedding characteristics of clastic rocks, and the volcanic epiclastic origin of some rocks of the Jim Pond Formation represent accumulations under different conditions: respectively, the volcanic-arc, continental-rise, and island-arc marine environments (Boudette and Boone, 1976). The Jim Pond Formation was probably accumulated close to its source from materials largely contributed by volcanic and intrusive rocks associated with the Boil Mountain Complex. Blue quartz in lapillite and in metagreywacke beds within the greenstone member was interpreted by Boudette (1970, p. C-9) to be from a tonalite facies of the complex. Some of the clastic material was probably contributed through subaerial weathering of emergent parts of the intrusion, as would occur in volcanic islands. The greenstone and the related mafic clastics appear, in part, to compose relict shield volcanoes that have been bevelled after tilting toward the southeast (present context) following burial by the clastic rocks of the metaquartzwacke member and the younger Hurricane Mountain and Dead River Formations (Boone, 1973).

Pillow structures throughout the entire section of the greenstone member indicate aquagene accumulation in a relatively quiet environment probably interrupted by explosive events which produced tephra. Though parts of the volcanic edifices associated with the greenstone member could have been subaerial, no known direct evidence is preserved. Finer grained clastic materials interlayered with the greenstone volcanics could have originated as shallow water cinder-rich blankets. Associated graywacke and quartzwacke could be derived simply from subaerial weathering of older, emergent volcanic rocks. Studies of vesicle diameters and abundance in amygdular greenstone indicate accumulation in water depths in excess of 2 km (Dale Katovich, oral commun., 1974).

The lower contact of the Jim Pond Formation is gradational with the plutonic rocks of the Boil Mountain Complex. Though this zone preserves the features of an intrusive contact at outcrop scale, it is interpreted as being the lithologic progression from the rocks within a pluton to the overlying, consanguineous, eruptive rocks.

The quartz clasts in the quartzwacke member and some rocks within the meta-greywacke member could represent a remote provenance. The immaturity of the quartz clasts in most of the metaquartzwacke, however, suggests a local source such as that of exposed and weathered tonalite within the Boil Mountain Complex. Finer grained quartz clasts and pelitic material in rocks of the metagraywacke member could be, in part, a pelagic contribution.

The iron-formation member is interpreted to represent a deposit accumulated under very quiet marine conditions. This condition was apparently reached immediately following the major pulses of emplacement of the metadacite member. The distribution and the limited stratigraphic and spatial extent of the iron-formation indicate that the accumulation occurred in localized, protected troughs close to the dacite. The excess silica and iron represented in the deposits were probably derived from the active volcanic center. Thus, these deposits are viewed as having an exhalative volcanic origin.

IMPLICATIONS OF OPHIOLITE EMPLACEMENT

The tectonic environment required to accommodate the emplacement of ophiolite probably involved a local metamorphic event. The emplacement of the ophiolite in the early Paleozoic is interpreted to be a precursor event of the Taconian orthotectonic interval. This emplacement is apparently related to northwestward (present context) thrust dislocation of oceanic crust against the Precambrian rocks during the Cambrian (or possibly Early Ordovician). The ophiolite sequence probably represents the final emplacement of oceanic crust on the Precambrian rocks in an area proximal to a spreading zone. It is postulated that the spreading zone was close to the site of subduction such as would occur in a back-arc or inter-arc environment.

The ophiolite-flysch succession must have been initially deformed as the oceanic crust overrode the Precambrian. The exact progression of deformation in Division II rocks, starting with disjuncture and terminating with the orthotectonic culmination of the Taconian event, is not known. It is only certain that deformation of rocks in Division II and in (at least) the lower part of III, by the end of the Taconian, progressed to the stage of near isoclinal folding and the development of flow cleavage essentially as they are seen today (Harwood, 1970). The main effects of Acadian deformation were probably broad warping and tightening of folds.

It was postulated in an earlier section that diapiric serpentinite apparently flowed along a fault plane in places tangential to the ophiolite. The ophiolite probably supplied the ultramafic material to the fault from its stratiform ultramafic component. The presence of diapiric serpentinite along faults subsidiary to the Squirtgun suggests that the tectonic plating of ultramafic material may have been even more extensive along the Squirtgun at one time.

The diamictic character of the Precambrian Helikian rocks of the Chain Lakes massif presents a contrast to the younger Hadrynian (Proterozoic Z) sequences of the Appalachians. We do not presently know whether the Chain Lakes massif is allochthonous, being a detached raft of ancient continental crust, or is essentially in place. If it is a raft of a continental segment, then the transport of the massif into place in central western Maine is interpreted to have occurred during the earliest part of Taconian-Caledonian time (Late Cambrian to Early Ordovician), when the proto-Atlantic Ocean (Iapetus) closed in an event that produced an ophiolitic mélange (Gansser, 1974) and suture along the Baie Verte-Brompton-Ottauquechee line (Williams, 1978). At the time of stabilization of the Chain Lakes raft (Division I), the ophiolite-flysch succession (Boil Mountain Complex and Division II) overrode the Chain Lakes massif and welded itself into place (Boudette, 1978, p. 193, 324). The rocks of central western Maine were probably subjected to metamorphism up to at least the epidote-pumpellyite facies at this time. This obduction event was probably preceded by the convergence and piling-up of much of the flysch-aquagene volcanic cover on the floor of Iapetus lower rise and abyssal plain by the process of dislocation. Prior to subduction the ocean floor acted as a conveyer belt, probably delivering the clastic accumulation into a narrow downwarp in front of the Chain Lakes continental crust segment. It is possible that a subduction surface existed beneath the Chain Lakes raft at this time which subsequently triggered the development of the obduction surface.

The obduction event was followed by tectonic quiescence while the Chain Lakes was gradually lowered into a relatively narrow marine basin marginal to paleo-Appalachia. At this time the Chain Lakes was covered mainly by lower Paleozoic sediments principally derived from rocks that were then being eroded into the basin that lay to the east.

Also during this quiescent period, minor thrusting began along the Squirtgun thrust fault. This fault, believed to be correlative with the Ammonoosuc fault of New Hampshire, developed more fully in a stage of brittle deformation upon its reactivated roots. Major transport on the Squirtgun begun in the Early Devonian, when rocks of the Moose River synclinorium as young as latest Early Devonian age were transported southeast over northwest. It is not known whether this transported segment of the synclinorium simply ended up at a higher structural position in the crust and was subsequently removed by erosion, or whether it was stacked along the international boundary in Quebec to the northwest as the Compton belt.

The third Paleozoic orthotectonic event closed out the Acadian orogeny probably no later than 380 Ma ago. At some time, probably either late Acadian or Hercynian (Pennsylvanian to Permian time), segments of the Acadian thrust fault system were steepened by rotation and selectively passed into right lateral motion. This motion was dissipated by complex refraction effects. During this fault modification, some of the serpentinite along the Squirtgun fault was remobilized, tectonically smeared, and passed into transverse pods, especially along northwest-striking fault planes.

Rocks of the Boil Mountain Complex in the western part of the ophiolite are partially contained within quartzwacke (Fig. 3). Farther to the west and southwest, the ophiolite becomes completely segmented. Such a change in emplacement habit could reflect simple tectonic smearing resulting in cold diapirism. More likely, however, the ophiolite was originally dismembered by detachment gliding and rafting due to gravitational instability. Such instability would occur on a trench margin. Thus, it is interpreted here that the mapped distribution in the region, especially of plutonic ophiolite, shows various degrees of tectonic dismemberment. If this interpretation is true, then it is proposed that the ophiolite passes to the northeast and southwest into ophiolitic mélange.

This conjectured line of ophiolite mélange and the ophiolite in central western Maine mark an important tectonic lineament. This lineament probably is the expression of a microplate closure which extends on strike in the orogenic tectonic grain throughout much of the northern Appalachians.

ACKNOWLEDGEMENTS

Field conferences and discussions with a substantial corps of concerned ophiolite collectors and Appalachian lineament watchers were vital to crystallizing my thoughts on the rocks of central western Maine.

I would like to thank all of these people who generously shared with me many excellent ideas and observations. I am especially grateful for the consultation and unselfish interest of J. B. Lyons and R. E. Stoiber of Dartmouth College; G. M. Boone of Syracuse University and the Maine Geological Survey; and L. R. Page, G. H. Espenshade, R. H. Moench, and T. P. Thayer of the U.S. Geological Survey. Help also came across the international boundary from Canada, to me by such memorable missionaries as Pierre St-Julien, M. J. Kennedy, Harold Williams, W. H. Poole, and W. R. Church.

Editions of the manuscript were constructively reviewed by J. B. Lyons, W. A. Bothner, B. R. Lipin, and G. M. Boone, who substantially improved my logic and the presentation. Significant contributions to field studies of the ophiolite were made by Woodrow Thompson, Jay Murray, Steven Goldthwaite, Timothy Targett, and D. S. Harwood.

REFERENCES

Billings, M.P., 1956, Bedrock geology, pt. 2 of The geology of New Hampshire: New Hampshire State Planning and Development Commission, Concord, 203 p.

Boone, G.M., 1973, Metamorphic stratigraphy, petrology, and structural geology in the Little Bigelow Mountain map area, western Maine: Maine Geol. Survey Bull. 24, 136 p.

Boone, G.M., Boudette, E.L. and Moench, R.H., 1970, Bedrock geology of the Rangeley Lakes-Dead River basin region, western Maine: in Boone, G.M., ed., New England Intercollegiate Geological Conference Guidebook for Field Trips in the Rangeley Lakes-Dead River Basin Region, Western Maine, p. 1-24.

Boucot, A.J., 1961, Stratigraphy of the Moose River synclinorium, Maine: United States Geol. Survey Bull. 1111-E., p. 153-188.

Boudette, E.L., 1970, Pre-Silurian rocks in the Boundary Mountains anticlinorium, northwestern Maine: in Boone, G.M., ed., New England Intercollegiate Geological Conference Guidebook for Field Trips in the Rangeley Lakes-Dead River Basin Region, Western Maine; p. C-1 to C-21.

_____, 1978, The stratigraphy and structure of the Kennebago Lake quadrangle, west central Maine: Ph.D. Thesis, Dartmouth College, 346 p.

Boudette, E.L. and Boone, G.M., 1976, Pre-Silurian stratigraphic succession in central western Maine: in Page, L. R., ed., Contributions to the Stratigraphy of New England: Geol. Soc. America Memoir 148, p. 79-96.

Church, W.R., 1972, Ophiolite, its definition, origin as oceanic crust, and mode of emplacement in orogenic belts, with special reference to the Appalachians: in Irving, E., ed., The Ancient Oceanic Lithosphere: Dept. Energy, Mines and Resources, Ottawa, Earth Physics Branch Publication, v. 42, no. 2, p. 71-85.

Coleman, R.G. and Peterman, Z.E., 1975, Oceanic plagiogranite: Jour. Geophys. Research, v. 80, p. 1099-1108.

Ekren, E.B. and Frischknecht, F.C., 1967, Geological and geophysical investigations of bedrock in the Island Falls quadrangle, Aroostook County, Maine: United States Geol. Survey Prof. Paper 527, 36 p.

Faul, Henry, Stern, T.W., Thomas, H.H. and Elmore, P.L.D., 1963, Ages of intrusion and metamorphism in the northern Appalachians: American Jour. Sci., v. 261, p. 1-19.

Faust, G.T. and Fahey, J.J., 1962, The serpentine-group minerals: United States Geol. Survey Prof. Paper 384-A, 92 p.

Gansser, A., 1974, The ophiolitic melange, a world-wide problem on Tethyan examples: Eclogae Geologicae Helvetiae, v. 67, p. 479-507.

Harwood, D.S., 1970, Nature of the Taconic orogeny in the Cupsuptic quadrangle, west-central Maine: in Boone, G.M., ed., New England Intercollegiate Geological Conference Guidebook for Field Trips in the Rangeley Lakes-Dead River Basin Region, Western Maine, p. H-1 to H-19.

_____, 1973, Geology of the Cupsuptic and Arnold Pond quadrangles, Maine: United States Geol. Survey Bull. 1346, 90 p.

Hurley, P.M. and Thompson, J.B., 1950, Airborne magnetometer and geological reconnaissance survey in northwestern Maine: Geol. Soc. America Bull., v. 61, p. 835-842.

Kennedy, M.J. and Phillips, W.E., 1971, Ultramafic rocks of Burlington Peninsula, Newfoundland: Geol. Assoc. Canada Proceedings, v. 24, p. 35-46.

Lamarche, R.Y., 1972, Ophiolites of southern Quebec: in Irving, E., ed., The ancient oceanic lithosphere: Department of Energy, Mines and Resources, Ottawa, Earth Physics Branch Publ., v. 42, no. 2, p. 71-85.

Laurent, R., 1975, Occurrence and origin of the ophiolites of southern Quebec, northern Appalachians: Canadian Journal of Earth Sciences, v. 12, p. 443-455.

_____, 1977, Ophiolites from the northern Appalachians of Quebec: in Coleman, R.E. and Irwin, W.P., eds., North American Ophiolites: Oregon Dept. Geology and Mineral Industries Bull. 95, p. 25-40.

Laurent, R. and Hébert, Y., 1977, Features of submarine volcanism in ophiolites from the Quebec Appalachians: in Baragar, W.R.A., Coleman, L.C. and Hall, J.M., eds., Volcanic Regimes in Canada: Geol. Assoc. Canada Spec. Paper 16, p. 91-109.

Leonardos, O.H., Jr. and Fyffe, W.S., 1974, Ultrametamorphism and melting of a continental margin: The Rio de Janeiro region, Brazil: Contributions of Mineralogy and Petrology, v. 46, p. 201-214.

Lyons, J.B. and Faul, H., 1968, Isotope geochronology of the northern Appalachians: in Zen, E-an et al., eds., Studies of Appalachian Geology: Northern and Maritime: New York, Interscience Publishers, p. 305-318.

Malpas, J., 1978, Two contrasting trondhjemite associations from transported ophiolites in western Newfoundland: Initial report: in Barker, Fred, ed., Trondhjemites, dacites, and related rocks: New York, Elsevier, p. 465-484.

_____, 1979, The dynamothermal aureole of the Bay of Islands ophiolite suite: Canadian Jour. Earth Sci., v. 16, p. 2086-2101.

Moench, R.H., 1969, The Quimby and Greenvale Cove Formations in Western Maine: United States Geol. Survey Bull. 1274-L, p. L1-L17.

Naylor, R.S., 1969, Age and origin of the Oliverian domes, central western New Hampshire: Geol. Soc. America Bull., v. 80, p. 405-428.

_____, 1975, Age provinces in the northern Appalachians: Earth and Planetary Sci. Annual Review, v. 3, p. 387-400.

Naylor, R.S., Boone, G.M., Boudette, E.L., Ashenden, D.O. and Robinson, P., 1973, Pre-Ordovician rocks in the Bronson Hill and Boundary Mountains anticlinoria, New England, USA (Abst.): EOS, American Geophys. Union Trans., v. 54, no. 4, p. 495.

Neale, E.R.W., Béland, J., Potter, R.R. and Poole, W.H., 1961, A preliminary tectonic map of the Canadian Appalachian region based on age of folding: Canadian Mining Metall. Bull., v. 54, no. 593, p. 687-694.

Page, N.J. and Coleman, R.G., 1967, Serpentine-mineral analyses and physical properties: in Geol. Survey Research, 1967: United States Geol. Survey Prof. Paper 575-B, p. B103-B107.

Poole, W.H., 1967, Tectonic evolution of Appalachian region of Canada: in Neale, E.R.W. and Williams, H., eds., Geology of the Atlantic Region: Geol. Assoc. Canada Special Paper 4, p. 9-51.

Upadhyay, H.D., Dewey, J.F. and Neale, E.R.W., 1971, The Betts Cove ophiolite complex, Newfoundland-Appalachian oceanic crust and mantle: Geol. Assoc. Canada Proceedings, v. 24, p. 27-34.

Williams, Harold, 1971, Mafic-ultramafic complexes in western Newfoundland Appalachians and the evidence for their transportation: a review and interim report: Geol. Assoc. Canada Proceedings, v. 24, p. 9-25.

_____, 1978, Tectonic-lithofacies map of the Appalachian orogen: Memorial University of Newfoundland, Map No. 1.

Williams, Harold, and Payne, J.G., 1975, The Twillingate Granite and nearby volcanic groups – An island arc complex in northeast Newfoundland: Canadian Jour. Earth Sciences, v. 12, p. 982-995.

Wing, L.A., 1951a, Asbestos and serpentine rocks of Maine: Maine Geol. Survey Report, State Geologist, 1949-1950, p. 35-46.

_____, 1951b, Summary report on Maine greenstones: Maine Geol. Survey Report, State Geologist, 1949-1950, p. 47-60.

Wing, L.A. and Dawson, A.S., 1949, Preliminary report on asbestos and associated rocks of northwestern Maine: Maine State Geol. Survey Report, State Geologist, 1947-1948, p. 30-62.

Zartman, R.E., Hurley, P.M., Krueger, H.W. and Gilletti, B.J., 1970, A Permian disturbance of K-Ar radiometric ages in New England: Its occurrence and cause: Geol. Soc. America Bull., v. 81, p. 3359-3374.

Manuscript Received June 15, 1980
Revised Manuscript Received April 10, 1981

Major Structural Zones and Faults of the Northern Appalachians, edited by
P. St-Julien and J. Béland, Geological Association of Canada Special Paper 24, 1982

THE DOVER-HERMITAGE BAY FAULT:
BOUNDARY BETWEEN THE GANDER AND AVALON
ZONES, EASTERN NEWFOUNDLAND

M. J. Kennedy
Department of Geology, University College, Belfield, Dublin 4.

R. F. Blackwood, S. P. Colman-Sadd,
C. F. O'Driscoll and W. L. Dickson
Mineral Development Division, Newfoundland Department of Mines and Energy,
P.O. Box 4750, St. John's Newfoundland A1C 5T7

ABSTRACT

The Dover-Hermitage Bay Fault extends 205 km from Bonavista Bay in northeast Newfoundland to Hermitage Bay on the south coast of the island. It separates plutonic and
metamorphic rocks of the Gander Zone to the northwest from weakly metamorphosed to
non-metamorphic rocks of the Avalon Zone to the southeast, with relatively few plutons. In the
Bonavista Bay region the fault is characterized by mylonites derived from rocks of both zones,
which have undergone subsequent local brecciation. In the Hermitage Bay region the fault is
characterized by a zone of brecciation, although mylonitic rocks occur locally associated with
the breccias. Here the fault is also intruded by a granite which post-dates the mylonitization but
has suffered the later brecciation. The fault is cut by a major Devonian batholith, the Ackley
Granite. No clear direct indications of the direction of movement on the fault have been seen. It
has had a complex movement history. Latest movement is pre-Late Devonian and post-Silurian
but evidence of earlier movement is also present. It is interpreted as a fundamental crustal
fracture within the continental crust of the southeast side of the Appalachian System of Newfoundland.

IUGS
UNESCO
IGCP Project 27
Caledonide
Orogen

Canadian
Contribution
No. 28

RÉSUMÉ

La Faille de Dover-Hermitage Bay s'étend sur une distance de 205 km depuis la baie de Bonavista dans le nord-est de Terreneuve jusqu'à la baie d'Hermitage sur la côte sud. D'un coté, au nord-ouest, se trouvent les roches plutoniques et métamorphiques de la Zone de Gander et de l'autre coté, au sud-est, les roches faiblement ou non métamorphosées de la Zone d'Avalon où les plutons sont plus clairsemés. Dans le secteur de la baie de Bonavista la faille comporte des mylonites affectant les roches des deux zones juxtaposées; une bréchification locale subséquente est aussi manifeste. Dans le secteur de la baie d'Hermitage la bréchification prédomine quoique des mylonites soient localement associées aux brèches. En outre, dans ce même secteur, la faille est recoupée par une intrusion de granite postérieure à la mylonitisation mais affectée par la bréchification tardive. Aucune indication claire, directe de la nature du mouvement n'est connue. Toutefois la reconstitution des évènements montre une histoire complexe. Le dernier déplacement se serait fait au Silurien mais avant le Dévonien supérieur; d'autres mouvements antérieurs se sont aussi produits. Il s'agirait d'une faille crustale fondamentale, dans la croûte continentale bordant au sud-est le système appalachien de Terre-Neuve.

INTRODUCTION

Subdivision of the Appalachian Orogen in Canada into tectonostratigraphic zones based upon contrasting Ordovician or earlier depositional and/or structural development was first proposed by Williams *et al.* (1972). This was a refinement of the older three-fold division of Newfoundland into a Western Platform, Central Mobile Belt and Avalon "Platform" (Fig. 1). Following a more detailed description and naming of these zones in Newfoundland (Williams *et al.*, 1974), Williams has simplified the zonal scheme to enable it to be better applied to the Appalachian Orogen as a whole (1978a and b). The boundaries between the zones in either scheme are generally faults across which lithologic correlation is difficult. This paper is concerned with the boundary between the Gander and Avalon Zones which has been termed the Dover-Hermitage Bay Fault. Lithostratigraphic correlation across it is impossible, thus in this respect it represents possibly the most striking of all the zone boundaries. The Hermitage Bay Fault was first recognized on the south coast of Newfoundland by Widmer (1950) and the Dover Fault was first recognized in the Bonavista Bay region by Younce (1970). These faults and the geology of the immediately adjacent terranes have subsequently been described in the context of the zonal subdivision of the Newfoundland Appalachian Orogen by Blackwood and Kennedy (1975), Blackwood (1976) and Blackwood and O'Driscoll (1976). The Dover and Hermitage Bay Faults form the extremities of what was once a continuous fault that has been subsequently intruded by granitoid plutons, of which the Ackley Granite (Fig. 2) in east-central Newfoundland (Dickson *et al.*, 1980) is the largest.

The Gander Zone, on the northwest side of the fault, is characterized by a variety of metamorphic rocks and granitoid plutons (Fig. 2). In the Bonavista Bay region, gneissic rocks consisting of psammitic and semi-pelitic paragneisses with amphibolites (Square Pond Gneiss) and tonalitic orthogneisses (Hare Bay Gneiss) containing xenoliths similar to the Square Pond Gneiss (Blackwood, 1977) are in contact with psammitic and semi-pelitic schists of the Gander Group farther to the west (Kennedy and McGonigal, 1972; Kennedy, 1976; Blackwood, 1976, 1977, 1978).

The gneisses are intruded by a variety of granitoid plutons of different ages (Fig. 3). The Gander Group has been interpreted to exhibit the effects of pre-Middle Ordovician deformation by Kennedy and McGonigal (1972). Kennedy (1976) suggested that this deformation was probably Late Precambrian. These conclusions have been challenged by some other workers. For example, Bell *et al.* (1977) fail to see any evidence of pre-Ordovician plutonic activity from Rb/Sr whole rock isochron studies of granitic plutons in eastern Newfoundland, and Pajari and Currie (1978) do not agree with pre-Middle Ordovician deformation and metamorphism of the Gander Group. The gneisses have been interpreted as basement to the Gander Group by Kennedy and McGonigal (1972), Kennedy (1976) and Blackwood (1977) but Blackwood (1978) has described them to represent the higher grade and migmatitic equivalents of rocks exposed farther west. This latter interpretation follows the earlier interpretation of Jenness (1963) for rocks of the Gander and Botwood Zones.

In the Hermitage Bay region, the Gander Zone is characterized by similar lithologies (Fig. 4). Semi-pelitic and psammitic paragneisses with amphibolites and tonalitic orthogneisses with xenoliths of these lithologies form the Little Passage Gneisses (Colman-Sadd, 1974, 1978, 1980). The gneisses contain structural and metamorphic features which predate those of the structurally overlying psammites

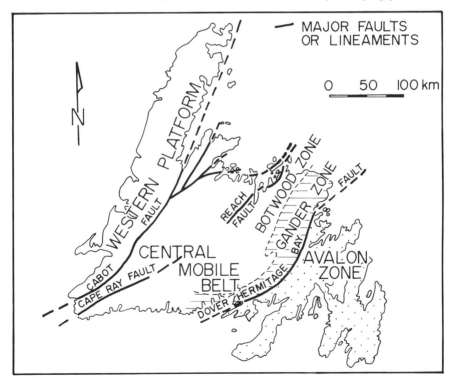

Figure 1. Map of Newfoundland to show the position of the Dover-Hermitage Bay Fault, other major faults and subdivisions of the Appalachian Orogen.

and semi-pelites of the Baie d'Espoir Group (Colman-Sadd, 1974). The Baie d'Espoir Group is correlated in part with the Gander Group farther to the northeast. Colman-Sadd (1980) has interpreted the deformation of the Baie d'Espoir Group as Acadian since he has found no evidence of a break between Middle Ordovician and older rocks of this group. The Little Passage Gneisses are also intruded by granitoid plutons of different ages.

The Avalon Zone adjacent to the Dover-Hermitage Bay Fault is characterized by Late Precambrian volcanic and sedimentary rocks (Fig. 2). Farther to the southeast these are overlain by Cambro-Ordovician and Devonian sedimentary sequences. The

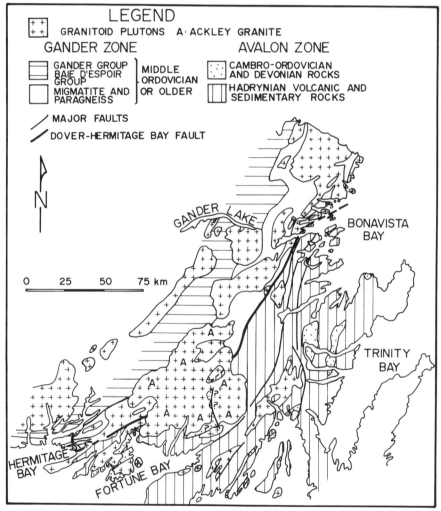

Figure 2. Map of eastern Newfoundland showing the general geology adjacent to the Dover-Hermitage Bay Fault.

Devonian sequences are in part post-Acadian in age. In the extreme north, rocks of the Hadrynian Love Cove Group (acid to intermediate tuffs, other pyroclastic rocks, flows and minor sandstones and conglomerates) occur in two fault-bounded belts (Fig. 3). One belt is bounded by the Dover Fault (Blackwood, 1977) and the other occurs farther to the southeast, separated by a belt of red and green sandstones, conglomerates and local volcanic rocks of the Musgravetown Group. The Musgravetown Group is overlain by Lower Cambrian sediments but the uppermost part of it may be of Early Cambrian age. In contrast to the Gander Zone, granitoid plutons are generally rare within this part of the Avalon Zone.

In the south (Fig. 4), the Avalon Zone is characterized by dominantly basic to acid tuffs and flows with minor sediments of the Connaigre Bay Group (Green and O'Driscoll. 1976; O'Driscoll and Strong, 1979) which are bounded by the Hermitage Bay Fault on the northwest. These rocks are probably equivalent to similar volcanic and sedimentary rocks of the Hadrynian Long Harbour Group that occurs farther to the southeast and is overlain by Lower Cambrian sediments. Devonian clastic rocks also occur in this region. The youngest unit (Great Bay de l'Eau Formation) postdates the main phase of the Acadian Orogeny (Williams, 1971). In contrast to the Bonavista Bay region, the Avalon Zone here contains an extensive variety of Devonian and older granitoid plutons. The Hermitage Bay Fault is also intruded by the Straddling Granite (Blackwood and O'Driscoll, 1976; O'Driscoll and Strong, 1979).

The main phase of movement on the Dover-Hermitage Bay Fault, which formed the mylonites, has been considered to have been Precambrian by Blackwood and Kennedy (1975) and Blackwood and O'Driscoll (1976) with later brecciation being probably of Acadian age. This interpretation was based upon the occurrence of clasts of foliated volcanic rocks (Love Cove Group) in late Precambrian conglomerates of the Musgravetown Group in the north and foliation of some volcanic rocks assigned to the Connaigre Bay Group close to the fault in the Hermitage Bay region, while other rocks of the Connaigre Bay Group adjacent to the fault were unaffected. The foliation of the Love Cove Group passes gradually into the mylonitic foliation of the Dover Fault Zone.

This paper will summarize the relevant features along the line of the Dover-Hermitage Bay Fault and will discuss their significance both in terms of the age of the fault and its position within the Appalachian-Caledonian Orogen.

RELATIONSHIPS ACROSS THE DOVER-HERMITAGE BAY FAULT

Lithostratigraphic units of the Gander Zone do not occur in the Avalon Zone. Ordovician sediments to the northwest of the fault are broadly time stratigraphic equivalents of Ordovician rocks of the Avalon Zone, but the facies are different, being generally turbidites in the Central Mobile Belt and shallow water clastics in the Avalon Zone. It is possible that some of the sediments of the Gander Group are time equivalents to Cambrian or Precambrian rocks of the Avalon Zone, but again the facies are different (Kennedy, 1976). Some granitoid intrusive rocks, however, do cross the Dover-Hermitage Bay Fault. The Ackley Granite was intruded across the fault in Devonian time and the Straddling Granite, although brecciated by the fault, occurs on both sides of it.

The Dover Fault

Relationships across the Dover Fault are best exposed in the northeast (Fig. 3) where the fault zone is transected by the coast (Blackwood and Kennedy, 1975; Blackwood, 1977). In the Hare Bay area, the fault is represented by a topographic depression, but along the southern shore of Lockers Bay, there is uninterrupted exposure from the Gander Zone to the Avalon Zone. South of Bonavista Bay, exposure is generally not as good but the line of the fault can be followed with little difficulty, even though it has no topographic expression. The fault is characterized by a 300 to 500 m wide zone of mylonites which in the northeast have been subsequently brecciated.

Cataclastic Deformation. Close to the Dover Fault, rocks of the Gander Zone are characterized by a strong cataclastic foliation which is sub-vertical and trends parallel to the trace of the fault. The foliation intensifies towards the southeast and passes into protomylonites and ultramylonites of granite protoliths closest to the fault itself. Cataclastic effects on this foliation decrease westwards. The mylonitic fabrics affect the Hare Bay and Square Pond Gneisses, the Lockers Bay Granite and equigranular granites termed Dover Fault Granite closest to the fault. Two mica, locally garnetiferous, granites that cut the gneisses as dikes and plutons are also affected by this cataclastic deformation. Gneisses, megacrystic granites and two mica, locally garnetiferous granites are also overprinted by cataclastic deformation south of Bonavista Bay. All granites that cut the gneisses of the Gander Zone postdate the gneissic foliations. Some granitic bodies are unaffected by cataclastic deformation and hence have been interpreted to post-date this cataclasis. The granitic mylonites of the Dover Fault Zone have been subjected to later deformation which has resulted in mesoscopic gently plunging upright folds of the mylonitic foliation.

The Lockers Bay Granite has yielded a Rb/Sr whole rock isochron age of 300 + 18 Ma (Bell and Blenkinsop, 1977) and an U-Pb zircon concordia age of 460 ± 20 Ma (O'Driscoll and Gibbons, 1980; Dallmeyer et al., 1981). The Middle Brook Granite, a largely undeformed body, has yielded a Rb/Sr whole rock isochron age of 420 ± 20 Ma (Bell and Blenkinsop, 1975; Bell et al., 1977). Megacrystic granite of the "Freshwater Bay Pluton" (Maccles Lake Granite of Blackwood, 1976) has yielded an age of 355 ± 10 Ma by Rb/Sr whole rock isochron (Bell and Blenkinsop, 1975; Bell et al., 1977). The Rb/Sr radiometric age of the Lockers Bay Granite is impossible to reconcile with field relationships. Small plutons of Dover Fault Granite which occur between the Hare Bay Gneiss and the Dover Fault itself are progressively mylonitized into the fault zone. It is these rocks that comprise the dominant protolith of the granitoid mylonites of the fault zone in the north. Variably mylonitized samples from one of these plutons has yielded a Rb/Sr whole rock isochron age of 400 ± 30 Ma (Blenkinsop et al., 1976).

Close examination of the microstructural features of the Dover Fault granitoid mylonites indicates that feldspar phenocrysts have been broken down into flattened polycrystalline lenses surrounded by fluxion lines. Some plagioclase porphyroclasts have either resisted this deformation or have grown as porphyroblasts during a later stage of the cataclastic deformation. They are themselves cataclastically deformed and contained in augen of the cataclastic foliation.

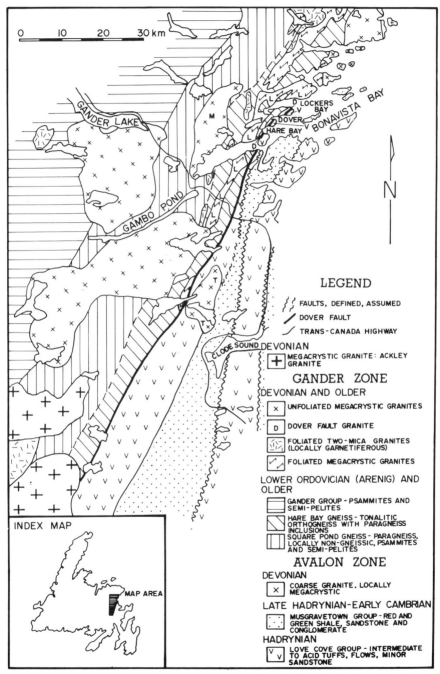

Figure 3. Geologic map of the area north of the Ackley Granite. L: Lockers Bay Granite, M: Middle Brook Granite, T: Terra Nova Granite.

On the southeast side of the Dover Fault, schistose volcanic rocks of the Love Cove Group became increasingly mylonitic as they are traced towards the fault. The mylonitic foliation merges into the regional foliation of the Love Cove Group in this region and no evidence has been seen that cataclasis either pre-dates or post-dates the regional deformation. Mylonitized granitic plutons or dykes and veins are absent, in strong contrast to the adjacent Gander Zone. The regional foliation in the Love Cove Group is subvertical and parallel to the trace of the Dover Fault. Mylonitization in the Love Cove Group has been interpreted as broadly contemporaneous with regional deformation of these rocks (Blackwood and Kennedy, 1975). The resulting foliation is defined by chlorite and sericite. It is locally refolded by a steep northeast-southwest crenulation cleavage or kink bands.

The Love Cove and Musgravetown rocks are intruded by a generally unfoliated granitoid pluton, the Terra Nova Pluton which has been dated by whole rock Rb/Sr isochron at 340 ± 10 Ma (Bell et al., 1977). The pluton presumably postdates deformation of both the Love Cove and Musgravetown Groups.

Mylonites of the Dover Fault are affected by later brecciation in the northeast. Here, brecciation is associated with fluidization (Reynolds, 1954) so that disoriented fragments of mylonite occur in a fine-grained matrix of similar composition. This brecciation is associated with alteration of feldspar to epidote and sericite and results in the fault being eroded to form a depression. Farther to the southeast brecciation is absent. No intrusive rocks are found associated with this brecciation.

The Hermitage Bay Fault

In contrast to most of the line of the Dover Fault, the Hermitage Bay Fault is marked by a linear depression that continues southwestwards to form Hermitage Bay. Exposure within the fault valley is poor, but local exposures here and on the sides of the valley make it possible to recognise the major characteristics of the fault itself. Rocks of the adjacent Gander and Avalon Zones are moderately to well exposed on both sides of the fault zone. Unlike the Dover Fault, the Hermitage Bay Fault is characterized by a 50 to 100 m wide breccia zone rather than mylonite, but mylonitic rocks do occur at a few localities along its length and cataclastic deformation is also evident in the adjacent Gander Zone. The fault is cut by the Straddling Granite that has been affected by brecciation but not mylonitized.

Cataclastic Deformation. Rocks of the Gander Zone are overprinted by a vertical to steeply northwest dipping foliation with a general trend parallel to the Hermitage Bay Fault. This foliation affects most of the granitic rocks of the adjacent Gander Zone (Fig. 4) which includes the Gaultois Granite, locally garnetiferous two mica granites, as well as the Little Passage Gneisses (Colman-Sadd, 1978). All these granitic rocks postdate the gneissic foliations. It does not affect some later granitoid intrusions, of which the Straddling Granite is the most important. The foliation is generally not associated with cataclasis, but locally breakdown of mineral grains into sub-grains and aggregates of smaller grains is evident close to the fault. Quartz is elongated in this foliation, which is also defined by chlorite and biotite.

In the Avalon Zone, dominantly volcanic rocks with minor red sandstones and conglomerates of the Hadrynian Connaigre Bay Group contain gently plunging folds

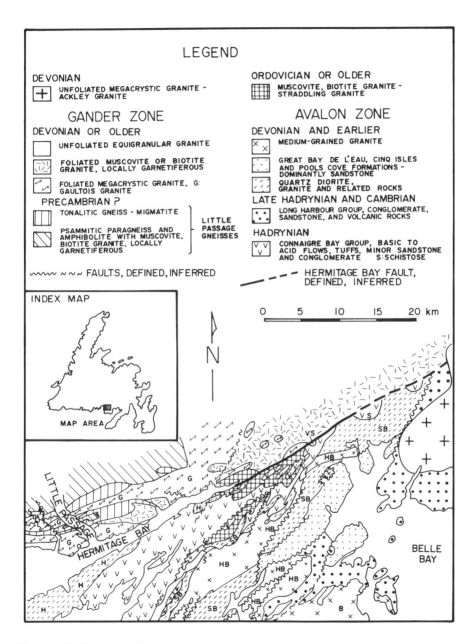

Figure 4. Geologic map of the area southeast of the Ackley Granite. H: Hermitage Complex, HB: Harbour Breton Granite, SB: Simmons Brook Batholith, B: Belleoram Granite.

with a weak inter-penetrative axial planar foliation. This foliation is steep with a northeast trend. There is no cataclastic component associated with this foliation. However, at two isolated localities close to the trace of the Hermitage Bay Fault, the volcanic rocks contain a fine-grained penetrative foliation of similar attitude to that developed elsewhere in the Connaigre Bay Group. This foliation has a clear cataclastic component at one of these localities where the rocks are also pre-tectonically intruded by granodiorite (not shown on Fig. 4). The plagioclase crystals in the granodiorite are strongly flattened into lenses and broken down into polycrystalline aggregates. Quartz and biotite form augen around these porphyroclasts (Blackwood and O'Driscoll, 1976). This cataclastic foliation is steep and parallel to the fault. Other granitic rocks (Fig. 4) of the Avalon Zone adjacent to the fault (i.e., Simmons Brook Batholith) are unaffected by this cataclastic foliation. The Simmons Brook Batholith is overlain nonconformably by the probable Devonian Pools Cove Formation. These relationships have led Blackwood and O'Driscoll (1976) to argue that mylonitization along the Hermitage Bay Fault is earlier than accumulation of most of the Connaigre Bay Group and hence is of Precambrian age. The regionally developed weak foliation and folding of the Connaigre Bay Group is interpreted to be of Acadian age.

The Hermitage Bay Fault is characterised by a zone of brecciation (Widmer, 1950; Blackwood and O'Driscoll, 1976). Fragments of Gander and Avalon Zone lithologies can be recognized within these breccias. Most of the fragments are sub-rounded to rounded, but angular fragments also occur. Gander Zone fragments can be distinguished by their disoriented internal pre-brecciation foliation and consist predominantly of granitic rocks. Avalon Zone fragments are represented by previously unfoliated fragments of volcanic and granitoid rocks. Intermixing of fragments from both zones is rare. Generally there is a sharp contact between breccias derived from rocks of the Gander Zone and those derived from rocks of the Avalon Zone. Much of the zone of brecciation is not exposed. Brecciation affects the Simmons Brook Batholith adjacent to the fault. This batholith is restricted to the Avalon Zone (Fig. 4).

The equigranular Straddling Granite intrudes rocks of both the Gander and Avalon Zones. In the Gander Zone it postdates the late foliation and in the Avalon Zone it intrudes rocks of the Connaigre Bay Group and the Hermitage Complex. It is not foliated, except for local shear zones, and probably predates the Acadian fabric of the Avalon Zone. It is cut by the Hermitage Bay Fault along which it is severely brecciated. The Straddling Granite has yielded a Rb/Sr whole rock isochron age of 490 ± 10 Ma (Blenkinsop et al., 1976). Blackwood and O'Driscoll (1976) have interpreted the main brecciation on the Hermitage Bay Fault as Acadian.

THE ACKLEY GRANITE

This granite occupies the largest granitoid pluton is eastern Newfoundland. It is an irregular body of massive coarse-grained biotite granite which is intrusive into both the Gander and Avalon Zones. Recent mapping (Dickson et al., 1980) has better differentiated between granitic rocks of the batholith and older granitoids of the adjacent Gander Zone. This work, and also that of O'Brien and Nunn (1980) has indicated that rocks in the southeast side of the batholith and pendants within it more probably belong to the Precambrian rather than the Paleozoic sequences. Most of the batholith is coarse-grained and porphyritic. Dickson *et al.* (1980) have recognized three separate lithologic units within it. The batholith truncates the Dover-Hermitage Bay Fault and is unaffected both by mylonitization and brecciation related to it. A possible roof pendant in the eastern part of the batholith may contain a remnant of the fault. Here, coarse-grained porphyroclastic K-feldspar, biotite granite to the northwest is apparently faulted against probable Hadrynian volcanic rocks (Fig. 2). The Ackley Granite Batholith has yielded an age of 345 ± 5 Ma by Rb/Sr whole rock isochron (Bell *et al.*, 1977). This age suggests that cataclasis on the Hermitage Bay Fault was completed by late Devonian time.

DISCUSSION

Although some of the radiometric ages from rocks associated with the Dover-Hermitage Bay Fault conflict with each other, it is possible to reconstruct the general history of movements along this fault zone. It is clear that the fault is a major crustal fracture system within the eastern part of the Newfoundland Appalachians. Correlatives may occur elsewhere in the Appalachian-Caledonian Orogen.

The Dover-Hermitage Bay Fault

Mylonitic fabrics within the fault zone do not show a strong linear element and hence estimates of the orientation of the maximum strain axis is not possible without detailed petrofabric studies. The mylonites appear to be the product of strong flattening with no dominant single direction of extension. It is thus not possible to interpret the kinematics of displacement during cataclastic deformation along the fault. Parallel faults within the Avalon Zone in the north are associated with fanglomerates in the Musgravetown Group, suggesting that these faults were active in the Late Hadrynian, probably as dip-slip fractures. These movements may be related to early movements on the Dover Fault.

Controls on the movement history of the Dover-Hermitage Bay Fault are limited and indirect. The unfoliated granites adjacent to the fault (with the exception of the Straddling Granite) give Rb/Sr dates which generally place a Late Devonian upper limit on the mylonitic deformation. Brecciation locally predates and postdates these intrusions. Age determinations on foliated granites (with the exception of the Lockers Bay Granite) suggest a Silurian lower limit for formation of regional cataclastic fabrics. Thus, the main cataclastic deformation and the associated Dover-Hermitage Bay Fault is best interpreted as Acadian in the light of geochronological studies (Dallmeyer *et al.*, 1981), but this interpretation conflicts with some of the stratigraphic and structural relationships.

Contrasts in the degree of deformation of the Love Cove and Musgravetown Groups and the presence of greenschist clasts in fanglomerates of the Musgravetown Group (Jenness, 1963; Hussey, 1979) have led to the suggestion that deformation of the Love Cove Group and the associated Dover-Hermitage Bay Fault was pre-Musgravetown Group and hence Precambrian (Blackwood and Kennedy, 1975; Blackwood and O'Driscoll, 1976).

Hussey (1978, 1979), however, has shown that around and south of Clode Sound the western belt of the Love Cove Group is conformable with the central belt of Musgravetown Group. The same strong deformation is developed in both groups in the contact area. He suggests that for most of the western Avalon Zone, high-angle faults that formed early in the history of the depositional basin have been continuously re-activated and now largely obscure the generally conformable nature of the Love Cove and Musgravetown Groups. This model readily explains local metamorphic detritus in fanglomerates (Jenness, 1963; Blackwood and Kennedy, 1975; presumably related to the active fault scarps) of the Musgravetown Group. The extent of early deformation is not known, but it was almost certainly localized around high-angle faults, including the Dover-Hermitage Bay Fault. The model would involve formation of schistose rocks in the fault zones at depth while sedimentation was continuing at the surface. Repeated displacement would eventually expose these schists on the fault scarps from which they would be eroded to contribute detritus to fanglomerates. Elsewhere rocks of both groups would be in conformable contact with each other. The Connecting Point Group of greywackes, siliceous siltstones and shales with minor volcanic rocks has generally been considered to be younger than the Love Cove Group and older than the Musgravetown Group, that rests upon it with local unconformity (Jenness, 1963; Kennedy, 1976). The Connecting Point Group (Hayes, 1948) is also locally strongly deformed in the western part of the Avalon Zone in contrast to adjacent rocks of the Musgravetown Group. It is possibly laterally equivalent to part of the Love Cove Group and locally affected by early deformation. Regionally developed foliations which affect the Love Cove, Connecting Point and Musgravetown Groups are interpreted to have formed during the Acadian Orogeny. These fabrics merge with the main cataclastic foliation along and adjacent to the Dover-Hermitage Bay Fault. Thus, although movement along the fault probably dates back to the Precambrian, the presently observable mylonitic fabrics correlate with the regional fabrics of the Avalon Zone and are Acadian. Foliated fragments of Love Cove Group lithologies in the conglomerates of the Musgravetown Group are the earliest manifestations of movement on the Dover-Hermitage Bay Fault and related faults at depth.

General considerations of the geology of the Gander and Avalon Zones in Newfoundland indicate that a deeper crustal level is exposed on the northwestern side of the Dover-Hermitage Bay Fault. This would support interpretation of the fault as a major dip-slip fracture, possibly a high-angle reverse fault. The absence of lenticles of extraneous rock types along the fault line, a characteristic of some strike-slip faults, would also support interpretation as a dip-slip fault. This is also supported by the

---

Figure 5. Map of the northern part of the Appalachian Orogen showing the extension of the Gander and Avalon Zones of the southeast, A: Precambrian and/or early Paleozoic rocks of the Avalon Zone, G: rocks correlated with those of the Gander Zone, FZ: Ordovician and Silurian rocks of the Fredericton Zone. K: Kennebecasis Deformed Zone. M: Meguma Zone, DH: Dover-Hermitage Bay Fault, C: Cobequid-Chedabucto Bay Fault, LB: Lubec-Belleisle Fault, F: Fredericton Fault, N: Norumbega Fault, CN: Clinton-Newbury Fault. Carboniferous and younger cover rocks are shown by dots, gneisses and amphibolite facies metasedimentary rocks that are in part basement and in part possibly metamorphosed equivalents of Paleozoic sequences are shown by stipple.

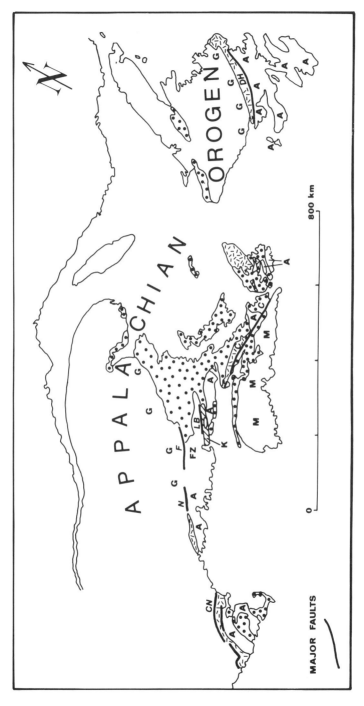

Figure 5

geology of the Gander Zone itself which has been invaded by considerable volumes of granitic magma. It has also been intensely deformed and hence the crust has probably been thickened. Bouguer anomalies over the Gander Zone would support this (Weaver, 1967). Uplift on the northwest side of the fault would have occurred in response to this crustal thickening. Overthrusting of probable ophiolite over the Gander Group (Kennedy, 1976) along the boundary between the Gander and Botwood Zones farther to the west may also be associated with movement on the Dover-Hermitage Bay Fault.

The fault shows some of the characteristics of structures referred to as deep faults by Soviet geologists (Beloussov, 1962, 1972). Such deep faults often form the boundaries of basins or other subdivisions within geosynclines and can exhibit a prolonged movement history which ceases following healing of the fracture at the end of the tectonic cycle. The Dover-Hermitage Bay Fault ceased movement with the Acadian Orogeny. Localization of equigranular granites (Dover Fault Granite) along the fault line in the north was probably controlled by a fracture system which was capable of tapping a source of granitic magma. The latest movements are associated with the Acadian Orogeny and the final stages of closure of the Iapetus Ocean in this part of the Appalachian-Caledonian Orogen.

Similar Faults in Comparable Positions Elsewhere in the Appalachian-Caledonian Orogen

Rocks of the Gander and Avalon Zones can be followed southwestwards into the Appalachians of mainland Canada and the United States (Rast, Kennedy and Blackwood, 1976; Williams, 1978a and b; Kennedy, 1980) and northeastwards into the Caledonides of western Europe (Rast, O'Brien and Wardle, 1976; Williams, 1978a; Kennedy, 1980). No fault analogous to the Dover-Hermitage Bay Fault is present, forming the boundary between rocks of the Avalon Zone to the southeast and rocks of the Gander Zone to the northwest. However major faults occur either within the extensions of the Avalon or Gander Zones and generally form the boundary between Paleozoic and late Precambrian or crystalline terranes. These crystalline rocks are locally recognized as basement to the Avalon Zone.

No contact between rocks of Avalon and Gander affinities has been recognized in Cape Breton Island, but this may possibly be the result of Mississippian and Pennsylvanian sequences which could have been deposited across the contact between the zones. In New Brunswick (Fig. 5) rocks of both zones are separated from each other by a 90 km wide belt of Ordovician and Silurian rocks, whcih has been referred to as the Fredericton Zone (Rast, Kennedy and Blackwood, 1976). Although this zone is represented by a depositional basin, it is not known if it is the product of some spreading as opposed to rifting with no spreading. Rast et al. (1976) have suggested that the Fredericton Zone has resulted from subsequent modification of an older situation that was analogous to the Dover-Hermitage Bay Fault. A broad mylonitic belt which is wider than but not unlike the Dover-Hermitage Bay Fault zone has been described as the Kennebecasis Deformed Zone by Ruitenberg et al. (1979). This zone (Fig. 5) is separated from deformed Paleozoic rocks to the northwest by the Lubec-Belleisle Fault and from Precambrian basement (Greenhead Group) and Early Paleozoic less deformed sequences to the southeast by another fault. Ruitenberg *et al.* (1979) suggest the mylonitization in this belt is pre-Acadian. Rast and Currie (1976) have described mylonitized granitic rocks in the same belt which they believe were mylonitized during the Avalonian Orogeny. It is probable that these mylonitic rocks represent a similar zone to the Dover-Hermitage Bay Fault although the situation has been modified by deposition of Early Paleozoic sequences.

Farther to the southeast, in New England, faults again apparently characterize the northwest boundary of the Avalonian terrane. Gneisses and amphibolite facies metasedimentary rocks (Casco Bay and Blackstone Groups) probably represent basement to the Avalon Zone. In Massachusetts, the Clinton-Newbury Fault forms the northwestern boundary of amphibolite facies rocks faulted against Paleozoic slates to the northwest of the Merrimack Synclinorium. Rocks of Avalonian affinities are restricted to the southeast side of this fault. These relationships suggest modification of a situation similar to that shown by the Dover-Hermitage Bay Fault of Newfoundland.

In the Caledonides of the British Isles, no direct correlations of the Dover Fault have been recognized. Rocks of Avalonian aspect rest unconformably on metamorphosed rocks of the late Precambrian Mona Complex, which is generally considered as the equivalent of the Gander Zone (Williams, 1978a; Kennedy, 1979). Major northeast trending faults are characteristic of this general region, similar to the western part of the Avalon Zone of Newfoundland, but the Paleozoic Welsh Basin intervenes between the Mona Complex with a local Precambrian Avalonian cover and the more extensive Precambrian Avalonian rocks of the Welsh Borderlands and the English Midlands.

ACKNOWLEDGEMENTS

Field and laboratory work upon which this paper is based has been supported by grants from the National Research Council and Natural Sciences and Engineering Research Council (M.J.K.) at Memorial University and Brock University and by the mapping program of the Newfoundland Mineral Development Division under an agreement with the Department of Regional Economic Expansion. The authors also wish to acknowledge much open and useful discussion with friends and colleagues in the Newfoundland Mineral Development Division and in the Department of Geology at Memorial University.

REFERENCES

Bell, K., and Blenkinsop, J., 1975, Geochronology of eastern Newfoundland: Nature, v. 254, p. 410-411.
_____, 1977, Geochronological evidence of Hercynian activity in Newfoundland: Nature, v. 265, p. 616-618.
Bell, K., Blenkinsop, J. and Strong, D.F., 1977, The geochronology of some granitic bodies from eastern Newfoundland and its bearing on Appalachian evolution: Canadian Jour. Earth Sci., v. 14, p. 456-476.
Beloussov, V.V., 1962, Basic problems in geotectonics: New York, McGraw-Hill.
_____, 1972, Basic trends in the evolution of the continents: in Ritsema, A.R., ed., The Upper Mantle: Tectonophysics, v. 13, p. 95-117.

Blackwood, R.F., 1976, The relationship between the Gander and Avalon Zones in the Bonavista Bay region, Newfoundland: M.Sc. Thesis, Memorial University of Newfoundland, St. John's, 156 p.

—————————, 1977, Geology of the east half of the Gambo (2D/16) map area and the northwest portion of the St. Brendan's (2C/13) map area, Newfoundland: Newfoundland Mineral Development Division, Rept. 77-5, 20 p.

—————————, 1978, Northeastern Gander Zone, Newfoundland: Newfoundland Mineral Development Division, Rept. 78-1, p. 72-79.

Blackwood, R.F. and Kennedy, M.J., 1975, The Dover Fault: western boundary of the Avalon Zone in northeastern Newfoundland: Canadian Jour. Earth Sci., v. 12, p. 320-325.

Blackwood, R.F. and O'Driscoll, C.F., 1976, The Gander Avalon Zone Boundary in southeastern Newfoundland: Canadian Jour. Earth Sci., v. 13, p. 1155-1159.

Blenkinsop, J., Cucman, P.F. and Bell, K. 1976, Age relationships along the Hermitage Bay-Dover Fault System, Newfoundland: Nature, v. 262, p. 337-378.

Colman-Sadd, S.P., 1974, The geologic development of the Bay d'Espoir area, southeastern Newfoundland: Ph.D. Thesis, Memorial University of Newfoundland, St. John's, 271 p.

—————————, 1978, Gaultois map area (1M/12), Newfoundland: Mineral Development Division, Rept. 78-1, p. 90-96.

—————————, 1980, Geology of south-central Newfoundland and evolution of the eastern margin of Iapetus: American Jour. Sci., v. 280, in press.

Dallmeyer, R.D., Blackwood, R.F. and Odom, A.L., 1981, Age and origin of the Dover Fault: tectonic boundary between the Gander and Avalon Zones of the northeastern Newfoundland Appalachians: Canadian Jour. Earth Sci., v. 18, p. 1431-1442.

Dickson, W.L., Elias, P. and Talkington, R.W., 1980, Geology and geochemistry of the Ackley Granite, southeast Newfoundland: Newfoundland Mineral Development Division, Rept. 80-1, p. 110-119.

Greene, B.A. and O'Driscoll, C.F., 1976, Gaultois map-area: Newfoundland Mineral Development Division, Rept. 76-1, p. 56-63.

Hayes, A.O., 1948, Geology of the area between Bonavista and Trinity Bays, eastern Newfoundland: Geol. Survey Newfoundland, Bull. 32, Pt. 1.

Hussey, E.M., 1978, Geology of the Sound Island map area (west half), Newfoundland: Newfoundland Mineral Development Division, Rept. 78-1, p. 110-115.

—————————, 1979, The Stratigraphy structure and petrochemistry of the Clode Sound map area, northwestern Avalon Zone, Newfoundland: M.Sc. Thesis, Memorial University of Newfoundland, St. John's, 312 p.

Jenness, S.E., 1963, Terra Nova and Bonavista map areas, Newfoundland: Geol. Survey Canada Memoir 327, 184 p.

Kennedy, M.J., 1976, Southeastern margin of the northeastern Appalachians: late Precambrian orogeny on a continental margin: Geol. Soc. America Bull., v. 87, p. 1317-1325.

—————————, 1979, The continuation of the Canadian Appalachians into the Caledonides of Britain and Ireland: in Harris, A.L., Holland, C.H. and Leake, B.E., eds., The Caledonides of the British Isles – Reviewed: Geol. Soc. London, Spec. Publ. No. 8, Scottish Academic Press, p. 33-64.

Kennedy, M.J. and McGonigal, M.H., 1972, The Gander Lake and Davidsville Groups of northeastern Newfoundland: new data and geotectonic implications: Canadian Jour. Earth Sci., v. 9, p. 452-459.

O'Brien, S.J. and Nunn, G.A.G., 1980, Terenceville (1M/10) and Gisborne Lake (1M/15) map areas, Newfoundland: Newfoundland Mineral Development Division, Rept. 80-1, p. 120-133.

O'Driscoll, C.F. and Strong, D.F., 1979, Geology and geochemistry of Late Precambrian volcanic and intrusive rocks of southwestern Avalon Zone in Newfoundland: Precambrian Research, v. 8, p. 19-48.

O'Driscoll, C.F. and Gibbons, R.V., eds., 1980, Geochronology report – Newfoundland and Labrador: Newfoundland Mineral Development Division Rept. 80-1, p. 143-146.

Pajari, G.E. and Currie, 1978, The Gander Lake and Davidsville Groups of northeastern Newfoundland: a re-examination: Canadian Jour. Earth Sci., v. 15, p. 708-714.

Rast, N. and Currie, K.L., 1976, On the position of the Variscan front in southern New Brunswick and its relation to Precambrian basement: Canadian Jour. Earth Sci., v. 13, p. 194-196.

Rast, N., Kennedy, M.J. and Blackwood, R.F., 1976, Comparison of some tectonostratigraphic zones in the Appalachians of Newfoundland and New Brunswick: Canadian Jour. Earth Sci., v. 13, p. 868-875.

Rast, N., O'Brien, B.H. and Wardle, R.J., 1976, Relationships between Precambrian and Lower Paleozoic rocks of the "Avalon Platform" in New Brunswick, the northeast Appalachians and the British Isles: Tectonophysics, v. 30, p. 315-338.

Reynolds, D.L., 1954, Fluidisation as a geological process and its bearing on the problem of intrusive granites: American Jour. Sci., v. 252, p. 577-614.

Ruitenberg, A.A., Giles, P.S., Venugopal, D.V., Buttimer, S.M., McCutcheon, S.R. and Chandra, J., 1979, Geology and Mineral Deposits, Caledonia area. Mineral Resources Branch, New Brunswick Dept. Natural Resources and Canada Dept. Regional Economic Expansion Memoir 1.

Weaver, D.F., 1967, A geological interpretation of the Bouguer anomaly field of Newfoundland: Dominion Observatory Canada Publ., 35, (5).

Widmer, K., 1950, Geology of the Hermitage Bay area, Newfoundland: Ph.D. Thesis, Princeton University, Princeton, N.J., 459 p.

Williams, H., 1971, Geology of Belleoram map-area, Newfoundland (1M/11): Geol. Survey Canada Paper 70-65, 39 p.

_____, 1978a, Geological Development of the northern Appalachians: its bearing on the evolution of the British Isles: in Bowes, D.R. and Leake, B.E., eds., Crustal evolution in northwestern Britain and adjacent regions: Geol. Jour. Special Issue 10, Seel House Press, Liverpool, p. 1-22.

_____, 1978b, Tectonic lithofacies map of the Appalachian Orogen: Memorial University of Newfoundland, 2 p.

Williams, H., Kennedy, M.J. and Neale, E.R.W., 1972, The Appalachian Structural Province: in Price, R.A. and Douglas, R.J.W., eds., Variations in Tectonic Styles in Canada: Geol. Assoc. Canada Spec. Paper 11, p. 181-261.

_____, 1974, The northeastward termination of the Appalachian orogen: in Nairn, A.E.M. and Stehli, F.G., eds., The Ocean Basins and Margins: Vol. 2: Plenum Press, New York, p. 79-123.

Younce, G.B., 1970, Structural geology and stratigraphy of the Bonavista Bay Region, Newfoundland: Ph.D. Thesis, Cornell University, Ithaca, New York.

Manuscript Received December 23, 1979
Revised Manuscript Received April 4, 1980

Major Structural Zones and Faults of the Northern Appalachians, edited by
P. St-Julien and J. Béland, Geological Association of Canada Special Paper 24, 1982

THE POCOLOGAN MYLONITE ZONE

N. Rast
Department of Geology, University of Kentucky, Lexington, KY 40506

W. L. Dickson
Mineral Development Division, Newfoundland Department of Mines and Energy,
P.O. Box 4750, St. John's, Newfoundland A1C 5T7

ABSTRACT

The edge of the Avalon Platform in New Brunswick lies near the Belleisle Fault. To the southeast of the fault Precambrian rocks, well-exposed rocks of the Greenhead and Coldbrook groups, are overlain by Cambro-Ordovician rocks succeeded by Upper Devonian and Carboniferous strata. Lenticles of these rocks exist within the Belleisle fault zone. To the southeast of Saint John City strongly deformed Carboniferous strata and intrusions are overthrust onto the Precambrian terrain. The Precambrian locally shows evidence of strong cataclastic deformation that cannot be identified in Phanerozoic rocks. At Pocologan a 2 km wide zone has been mapped and is attributed to late Precambrian movements. This deformation is earlier than that associated with the Belleisle Fault or Carboniferous thrusting and is in places overprinted and retrogressed by both. The mylonite zone has polyphase minor structures and is characterized by a Late Precambrian post-tectonic metamorphism succeeded by a later retrogressive event that is attributed to Taconian-Acadian events.

It is proposed that the continental separation during the Late Precambrian had started along a major shear zone and only later was involved in tensional splitting.

RÉSUMÉ

Au Nouveau-Brunswick, la bordure de la plateforme d'Avalon se trouve près de la faille de Belleisle. Au sud-est de la faille, les roches précambriennes des Groupes de Greenhead et Coldbrook, sont bien exposées. Elles sont recouvertes par des roches cambro-ordoviciennes que surmontent les strates du Dévonien Supérieur et du Carbonifère. Ces dernières sont aussi en lentilles dans la zone de faille de Belleisle. Au sud-est de la ville de St-Jean des strates très déformées du Carbonifère et des intrusions chevauchent les terrains du Précambrien. Localement le Précambrien montre des évidences d'une déformation cataclastique poussée laquelle est

absente dans les roches du Phanérozoïque. A Pocologan une zone de mylonite de 2 km de largeur est attribuée à des mouvements datant du Précambrien tardif. Cette déformation est antérieure à celle qui est reliée à la faille de Belleisle ou au chevauchement du Carbonifère et, en certains endroits, ces deux déformations tardives lui sont superposées avec régression du métamorphisme. La zone de mylonite montre des structures mineures polyphasées et elle est caractérisée par un métamorphisme rétrograde attribué à des évènements taconiens-acadiens.

Il est proposé que la rupture du continent s'est produite à partir de fractures de tension développées sur le site d'une ancienne zone de cisaillement datant de la fin du Précambrien.

INTRODUCTION

The general geology of the Avalon Platform of New Brunswick has been recently described by Rast *et al.* (1976) and Ruitenberg *et al,* (1977) and the extent of it is suggested from geophysical data by Haworth and Lefort (1979) who have indicated the edge of the Avalon Platform in New Brunswick at the Belleisle Fault. The general succession has been established at Saint John by Hayes and Howell (1937), who recognized the earlier Greenhead Group of carbonates, quartzites and argillites, overlain by a later, principally volcanic, succession (the Coldbrook Group), as Precambrian basement rocks. The Proterozoic sediments and volcanics have infolds of deformed and weakly metamorphosed Cambro-Ordovician quartzites and greywackes and the whole is unconformably overlain by Late Devonian-Carboniferous conglomerates, sandstones and sandy shales. The Avalon Platform of southern New Brunswick is, thus, characterized by the absence of Silurian-Lower Devonian formations, which exist immediately to the northwest of the Belleisle Fault, the trace of which has been discussed by Garnett and Brown (1973). Rast and Currie (1976) have suggested that the Precambrian rocks of the Avalon Platform actually extend for a few miles beyond the Belleisle Fault, to the northwest of it although Howarth and Lefort (1979) do not so interpret this region. Their plot (Howarth and Lefort, Fig. 6b) of magnetic anomalies suggests that Precambrian volcanic and plutonic rocks only just continue across the Belleisle Fault, and therefore it can be assumed that the Central Mobile Belt of New Brunswick is devoid of Avalonian basement.

Lenticles of Phanerozoic rocks have been mapped along the trace of the Belleisle Fault by Helmstaedt (1968) who recognised that within the fault zone both Cambrian and Carboniferous strata are present. He also suggested that there are lenticles of Silurian rocks in Beaver Harbour (Fig. 1) and although smears of these may be present, the main lenticle so mapped by him has volcanic lapilli tuffs indistinguishable from fossil-dated Cambrian volcanic rocks and carbonates. Detailed mapping by Rast and Grant (in prep.) establishes the lateral passage and the lithologic identity of the volcanic rocks assumed as Silurian and proven Cambrian volcaniclastic strata. The lenticle with the Cambrian strata in the Beaver Harbour has an adjacent lenticle of Carboniferous rocks and to the southeast of it lie Precambrian volcanic and volcaniclastic rocks and occasional sediments of the Coldbrook Group intruded by granitic and gabbroic plutons and by numerous dykes of amphibolitized diabases described by Rast (1979). Wardle (1978) has shown that the Cambro-Ordovician succession of Saint John City is not invaded by such intrusions. Moreover, the aforementioned amphibolite dyke swarm cuts the plutons intrusive into Greenhead and Colbrook Groups (Helmstaedt, 1968; Rast and Currie, 1976; Rast, 1979), but only a few, virtually unmetamorphosed, diabase dykes are found in Cambro-Ordovician strata of Saint John City. The Coldbrook Group to the southeast of Cambrian volcanic and sedimentary rocks in the Beaver Harbour area (Fig. 1) has been traced continuously and is highly deformed along a belt that extends from Negro Harbour northeast through Saint John to the vicinity of Moncton a distance of about 200 km.

Intersecting the variably deformed Coldbrook rocks and associated intrusions is a well developed cataclastic zone that is especially well exposed in the vicinity of the village of Pocologan. This zone has already been briefly alluded to by Rast and Currie (1976). The width of the cataclastically deformed zone of the Coldbrook rocks is approximately 2 km. In our area (Fig. 1), situated 40 km southwest of Saint John, New Brunswick, it is bounded to the southeast by the New River Beach Fault which is continuous with the Kennebecasis Fault to the north of Saint John (Potter *et al.*, 1968). The fault juxtaposes the zone of mylonites and cataclastics against the essentially massive gabbros and diorites referred to as the Golden Grove intrusives from near Saint John (Hayes and Howell, 1937). To the southeast of the latter lies the zone of variably deformed rocks of the Caledonian Mountains of southeastern New Brunswick. These three zones recognized here, essentially corresponded to 5A, 5B and 5C of Ruitenberg *et al.* (1977).

THE LITHOLOGY AND STATE OF DEFORMATION

In this paper only the deformed zone is being considered since it contains a 2 km-wide zone of mylonites that can be traced at least as far as Kingston Penninsula just north of the city of Saint John. The zone is best exposed in the area of New River Beach and Pocologan Harbour (Rast and Currie, 1976) and will be here referred to as the Pocologan mylonite zone. It affects only Precambrian rocks and while its precise age has not directly been determined it is in all probability Late Precambrian since the amphibolites of the aforementioned dyke swarms are at least in part later than the most intense mylonitization. To the southeast the mylonite zone (Fig. 1) is margined by the New River Beach Fault and to the northwest it partly overlaps the belt of dykes, affecting some and being cut by others. Deformation associated with the mylonite zone gradually becomes less intense towards the Belleisle Fault. Therefore, the Belleisle Fault as such has not been responsible for mylonitization.

In the mylonite zone, which is generally characterized by a pervasive schistosity, the following rocks are involved: a) Coldbrook volcanics and volcaniclastic horizons, b) plutons varying from gabbro to granite with granodiorite and quartz diorite being most significant, c) granophyres and prophyries, d) basic dykes and sills. It would have been impossible to identify all these rock types were it not that within the mylonite zone there are large and small tectonic inclusions of original rocks that are internally very little deformed.

Cataclastic Coldbrook Horizons. The Coldbrook rocks tend to become deformed into finely laminated mylonites in which the fine grain size of the matrix has a few phenoclasts as well as freshly grown mica and chlorite. However, where the rocks have been primarily agglomeratic, larger fragments maintain their identity and therefore the rock has numerous lenticular fragments exhibiting preferred dimensional orientation. Such rocks are common along the northwest margin of the mylonite zone. Some laminated horizons, however, were possibly deformed granitic rocks, since they contain coarse remnants of original quartz crystals as phenoclasts, and there are also occasional larger feldspar grains that still are almost euhedral and have evidently been phenocrysts of the original granitic rock. Occasionally, original sedimentary and tuffaceous horizons can be recognized by strong colour banding with varying grain size and texture from lamina to lamina.

Plutonic Rocks. The deformed and mylonitized plutonic rocks are often recognizable by the preservation of protolithic grains, fragments and blocks (Fig. 2) of varying sizes. In places contacts between plutonic rocks and Coldbrook horizons can be detected. The protolithic grains not uncommonly consist of hypidiomorphic intergrown quartz and feldspar, while, in larger fragments, deformed granitic texture is detectable. The largest block so far found is 40 m wide and occurs at New River Bridge (Fig. 1). Here the plutonic rock (hornblende granodiorite) has primary igneous

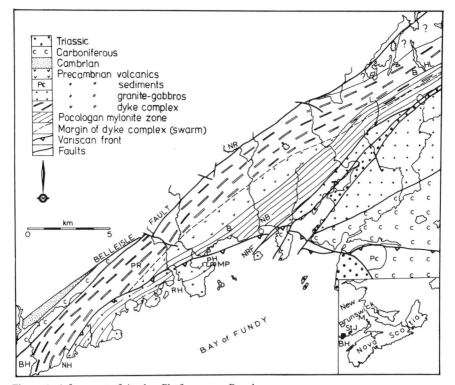

Figure 1. A fragment of Avalon Platform near Pocologan.
BH – Beaver Harbour; B – biotite locations; HL – Hanson Lake; L – Lepreau Village; MP – McCarthy's Point; NB – New River Bridge; NH – Negro Harbour; NR – New River; NRF – New River fault; O – Oland's Beach; P – Pocologan; PH – Pocologan Harbour; PR – Penfield ridge; RH – Redhead. Insert – BH – Beaver Harbour, M – Moncton, StJ – Saint John. The dykes cross variable rocks not distinguished on the map.

banding with axiolitic harristic amphiboles. We interpret the banding to have been originally subhorizontal. At present it is inclined at about 35° to the southeast. The plutonic rocks are in places cut by steeply inclined dykes of leucocratic, sodic granite (trondhjemite). The junctions of the block against the strongly deformed rocks of the surrounding mylonites are very sharp although near to the junction the mylonite is recognizably the same rock as the protolith. The banding within the protoliths gets completely destroyed in mylonitization, but the mylonite often acquires a new mylonitic banding which does not appear to have any connection with banding in protoliths.

Granophyres and Quartz and Feldspar Porphyries. These rocks outcrop mainly at the northwest margin of the mylonite zone and represent minor acid intrusions. Some such intrusions are almost undeformed and the original porphyritic feldspar occurs as still almost euhedral crystals of plagioclase (albite) crowded with fairly large epidote grains and mica flakes. The mineral has been clearly entirely recrystallized. In some locations there are deformed phenoclasts of microcline, but potash feldspar occurs generally in the groundmass where it is intergrown with quartz. In some thin sections a disrupted and recrystallized granophyric texture can be recognized around glomeroporphyritic quartz and feldspar aggregates. Some of the glomeroporphyries are pulled apart and boudinaged.

Basic Minor Intrusions. There are two varieties of old (Precambrian) basic minor intrusions among the volcanic and plutonic rocks, a) the tholeiitic amphibolites, and b) the xenolith-bearing intrusions. The tholeiitic amphibolites vary in composition from low K_2O to normal K_2O basalts to andesite, while the xenolithic intrusions vary considerably and have in general alkalic lamprophyric affinities. The xenoliths in these intrusions often lie parallel or subparallel to the foliation and are severely deformed. The xenolithic intrusions by virtue of their high deformation appear to be earlier than the essentially subvertical dykes of the dyke swarm. A set of cross-cutting, undeformed but metamorphosed mafic intrusions seems to be the latest in the local sequence of minor intrusive activity, but these are probably Phanerozoic and may be Siluro-Devonian since they are indistinguishable from dykes and sills that are emplaced into fossil-dated Silurian strata of the Mascarene Peninsula (Donohoe, 1978).

Minor Structures. The igneous and tuffaceous rocks involved in the mylonite zone bear evidence of polyphase deformation. All the rock types within the zone are generally very strongly foliated and the foliation to which we refer as mylonitic schistosity is very pervasive. Generally two cataclastic styles are present. In one, the rock consists of a series of lenticular bodies with a strong parallel orientation in the plane of the schistosity but lacking pronounced linear orientation. In the other a very fine laminar structure with a strong lineation, at times marked by oriented crystals, occurs. At the edges of the mylonite belt, as for instance on McCarthy's Point (Fig. 1), individual shear zones show transitions from one style to the other. This schistosity, to which one can refer to as Sp, is the most obvious foliation and contains intrafolial folds of original bands.

Within the mylonite zone, Sp can be often recognized as a composite two phase structure involving a relatively slight protomylonitic schistosity Sp_1 later crossed by a very pronounced Sp_2 schistosity. The two schistosities can be particularly well recognized by the development of peculiar kink band-like structures (Fig. 3), where the rock as a whole is more deformed than in the kink band. In places it is possible to demonstrate that the rocks have been further deformed since the schistose amphibolite sills gently transgress the 'kink bands'. The kink bands themselves are less deformed than the bulk of the mylonitized rock and therefore the term pseudo-kink band can be applied to them. The origin of these structures will be discussed elsewhere.

The composite schistosity is commonly tightly folded, sometimes without developing an obvious axial planar structure. These folds have normally almost completely horizontal axes and where they refold Sp a crenulation schistosity (Sp$_3$) can be detected (Fig. 4, Pocologan Harbour). The only later structure that has been recognized consists of variably spaced fracture or crenulation cleavage generally dipping southward or southeastward at angles between 20-45°. This structure, as a result of intersection with Sp or Sp$_3$, produces variably plunging intersection lineation and is locally referred to as Sc. This foliation cross-cuts all the previous structures and everywhere intensifies towards Carboniferous thrust planes; it is therefore inferred to be of Carboniferous age. The Sc planar structure is commonly associated with, and is often isolated by, very thin mylonitic bands containing fragments of previously mylonitized rocks. We interpret the bands as the product of Late Carboniferous deformation. In parts of the section (Red Head Harbour) up to 5 cm-thin bands of Carboniferous mylonitic material abruptly cut across the Precambrian. Nowhere in the section is there evidence for a Phanerozoic deformation other than Carboniferous while the Sc schistosity further to the southeast is found in very deformed Carboniferous rocks lying to the south of the Variscan front (Rast et al., 1978).

THE METAMORPHISM AND ITS RELATION TO STRUCTURE IN THE MYLONITIC ROCKS AT POCOLOGAN

Within the Pocologan mylonite zone a great variety of deformation styles can be detected resulting in extensive and variable cataclastic metamorphism. In general strongly deformed quartz can be recognized either as intensely mortared crystals or as bands of minute domains with diversely oriented optic axes. This type of texture has been described by Phillips (1965) and Higgins (1971). At borders of protoliths (Fig. 2) the mylonites pass into less deformed partially mylonitized rocks in which only a proportion of quartz crystals are thus affected and the texture still is recognizably hypidiomorphic. So far little petrofabric analysis has been carried out since the deformation is polyphase and effects of metamorphism (neocrystalline growth fabrics and new minerals) are variable in the zone. In the simplest cases the texture has a flaser aspect with granulitized quartz lenticles being dominant. In protomylonites occurring at the borders of large protoliths the quartz shows well developed lattice distortion resulting in deformation bands and lamellae, commonly so well-developed that quartz can be mistaken for microcline. Granulitization often concentrates at the edges of such crystals giving rise to mortar texture. The mylonitic fabric thus produced Sp$_1$ - S$_2$ leading to a partial mylonitization accompanied by syntectonic metamorphism referred to as M$_1$.

The feldspar crystals and any original ferromagnesian minerals throughout the mylonite zone are largely converted into a fine grained sericite, chlorite and epidote

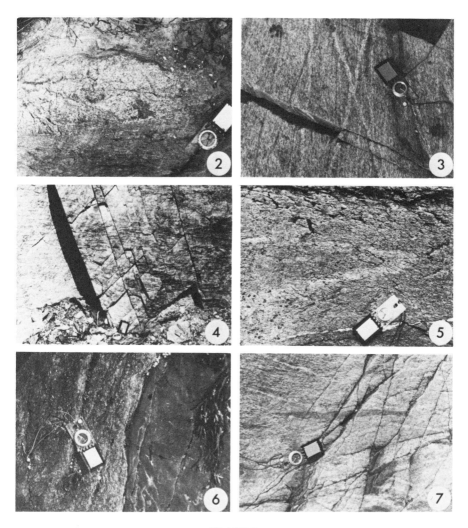

PLATE I

Figure 2. A protolith of granite in the mylonite (near Pocologan).

Figure 3. Mylonite schistosity crossed by kink band type of structures at loc. 0 (Fig. 1).

Figure 4. Subhorizontal crenulation Sp_3 (Pocologan).

Figure 5. Subisoclinally folded aplite with a tight closure in granite mylonite. Faint cross-cutting closely spaced fractures reflect S_c.

Figure 6. Latest basic intrusions cross-cutting the mylonite schistosity. Loc. 0 (Fig. 1).

Figure 7. Strongly deformed basic xenoliths in mylonitized granite. Loc. 0 (Fig. 1).

groundmass in which, nevertheless, saussuritized plagioclase phenoclasts are frequently encountered. In places (Fig. 1), fairly coarse metamorphic biotite is found to overgrow the cataclastically deformed matrix. The biotite is principally post-tectonic with respect to the composite Sp_{1-2} foliation indicating the effects of a post-tectonic M_2 metamorphism. This observation is at odds with the normally held view that the cataclastically deformed quartz recrystallizes as a result of annealing associated with post-tectonic metamorphism (Voll, 1960; Rast, 1965). Where the main schistosity, to be referred to as the Sp_{1-2}, is folded by later folds producing crenulation and a slight schistosity (Sp_3) as is the case in Pocologan Village (Fig. 1), the metamorphic biotite is replaced by chlorite and some surfaces of crenulated composite foliation are completely covered by chlorite and muscovite imparting highly lustrous aspects to other surfaces as for instance in rocks shown in Figure 4. This metamorphic phase of retrogression (M_3) thus is later than the progressive phase that produced biotite. The retrogressive phase is clearly associated with Sp_3.

In the Pocologan district the mylonite zone in the vicinity of the Variscan Front (Rast and Currie, 1976) has developed even a later foliation of Carboniferous age Sc, which occurs as either fracture (Fig. 5) or crenulation cleavage that crosscuts the pre-existing foliation at an appreciable angle. Within the crenulation cleavage, chlorite and sericite are the dominant minerals. In the Pocologan area no higher grade minerals associated with Sc structures have been found, but further to the southwest among deformed and in part mylonitized Carboniferous rocks (Rast *et al.*, 1978), biotite, chloritoid and almandine garnet, post-tectonic to Sc_{1-3}, have been recognized. The Carboniferous mylonitization, however, is entirely later than the Pocologan mylonite zone.

To the northeast of Pocologan area (Fig. 1, loc. HL) the mylonitized rocks are in parts strongly recrystallized and converted into blastomylonites which in thin sections show little evidence of former internal strain but a very pronounced orientation of neocrystalline quartz and feldspar that generally show an LS tectonic fabric that deflects around larger porphyroclasts of feldspar. The blastomylonites often pass into augengneisses that have been reported by Garnett (1973). It is suggested that blastomylonites have been produced during the post Sp_2 progressive metamorphic event.

Thus, in the Pocologan area, one can recognize at least three phases of metamorphism on the basis of relationships between minor structures in mylonitized granites and phases of mineral growth. There is, however, another method of estimating the number and continuity of metamorphic events and that is based on the relationship of basic dykes to the Pocologan mylonite zone.

In the Pocologan mylonite zone and immediately to the north of it four phases of basic dyke intrusion have been recognized as follows:

a) very strongly deformed basic lenticular bodies which antedate the mylonitization (Fig. 1, loc. 0); dk_1 episode;

b) xenolith-bearing dykes that lie parallel to Sp_{1-2} (Fig. 1, loc. NB) but are pre-Sp3; dk2 episode;

c) variably deformed dykes of the swarm (Fig. 1) that generally parallel and slightly overprint the Pocologan zone (Rast, 1979); dk_3 episode;

d) very slightly deformed transgressive dykes, sills and sheets (Fig. 1, loc. 0) that appear to postdate (Fig. 6) all interpenetrative structures in the mylonite zone; dk_4 episode.

The dykes and sheets belonging to episodes dk_1, dk_2 and dk_3 are all interpreted as Precambrian, since all are highly amphibolitized with the main amphibole being yellow-green hornblende and the recrystallized feldspar being oligoclase. Amphibolites of this type have not been found among any Phanerozoic rocks of the Beaver Harbour-Pocologan area. Wardle (1978, after O'Brien, 1976) refers to two sets of Precambrian dykes from the Saint John Area, the first being largely amphibolitized and the second only partly so. Our observations indicate that even in Saint John there are still earlier amphibolite dykes, partly or completely converted into gneisses. These three sets of dykes we correlate with dk_1, dk_2 and dk_3 of the Pocologan area. The dk_1 dykes which predate the mylonitic schistosity, but which in Saint John are rendered gneissose, contribute xenoliths to the granite that have been mylonitized and the xenoliths are strongly flattened and usually elongated parallel to the prevailing lineation in Sp_{1-2} composite schistosity. The xenoliths in three dimensions often show a slight discordance with this schistosity (Fig. 7) so that the apparent direction of movement in the mylonite zone could be determined in relation to individual xenoliths by assuming simple shear. However, further investigation at loc. 0 (Fig. 1) where the xenoliths are especially abundant, has shown that these are strike zones of xenoliths which have alternately clockwise and anticlockwise orientation relative to Sp_{1-2} (Fig. 8). This indicates that the originally deformed xenoliths have been de-

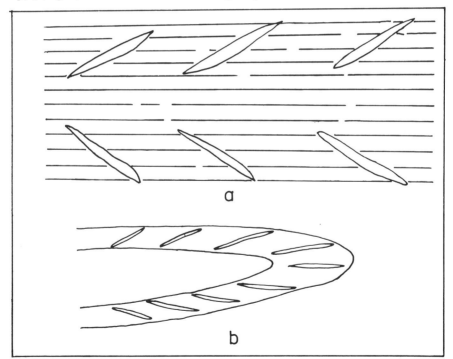

Figure 8. Diagrammatic relationship of xenoliths to mylonitic schistosity in granite (a) and the geometric interpretation (b) of the relationship. Xenoliths are represented as lenticles, schistosity as lines.

formed on the limbs of isoclinal folds, which are only seen on rare occasions (Fig. 5). The fairly uniform axial rations of xenoliths (1 : 10 : 20) indicate that they were originally of similar shape and that they have undergone considerable deformation. On the assumption that the xenoliths were originally nearly spherical, the deformation in the mylonite zone would involve horizontal distension by as much as 600 per cent and vertical distension by 300 per cent. Since, however, throughout the area in the non-mylonitic granitoids similar basic xenoliths are flow-oriented subparallel to the strike and have average ratios of 4 : 6 : 8, it means that the elongation in the xenoliths was of the order of 250 per cent horizontally and 160 per cent vertically. All the deformed xenoliths in the Pocologan mylonite zone have metamorphic biotite or chlorite after biotite and therefore together with the mylonite zone have suffered the M_1 and M_2 metamorphism, the latter reaching upper greenschist facies.

The xenolith-bearing dk_2 basic intrusions that are especially well developed in the mylonite zone are, in the area, characteristic of it and have not been seen elsewhere: they crosscut both the main nylonitic schistosity and the kink bands within it, but are metamorphosed by the mylonite zone as a whole. The intrusions were emplaced prior to the post-tectonic M_3 metamorphism. The dk_3 basic dykes form a pronounced dyke swarm (Fig. 1) in which the earliest dykes are deformed within the mylonitized zone and the latest are entirely post-tectonic, but are affected by M_2 metamorphism which causes their partial amphibolitization and conversion of prophyritic crystals of plagioclase into oligoclase with included crystals of epidote and/or zoisite. Occasional rosettes and large individual crystals of chlorite and blue-green amphibolite associated with epidote and ores are attributed to M_3 metamorphism, but the M_2 metamorphism is much more widely pervasive and all the dykes of the complex are affected by it.

The essentially undeformed basic intrusions (dk_4), which are nevertheless metamorphosed, uralitized and chloritized, we attribute to the Lower Paleozoic (pre-Acadian) episode since lithologically and metamorphically they are identical with the basic intrusions that intrude nearby Silurian formations of the Mascarene peninsula (Donohoe, 1978, p. 71-81). To the west of Pocologan, the only recognized low grade metamorphic effects are attributable to M_3 metamorphism.

RELATIONSHIPS OF THE POCOLOGAN MYLONITE ZONE TO THE DYKE SWARM

The Pocologan mylonite zone partly overlaps the dyke swarm (Fig. 1) in space and in time. The intrusion of the latest of dk_3 dykes, however, took place subsequent to the formation of the mylonite zone, since undeformed dykes cut the cataclastically deformed rocks. Yet, even a limited coexistence of the mylonite zone and dyke swarm in time implies a connection between the tectonic conditions necessary for the intrusion of dykes and the formation of the mylonite zone. The latter in all probability involved transcurrent movements since strain indices such as the xenoliths are elongated within the strike of mylonitic foliation and other linear structures, such as elongated phenoclasts, have also a subhorizontal orientation. Where, as north of loc. L (Fig. 1), there is an overlap between the dykes and the mylonite zone, hornblende lineation in the dykes is also subhorizontal. On the other hand where, as at locality PR (Fig. 1) the dykes lie outside the mylonite zone the hornblende lineation is sub-

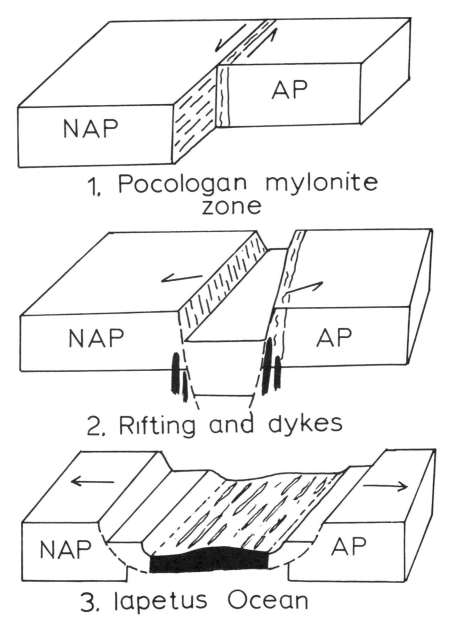

Figure 9. Stage by stage (1 to 3) evolution of the relationships of the mylonite zone dyke swarm and the opening of the Iapetus Ocean.
AP – Avalon plate; NAP – North America Plate.
The mylonite zone ~, the ocean lithosphere and dykes in black; arrows indicate the inferred direction of relative motion of the plates. The left lateral direction is 1 is assumed and at present the evidence is inconclusive.

vertical. In both cases the assumption is that the lineation is in a and corresponds to the principal extensional direction. Thus at partly overlapping times the tectonic setting of the region was affected by both vertical and horizontal compressional movements alternating with tensional episodes that permitted the emplacement of dykes. In a previous paper (Rast, 1979) it has been suggested that the dyke swarm represents the opening of the Iapetus Ocean. If this is so the dykes were presumably coast-parallel and the mylonite zone must have been also coast-parallel. It seems plausible to suggest that the existence of a coast-parallel shear zone implies that the opening of the Iapetus Ocean was nucleated on a major shear zone that originated at the border of Paleoamerica and the Avalonian plate (Fig. 9).

The timing of these events at present can only be referred to the Late Precambrian (700-600 Ma). The mylonite zone affected both volcanic and plutonic rocks and was obviously formed later than the Coldbrook volcanicity and intrusive activity. The deformation was episodal giving rise to a polyphase sequence within the zone, but the episodes were clearly continuous. The opening of the Iapetus Ocean and the intrusion of dykes began at a time when the mylonite zone was well developed. The fact that it often separates rhyolitic and mafic plutonic rocks suggests that its development involved a transcurrent dislocation. The existence of the vertical hornblende lineation in the dyke swarm suggests that the dykes were deformed and metamorphosed at the time of their injection (Rast, 1979). Furthermore the dykes affected by the post-tectonic M_2 metamorphism indicate that metamorphic conditions were prevalent during and after their emplacement. The M_2 metamorphism caused an extensive recrystallization of the latest diabases converting them often into partial amphibolites. These metamorphic conditions cannot be detected in the local Cambrian rocks and therefore the M_2 metamorphism was the latest major Precambrian event so far determined. The dating of the Sp_3 crenulation affecting the Sp_{1-2} schistosity and the foliation is at present imprecise. Firstly, these structures are localized, as indeed are the effect of the retrogressive M_3 metamorphism. It is possible that both these effects are Early to Middle-Paleozoic and are related to Taconian or Acadian movements but at present there is no conclusive evidence in the area examined by us. The effects of M_2 metamorphism on the mylonite zone indicate that deformation along it had stopped by Late Precambrian times.

The above chronological sequence, on a major scale, implies that prior to and in the early stages of the opening of the Iapetus Ocean the Avalonian plate started separating from the North American plate by strike-slip faulting of at present unknown sense that gave rise to the Pocologan mylonite zone. Extensional and vertical movements were produced prior, during, and later in the process, which was accompanied by deformational and metamorphic effects. Still later however, Paleozoic (?) and ultimately Upper Paleozoic deformations are clearly post-mylonitization.

ACKNOWLEDGEMENTS

The authors wish to thank Sandi Guptill for typing and other work related to the preparation of the paper. Bob McCulloch has done the photography and a map prepared by John Foran for an undergraduate thesis has been useful in the preparation. Natural Science and Engineering Research Council, Canada grant A1896 and a grant from the Graduate School, University of New Brunswick, are gratefully acknowledged. Final stages of preparation were financed from Hudnall Endowment, University of Kentucky, USA.

REFERENCES

Donohoe, H.V., 1978, Stratigraphy and structure of the Silurian and Lower Devonian rocks of the St. George-Mascarene Area, New Brunswick: in Ludman, A. ed., New England Intercollegiate Geological Conference Guidebook for Field Trips in Southeastern Main and Southwestern New Brunswick.

Garnett, J.A., 1973, Structural analysis of part of the Lubec-Belleisle Fault zone, Southwestern New Brunswick: Ph.D. Thesis, University of New Brunswick, Fredericton.

Garnett, J.A. and Brown, R.L., 1973, Fabric Variation in the Lubec-Belleisle Zone of Southern New Brunswick: Canadian Jour. Sci., v. 10, p. 1591-1599.

Haworth, R.T. and Lefort, J.P., 1979, Geophysical evidence for the extent of the Avalon zone in Atlantic Canada: Canadian Jour. Sci., v. 16, p. 552-567.

Hayes, A.O. and Howell, B.F., 1937, Geology of Saint John, New Brunswick: Geol. Soc. America Spec. Paper 5, p. 146.

Helmsteadt, H., 1968, Structural analysis of the Beaver Harbour Area, Charlotte County, New Brunswick: Ph.D. Thesis, University of New Brunswick, Fredericton.

Higgins, M.W., 1971, Cataclastic Rocks: United States Geol. Survey Prof. Paper 687.

O'Brien, B.H., 1976, The Geology of parts of the Coldbrook Group, Southern New Brunswick: M.S.c. Thesis, University of New Brunswick, Fredericton

Phillips, W.J., 1965, The Deformation of Quartz in a Granite: Geol. Jour., v. 4, part 2, p. 391.

Potter, R.R., Jackson, E.V. and Davies, J.L., 1968, Geological map of New Brunswick, 1:50,000: New Brunswick Dept. Natural Resources.

Rast, N., 1965, Nucleation and growth of metamorphic minerals: in Pitcher, W.S. and Flinn, G., eds, Controls of Metamorphism: Oliver and Boyd, London, p. 73-102

————————, 1979, Precambrian meta-diabases of Southern New Brunswick-the opening of the Iapetus Ocean?: Tectonophysics, v. 59, p. 127-137.

Rast, N. and Currie, K.L., 1976, On the position of the Variscan front in southern New Brunswick and its relation to Precambrian basement: Canadian Jour. Earth Sci., v. 113, p. 194-196.

Rast, N., O'Brien, B.H. and Wardle, 'R.J., 1976, Relationships between Precambrian and Lower Paleozoic rocks of the 'Avalon Platform' in New Brunswick, the Northern Appalachians and the British Isles: Tectonophysics, v. 30, p. 315-338.

Rast, N., Grant, R.H., Parker, J.S.D. and Teng, H.C., 1978, The Carboniferous deformed rocks west of Saint John, New Brunswick: in Ludman, A., ed., New England Intercollegiate Geological Conference Guidebook for Field Trips in Southeastern Maine and Southwestern New Brunswick.

Ruitenberg, A.A., Fyffe, L.R., McCutcheon, S.R., St. Peter, C.J., Irrinki, R.R. and Venugopal, D.V., 1977, Evolution of pre-Carboniferous Tectonostratigraphic zones: Geosci. Canada, v. 4, p. 171-181.

Voll, G., 1960, New Work on Petrofabrics: Liverpool Manchester Geol. Jour., v. 2, p. 503.

Wardle, R.J., 1978, The stratigraphy and Tectonics of the Greenhead Group: its relationship to Hadrynian and Paleozoic rocks, Southern New Brunswick: Ph.D. Thesis, University of New Brunswick.

Manuscript Received August 17, 1979
Revised Manuscript Received April 14, 1981

Major Structural Zones and Faults of the Northern Appalachians, edited by
P. St-Julien and J. Béland, Geological Association of Canada Special Paper 24, 1982

THE MINAS GEOFRACTURE

J. Duncan Keppie
Department of Mines and Energy, Halifax, Nova Scotia, B3J 2X1

ABSTRACT

The Minas Geofracture is defined as the east-west boundary between the Meguma and Avalon Zones in and adjacent to Nova Scotia. It is a deep crustal fault that separates dissimilar Precambrian and Lower Paleozoic rocks. It is inferred to have originated during the Middle Devonian Acadian Orogeny. Prior to that time, The Avalon Zone from Boston through coastal Maine and southern New Brunswick to near the Magdalen Islands is interpreted as a south-westerly strike continuation of Cape Breton Island. This palinspastic reconstruction is based upon correlation of Precambrian-Early Devonian lithotectonic and paleogeographic belts. During the Acadian Orogeny a dextral movement of at least 370 to 475 km took place on the Minas Geofracture, followed, in Mid-Late Devonian times, by a dextral movement of at least 280 km upon a fault parallel to Chignecto Bay. Some vertical relief on the Geofracture is indicated by the presence of Mid-Late Devonian molasse only on its northern side. As a consequence of these dextral movements the Magdalen Basin formed as a complex rift where the Appalachians swing through a Z-shaped bend. During Early Carboniferous times the Minas Geofracture was generally inactive; however, initial paleomagnetic results suggest that an internally coherent block, including the Avalon and Meguma Zones, moved 1500 to 2000 km northwards relative to cratonic North America. During the Late Westphalian to Permian Hercynian Orogeny, a dextral movement of approximately 165 km is inferred to have taken place on the Minas Geofracture and a dextral movement of some 95 km along Chignecto Bay. Finally, during Early Mesozoic times, a sinistral movement of approximately 75 km took place on the Minas Geofracture as the Bay of Fundy Rift formed in response to the opening of the Atlantic Ocean.

IUGS
UNESCO   *IGCP Project 27*   *Canadian*
          Caledonide   *Contribution*
          Orogen   *No. 29*

RÉSUMÉ

La géofracture de Minas est par définition la limite de direction est-ouest qui sépare les zones d'Avalon et de Méguma en Nouvelle-Ecosse et le secteur limitrophe. C'est une faille crustale profonde jouxtant des roches différentes rattachées au Précambrien et au Paléozoïque inférieur, respectivement. Elle est apparue au cours de l'orogénèse acadienne au Dévonien moyen. Auparavant, la zone d'Avalon depuis Boston en incluant la côte du Maine, le sud du Nouveau-Brunswick et jusqu'au voisinage des Iles-de-la-Madeleine constituait l'extension sud-ouest de l'Ile-du-Cap-Breton. Cette reconstitution palinspatique s'appuie sur une corrélation de zones lithotectoniques et paléogéographiques allant du Précambrien au Dévonien inférieur. Au cours de l'orogénèse acadienne, la géofracture de Minas a subi un mouvement dextre d'au moins 370 à 475 km, suivi au Dévonien supérieur moyen, par un déplacement dextre d'au moins 280 km le long d'une faille parallèle à la Baie de Chignecto. Une molasse du Dévonien supérieur moyen indique que le côté nord seulement de la géofracture manifestait un certain relief. De ces mouvements dextres est résulté le bassin des Iles-de-la-Madeleine formant un complexe de rift au lieu où les Appalaches s'incurvent en forme de Z. Au cours du Carbonifère inférieur, la géofracture de Minas est demeurée à peu près inactive; toutefois les premiers résultats d'une étude paléomagnétique suggèrent qu'un bloc cohérent englobant les zones d'Avalon et de Méguma s'est déplacé de 1500 à 2000 km vers le nord relativement au craton nord-américain. Au cours de l'orogénèse hercynienne allant du Westphalien supérieur au Permien, un mouvement dextre d'environ 165 km s'est manifesté le long de la géofracture de Minas en même temps qu'un déplacement dextre de 95 km se produisait à la Baie de Chignecto. Finalement, au Mésozoïque inférieur, un mouvement sénestre d'environ 75 km se produisait à la géofracture de Minas au moment de la création du rift de la Baie de Fundy reliée à l'ouverture de l'Océan Atlantique.

INTRODUCTION

The northern Appalachians have been divided into five zones by Williams (1979). From west to east, these are the Humber, Dunnage, Gander, Avalon and Meguma Zones. The Humber Zone represented during the Late Precambrian-Early Paleozoic, the ancient continental margin of eastern North America; the Dunnage Zone contains the vestiges of Iapetus; the Gander Zone records the development and destruction of a continental margin; the Avalon Zone was an epi-cratonic platform built upon a micro-continent and the Meguma Zone represented either a continental rise or an intra-cratonic trough on the southern margin of the Avalon Zone. Within this context, it is remarkable to find that all of these zones swing through a Z-shaped bend in the Quebec Reentrant (Williams, 1978). In the Humber Zone, this is attributed to the original curvature in the North American craton bordering Iapetus (Williams, 1979). However, it seems fortuitous that the eastern border of Iapetus should exactly parallel the western margin. In this connection, the significance of the major fault running from the Minas Basin into Chedabucto Bay, here named the Minas Geofracture (Fig. 1) is pertinent because it cuts obliquely across the general northeasterly trend of the Appalachians. It is the thesis of this paper that the Avalon and Meguma Zones were originally straight in this area and have been offset in the Quebec Reentrant along the Minas Geofracture.

DEFINITIONS

The Minas Geofracture is a term introduced here and defined as the east-west boundary between the Meguma and Avalon Zones in and adjacent to Nova Scotia (Fig. 1). As such, it is a deep crustal fault or megashear that separates dissimilar Precambrian and Cambro-Ordovician rocks. Although the Precambrian basement beneath the Meguma Zone is not exposed, its seismic compressional wave velocity of 5.4 km/sec on top of 6.25 km/sec contrasts with the 5.7 to 6.3 km/sec on top of 7.4 km/sec recorded in the Avalon Zone (Dainty *et al.*, 1966) exposed as gneiss, metasediments and metavolcanic rocks. The Cambro-Ordovician turbidites of the Meguma Group are distinctly different from the shelf deposits of the Avalon Zone. Nowhere along the Minas Geofracture can Precambrain or Early Paleozoic rocks be observed in contact because a system of faults parallel to the Minas Geofracture downthrows younger rocks between them. The surface projection of the Minas Geofracture runs from the Minas Basin to Chedabucto Bay. Its eastward continuation may be defined by the southern margin of the Collector Magnetic Anomaly (Haworth and Lefort, 1979). It should be noted that this definition of the Minas Geofracture differs from that of the Glooscap Fault System (King *et al.*, 1975). Thus, although the Glooscap Fault System follows the Minas Geofracture through Nova Scotia it departs from it at both ends where the former follows the Fundian Fault System in the west and the Newfoundland Fracture Zone in the east.

Faults parallel to the Minas Geofracture include the Cobequid Fault, Parrsboro Fault, Economy Fault, Riversdale Fault, North River Fault, Watervale Fault, West River St. Mary's Fault, Chedabucto Fault, Guysborough County Fault, and Guys-

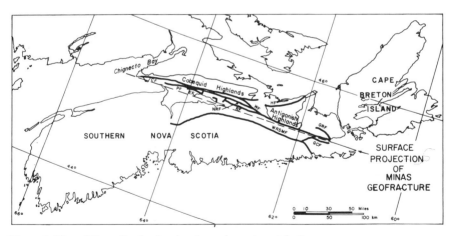

Figure 1. Map of Nova Scotia showing the main east-west faults.
CF = Cobequid Fault, CHF = Chedabucto Fault,
EF = Economy Fault, GCF = Guysborough County Fault,
GRF = Guysborough River Fault, HF = Hollow Fault
NRF = North River Fault, PF = Parrsboro Fault,
RF = Riversdale Fault, WF = Watervale Fault,
WRSMF = West River St. Mary's Fault.

borough River Fault. This has led to the use of the collective terms such as the Cobequid-Chedabucto Fault Zone or System (Webb, 1969) and the Cobequid-Chedabucto-Scatarie Fault System (Lefort and Haworth, 1978). Of all these faults, the Guysborough County Fault most closely approximates the surface projection of the deep crustal Minas Geofracture and all the other faults are merely sub-parallel subsidiaries. It is suggested that the collective term, Minas Fault System, replace the other more cumbersome terms. Movements on the Guysborough County Fault have intensely sheared rocks of the Meguma Group and granitoid plutons many of which exhibit horizontal slickensides. Movements along other individual faults have been variously interpreted to be dextral, sinistral, normal and reverse (Eisbacher, 1969; Benson, 1967, 1974). As such, individual movements probably record only a portion of the complex movement history of the Minas Geofracture.

PALINSPASTIC RECONSTRUCTIONS

Palinspastic reconstructions across major fault zones require the recognition of the same geological features on both sides of the fault. This is possible for faults crossing the Appalachian trend, such as the Minas Geofracture, but is often impossible for strike faults. Reconstructions across the Minas Geofracture can only be made with any degree of certainty for Late Hadrynian and Middle Devonian times using lithotectonic and metamorphic boundaries. Reconstructions for other times are more tentative because they are based upon palaeogeographic lines. Apparent offsets of such lines may be due to either fault displacement or palaeotopographic relief on the fault. For instance, as will be shown later, palaeotopographic relief of the Minas Geofracture Zone is probably the main factor in controlling the distribution of the Late Devonian – Early Carboniferous Horton Group. This renders Webb's (1969) latest Devonian palinspastic map invalid. The palinspastic maps presented here are not strictly palinspastic because they make no adjustments for folding or internal strain, and fault movements are summed on the Minas Geofracture and a fault parallel to Chignecto Bay.

Late Hadrynian

The Precambrian rocks of the Avalon Zone in northern Nova Scotia and southern New Brunswick may be subdivided into four northeast-southwest trending lithotectonic subzones (Fig. 2), all showing varying degrees of Late Hadrynian polyphase deformation (Keppie, 1979c).

The Cape Breton Island subzone is characterized by a gneissic basement of unknown age unconformably overlain by miogeoclinal rocks of the George River Group, which towards the top, interdigitate with volcanic rocks correlated with the Late Hadrynian volcanic rocks of the Fourchu Group. Spatially, gneiss and rocks of the George River Group lie to the northwest of a domain in which only volcanic rocks of the Fourchu Group crop out. The metamorphic grade increases from greenschist facies in the southeast to amphibolite facies in the northwest. Associated Late Hadrynian-Cambrian plutons are widespread in this subzone, where they are predominantly dioritoid and granitoid in composition. Tectonic interpretation of the Cape Breton Island subzone (Keppie, 1979a) suggests that it was part of an ensialic volcanic island arc. Polarity of the arc deduced from the spatial distribution of

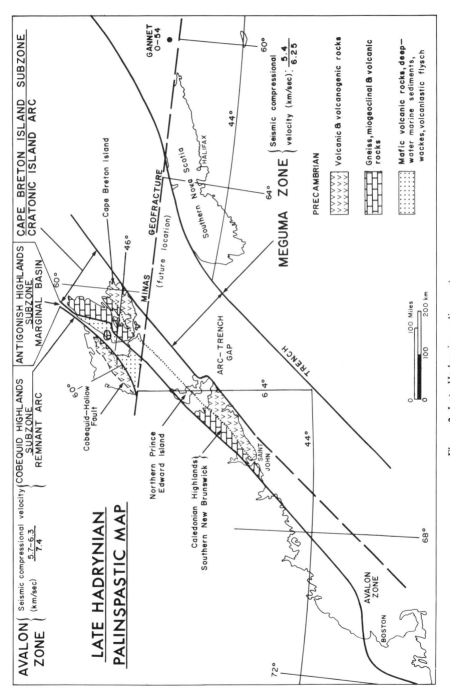

Figure 2. Late Hadrynian palinspastic map.

lithologies and metamorphism, and K_2O variations in the volcanic rocks suggest a northwesterly dipping palaeo-Benioff zone.

The Antigonish Highlands subzone is characterized by submarine mafic volcanic rocks, deep-water marine sediments, wackes and volcaniclastic flysch of the Georgeville Group (Keppie, 1978; Murphy *et al.*, 1979), which are inferred to be underlain by oceanic crust. Metamorphism is typically in the greenschist facies and plutonism is sporadic and mainly gabbroid in composition. The Antigonish Highlands subzone is inferred to represent part of a deformed marginal or back-arc basin (Keppie, 1979a).

The Cobequid Highlands subzone is characterized in the south by amphibolite facies gneisses, metavolcanic rocks, quartzite and schist of the Mt. Thom and Bass River Complexes, whereas in the north greenschist facies metavolcanic rocks with minor marble and siltstone of the Jeffers and Warwick Mountain Formations crop out (Donohoe and Wallace, 1978). Associated plutonism is limited to a few isolated dioritoid and granitoid plutons. The Precambrian rocks of the Cobequid Highlands are interpreted in terms of a remnant volcanic arc (Keppie, 1979a).

The Caledonian Highlands subzone (equivalent to Zone 5 or Ruitenberg *et al.*, 1977) is characterized by a southeastern belt of greenschist facies metavolcanics and volcanogenic rocks of the Coldbrook Group and a northwestern belt exposing gneiss, marble, quartzite and pelitic rocks of the Greenhead Group and metavolcanic rocks of the Coldbrook Group metamorphosed in the amphibolite and greenschist facies. Associated Late Hadrynian-Cambrian plutonism is mainly dioritoid and granitoid. The Caledonian Highlands subzone has been interpreted as either an intracratonic basin or an ensialic island arc (Rast *et al.*, 1976; Giles and Ruitenberg, 1977).

The Minas Geofracture Zone truncates the general northeast-southwest trend of all of the subzones in northern Nova Scotia and yet the Caledonian Highlands subzone may be traced northeastwards to the Magdalen Islands using gravity and magnetic anomalies (Haworth and Lefort, 1979) and southwestwards as far as the Boston area. This leads to speculations about the former southwestard continuation of the three subzones in northern Nova Scotia. A sinistral displacement on the Minas Geofracture is not likely because rocks of the Meguma Zone are not found elsewhere in the northern Appalachians. On the other hand, a dextral displacement along the Minas Geofracture implies that correlatives of the Cape Breton Island, Antigonish Highlands and Cobequid Highlands subzones may be found in the Avalon Zone of New Brunswick and New England. Comparisons suggest that the Cape Breton Island and Caledonian Highlands subzones are correlatives and share the following points in common: (a) a southeastern belt of volcanic and volcanogenic rocks metamorphosed in the greenschist facies, i.e., the Coldbrook and Fourchu Groups are equivalent; (b) a northwestern belt made up of diverse Precambrian lithologies including gneiss, quartzite, marble, pelites and metavolcanic rocks, metamorphosed in the amphibolite and greenschist facies, i.e., the Greenhead and George River Groups are correlatives; (c) widespread Late Hadrynian-Cambrian plutonism; (d) polyphase Late Hadrynian deformation; (e) tectonic interpretation of the volcanic rocks as part of the same ensialic island arc (Keppie, 1979a). Palinspastic restoration for Late Hadrynian times implies that a net dextral displacement of 370 km has taken place on the Minas Geofracture and dextral displacement of about 375 km on a major fault running northeast-southwest through Chignecto Bay since Precambrian times. The Cobequid Highlands were arbitarily restored by moving them along the Cobequid-Hollow Fault to a position west of Cape Breton Island where the trend of the Precambrian structures in the Cobequid Highlands becomes parallel to the Late Precambrian trends in Cape Breton Island (Fig. 2). It should be noted that this Late Hadrynian reconstruction does not include any displacements parallel to the Appalachian trend. Thus, although the Meguma Zone lay on the southeast side of the Avalon Zone it may have lain northeast or southwest along strike of the position shown in Figure 2.

As a consequence of such reconstructions, several conclusions follow. Firstly, the east-west line near the Magdalen Islands terminating the northeastern continuation of the Caledonian Highlands subzone is the former westward continuation of the Minas Geofracture. Secondly, the palinspastic map predicts that the Meguma Zone must have continued to the north somewhere east of Cape Breton Island. On the Grand Banks, schist, greywacke, phyllite, quartzite and granite dated at 376 Ma using K-Ar were encountered at the bottom of the following wells: Gannet D-54, Bittern M-62, and Jaeger A-49 (Jansa and Wade, 1974). This granite body lies well to the east of the easternmost Devono-Carboniferous granite in the Burin Peninsula of Newfoundland suggesting it may be linked with those of the Meguma Zone. Although the age of the other rocks is unknown, it is possible that they could be Meguma Zone rocks. This perhaps finds support in the seismic velocities of 5.1 to 6.0 km/sec recorded in the vicinity of these wells which are more similar to those in the Meguma Zone than those in the Avalon Zone (Sheridan and Drake, 1968). These data provide some constraints on the Late Hadrynian position of the Meguma Zone. Finally, the Late Hadrynian palinspastic map suggests that a back-arc basin equivalent to the Antigonish Highlands subzone should lie on the western side of the Caledonian Highlands subzone, a proposal also suggested by Rast *et al.* (1976).

Cambrian-Early Ordovician

The Cambro-Ordovician rocks of southern New Brunswick and Nova Scotia fall into two general facies belts which coincide with the Avalon and Meguma Zones (Fig. 3). The Cambro-Ordovician rocks of southern Cape Breton Island (Weeks, 1954; Hutchinson, 1952; Smith, 1978) and very similar to those of the Saint John Group in southern New Brunswick (Alcock, 1938; Patel, 1973). Thus, both successions begin with nonmarine rocks followed by shallow marine Middle to Late Cambrian deposits overlain in turn by deeper water Late Cambrian-Early Ordovician black shales. In both areas, the transgression took place from south to north. The recently identified Cambrian rocks in the northern Antigonish Highlands (Murphy *et al.*, 1979) appear to fit into the same general facies belt. These Cambro-Ordovician rocks of the Avalon Zone are markedly different from the thick turbidites of the Meguma Group in the Meguma Zone. The only point of similarity is the occurrence of Tremadocian black shales in both zones.

Unfortunately, the Cambro-Ordovician rocks on the Avalon Zone occur in small isolated areas. Thus, these data cannot be used to unequivocally determine whether the Cambro-Ordovician belts were originally curved or straight and subsequently displaced by faults. The palinspastic map for the Cambro-Ordovician (Fig. 3) relies on data from the Silurian and Early Devonian which suggests that there were no position changes from the Late Hadrynian reconstruction (Fig. 2).

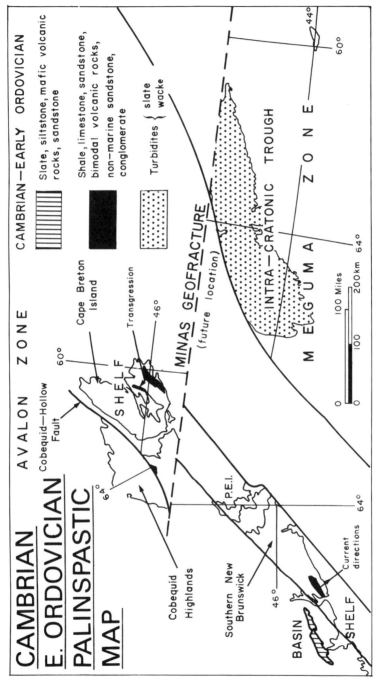

Figure 3. Cambrian-Early Ordovician palinspastic map.

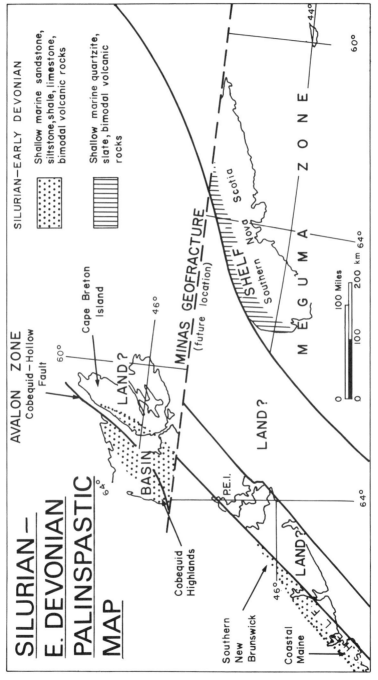

Figure 4. Silurian-Early Devonian palinspastic map.

Silurian – Early Devonian

The Silurian and Early Devonian rocks of southern New Brunswick and Nova Scotia may be divided into two broad facies belts separated by an inferred land area (Fig. 4). In the easternmost facies belt are the shallow marine Silurian-Early Devonian rocks along the northwestern side of the Meguma Zone. Sedimentological studies (Lane, 1975) indicate that the White Rock Formation changes from a shoreface facies on the northwestern side to a shelf mud facies to the southeast. The Early Devonian Torbrook Formation contains a fauna with Rhenish affinities (Boucot, 1960).

Silurian-Early Devonian rocks are absent in the Avalon Zone in southern New Brunswick and in Cape Breton Island except for an isolated occurrence near Mabou (Wait, 1959). This could be due to either nondeposition or deposition and subsequent erosion. However, the proximity of a shoreline on the northwestern side of the White Rock Formation and the northwesterly trending current directions in the latest Silurian-Early Devonian rocks in the Arisaig area (Boucot et al., 1974) suggest the presence of an intervening land area.

In the second facies belt, from the Antigonish and Cobequid Highlands and along the Coastal Volcanic Belt in Maine to Newburyport near Boston, Silurian-Early Devonian rocks are relatively common. Except for the more common occurrence of volcanic rocks from Jones Creek to Newburyport, the sediments and brachiopod communities are remarkably similar (Boucot et al., 1974; Watkins and Boucot, 1975; Cant, 1980), and indicate deep to shallow marine shelf and non-marine environments of deposition. It is significant that the Silurian-Early Devonian rocks in the southern Antigonish Highlands show none of the stratigraphic changes one would expect in approaching a large fault zone (Benson, 1967, 1974). This could be explained by assuming that the Minas Geofracture had no paleotopographic expression. A more palatable alternative is that the Minas Geofracture had not yet come into existence. Thus, the paleogeographic belts outlined above were originally continuous across the Minas Geofracture and have been displaced by subsequent faulting.

Middle Devonian

Coarse, molassic, mid-late Devonian rocks occur only to the north of the Minas Geofracture in Nova Scotia (Keppie, 1979b). In southern Nova Scotia, Early Carboniferous rocks rest unconformably upon Cambrian to Early Devonian rocks intruded by Devono-Carboniferous granitoid plutons. These granitoid plutons postdate the regional low pressure, high temperature regional metamorphism which in turn, post-dates the major Acadian folds (Keppie, 1977). Thus, the regional metamorphism is Middle-Late Devonian in age. Two areas of high temperature regional metamorphism occur around Shelburne and around Canso (Fig. 5). The Canso metamorphic high is bounded offshore to the west of the Fox I-22, Argo F-38, Crow F-52 and Wyandot E-53 wells, which bottomed in phyllite, sericite schist, chlorite-biotite metaquartzite and granite (Jansa and Wade, 1974). At present, this metamorphic high is truncated on the northern side by the Minas Geofracture (Keppie and Muecke, 1979). However, amphibolite facies metamorphism on Green Island, off the southern coast of Cape Breton Island, has recently been dated using $^{39}AR/^{40}$ Ar

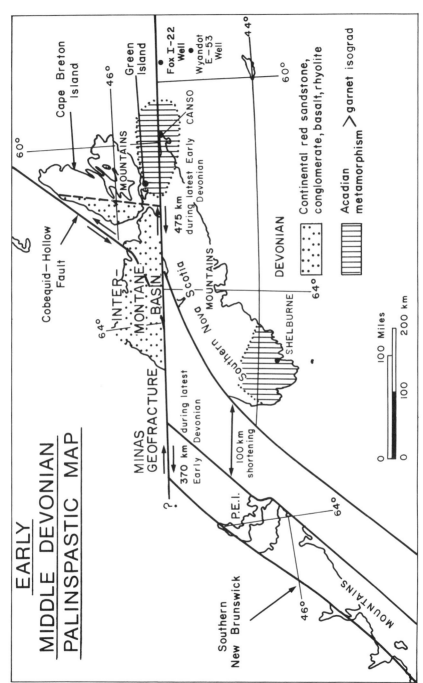

Figure 5. Middle Devonian palinspastic map.

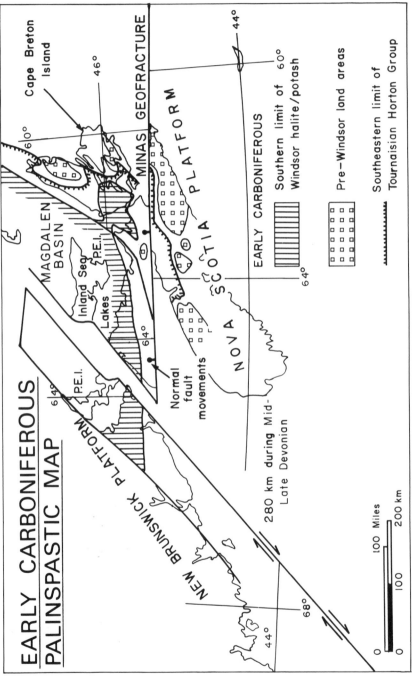

Figure 6. Early Carboniferous palinspastic map.

yielding 353 ± 2 Ma on biotite and 357 ± 4 Ma with a higher plateau at 391 ± 4 Ma on hornblende. The younger ages representing cooling ages whereas the older age probably represents the age of the regional metamorphic event, i.e., early Middle Devonian (Reynolds *et al.*, 1973; Reynolds and Muecke, 1978). The inferred offshore extent of this amphibolite facies metamorphism south of Cape Breton Island is shown in Figure 5. Palinspastic reconstructions across the Minas Geofracture for the Middle Devonian was accomplished by matching these metamorphic highs on either side of the Geofracture.

Carboniferous

Evaluation of data which may be used in palinspastic reconstructions across the Minas Geofracture provide no clear evidence. The Early Carboniferous age of most of the rocks south of the Minas Geofracture limits comparisons to rocks of this age. A line delimiting the southeastern limit of the Horton Group in Nova Scotia has a north-south to northeast-southwest trend except in the vicinity of the Minas Geofracture where it swings east-west (Fig. 6). This suggests that the Minas Geofracture had a paleotopographic expression during the deposition of the Horton Group. This was inherited from the Middle-Late Devonian at which time deposition was limited to the area north of the Minas Geofracture. Consequently, these data do not allow any firm palinspastic reconstructions to be based upon them (Webb, 1969). Also, facies data for the Horton Group are not defined clearly enough to be used.

Facies data for the Windsor Group on either side of the Minas Geofracture exhibit complex distribution patterns (Fig. 6) and do not allow any definite reconstructions to be made. Thus, in the absence of any clear data to the contrary, the Middle Devonian palinspastic reconstruction is applied to the Early Carboniferous. The generally quiescent tectonic conditions inferred during the deposition of the Horton and Windsor Groups lends some support to such a proposal. The position of southern New Brunswick relative to Nova Scotia (Fig. 6) assumes that the southern limit of halite and potash was originally a straight line across Chignecto Bay (P.S. Giles, pers. commun.).

Permo-Triassic

There appears to be no available data relevant to palinspastic reconstructions across the Minas Geofracture. The palinspastic map for Permian times (Fig. 7) is based upon tectonic criteria discussed below.

DISCUSSION

The palinspastic maps prepared for various times from Late Hadrynian to Permian times for southern New Brunswick and Nova Scotia provide some constraints on tectonic movements:

(i) The Late Hadrynian palinspastic map (Fig. 2) implies that the Caledonian Highlands was formerly a southwesterly continuation of Cape Breton Island. Data from the Cambro-Ordovician and especially the Silurian-Early Devonian suggest that this reconstruction persisted throughout the Early Paleozoic and into the Early Devonian. Thus, the Minas Geofracture did not exist until after the Early Devonian.

(ii) During the initial compressive phase of the Acadian Orogeny, the Minas Geofracture saw 370 km dextral movement between southern New Brunswick and Cape Breton Island and 475 km dextral movement of southern Nova Scotia (Figs. 2 to 5). The difference is inferred to have been taken up in folding and compressive strain in the area between southern Nova Scotia and the Caledonian Highlands subzone. These movements took place between early Emsian time (the age of the youngest rocks deformed by the compressive phase of the Acadian Orogeny) and the regional metamorphism at 391 ± 4 Ma, i.e., about 10 Ma. This gives a movement rate of 3.7 to 4.75 cm per year.

(iii) Negative evidence suggests that not much relative movement took place on the Minas Geofracture between late Middle Devonian and Early Carboniferous times. However, about 280 km dextral movement took place upon a fault or faults parallel to Chignecto Bay. It is inferred that as a consequence of these dextral movements the Magdalen Basin formed as a pull-apart basin where the Appalachians swing east-west through the Gulf of St. Lawrence. The Middle and Late Devonian volcanism is inferred to indicate the time during which rifting was active (Keppie and Dostal, 1980). Continued subsidence of the Magdalen Basin during the Carboniferous is inferred to be largely the result of isostatic adjustment due to sedimentary loading.

(iv) In a number of recent papers, various authors (Kent and Opdyke, 1978; Irving, 1979; Van der Voo et al., 1979) have inferred from limited paleomagnetic data that the eastern part of New England and the Maritime Provinces has moved about

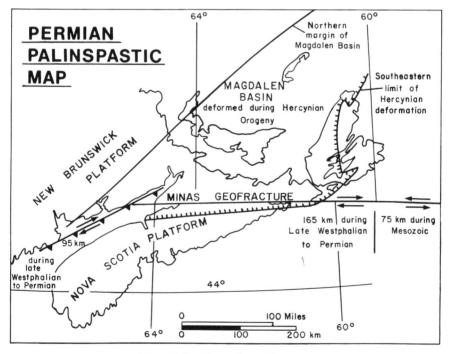

Figure 7. Permian palinspastic map.

1500 to 2000 km northeastwards relative to the rest of North America during the Early Carboniferous. They infer that these movements took place on a sinistral fault zone extending from New York City to Chaleur Bay. If movements of such a magnitude had occurred, intense compressive deformation would have occurred in the Carboniferous Magdalen Basin where the Appalachians swing through an east-west trending Gulf of St. Lawrence segment. Although there is some minor deformation of sedimentary rocks in the Magdalen Basin, it is mainly evident near its margins and is virtually absent in the centre. Thus, such movements are not in agreement with the geological data. Re-examination of the paleomagnetic data reveals that the sampling sites in New England and the Maritimes lie close to the coast of the Bay of Fundy and the Gulf of Maine. It is possible to accommodate major sinistral movements on major faults lying further south such as the Lubec-Belleisle Fault and Norembega-Fredericton Fault. These faults may be connected with the Cabot Fault in Newfoundland with only minor deflection in the Gulf of St. Lawrence, and it is possible that such deflection as there is was produced during the Late Carboniferous-Permian Hercynian Orogeny. If these faults were approximately straight through the Gulf of St. Lawrence and eastern New England, and the southern Maritimes and eastern Newfoundland moved as an internally coherent block, then major sinistral movement could be accommodated without producing intense deformation of the Magdalen Basin. Correlation of the halite-potash facies boundary in the Visean Windsor Group between southern New Brunswick and Nova Scotia (Fig. 6) would support such internal coherency, unless these major sinistral movements were completed by Visean times. More paleomagnetic data would indubitably provide better constraints on the time and location of these large sinistral movements.

(v) The Magdalen Basin was deformed between Early Westphalian and Late Triassic times by movements associated with the Hercynian Orogeny. During this time about 165 km dextral movement is inferred to have taken place on the Minas Geofracture. This figure is controlled by considering the magnitude of movements that have taken place during the Mesozoic to open the Bay of Fundy Rift (Keppie, 1977). These movements are consistent with the calc-alkaline nature of the Late Carboniferous volcanism at Saint John, New Brunswick, which implies the closing of a small ocean basin between Nova Scotia and New Brunswick (Strong et al., 1979). Contemporaneously, thrusting and about 95 km dextral displacement took place along northeast-southwest faults along the coast of southern New Brunswick. Deformation of the sediments in the Magdalen Basin at this time was most intense along its margins (Keppie, 1979c)

(vi) Rifting associated with the opening of the Bay of Fundy during Triassic, Jurassic and Cretaceous times was of limited extent because there appears to be no axial dyke. These rift movements are translated into sinistral movements on the Minas Geofracture and are inferred to be in the order of 75 km.

These data show that the Minas Geofracture has a complex history of repeated movement since it originated during the Devonian Acadian Orogeny. It is interesting to note that the deformation during both the Acadian and Hercynian Orogenies consists of folding and dextral movement upon the Minas Geofracture and the fault parallel to Chignecto Bay. In each case, this is interpreted in terms of dextral transpression caused by oblique collision. Thus, the initial collision was followed by dextral transcurrent faulting parallel to the strike of the orogen. The 1500 to 2000 km northward movement of the Avalon and Meguma Zones during the interval between the Acadian and Hercynian Orogenies is analogous to the rapid motion of microplates during the dying stages of the Tethys Ocean.

ACKNOWLEDGEMENTS

I would like to thank P. S. Giles for critically reading the manuscript, and for much useful discussion. Funds were supplied by the Nova Scotia Department of Mines and Energy and the Department of Regional Economic Expansion, Ottawa.

REFERENCES

Alcock, F. J., 1938, Geology of Saint John region, New Brunswick: Geol. Survey Canada Memoir 216.

Benson, D.G., 1967, Geology of Hopewell Map-area, Nova Scotia: Geol. Survey Canada Memoir 343.

Benson, D.G., 1974, Geology of the Antigonish Highlands, Nova Scotia: Geol. Survey Canada Memoir 376.

Boucot, A.J., 1960, Implications of Rhenish Lower Devonian brachiopods from Nova Scotia: 21st International Geol. Congress Rept., Part 12, p. 129-137.

Boucot, A.J. et al., 1974, Geology of the Arisaig area, Antigonish County, Nova Scotia: Geol. Soc. America Spec. Paper 139, 191 p.

Cant, D.J., 1980, Storm-dominated shallow marine sediments of the Arisaig Group (Silurian-Devonian) of Nova Scotia: Canadian Jour. Earth Sci., v. 17, p. 120-131.

Dainty, A.M., Keen, C.E., Keen, M.J. and Blanchard, J.E., 1966, Review of geophysical evidence on crust and upper mantle structure on the Eastern Seaboard of Canada: American Geophys. Union Monograph 10, p. 349-369.

Donohoe, H.V., Jr. and Wallace, P.I., 1978, Geology map of the Cobequid Highlands: Nova Scotia Dept. Mines Preliminary Map 78-1.

Eisbacher, G.H., 1969, Displacement and stress field along part of the Cobequid Fault, Nova Scotia: Canadian Jour. Earth Sci., v. 6, p. 1095-1104.

Giles, P.S. and Ruitenberg, A.A., 1977, Stratigraphy, paleogeography and tectonic setting of the Coldbrook Group, Caledonian Highlands, southern New Brunswick: Canadian Jour. Earth Sci., v. 14, p. 1263-1275.

Haworth, R.J. and Lefort, J.P., 1979, Geophysical evidence for the extent of the Avalon Zone in Atlantic Canada: Canadian Jour. Earth Sci., v. 16, p. 552-567.

Hutchinson, R.D., 1952, The stratigraphy and trilobite faunas of the Cambrain sedimentary rocks of Cape Breton Island, Nova Scotia: Geol. Survey Canada Memoir 263.

Irving, E., 1979, Paleopoles and paleolatitudes of North America and speculations about displaced terrains: Canadian Jour. Earth Sci., v. 16, p. 669-694.

Jansa, L.F. and Wade, J.A., 1974, Geology of the continental margin off Nova Scotia and Newfoundland: in Van der Linden, W.J.M., Wade, J.A., eds., Offshore geology of Eastern Canada: Geol. Survey Canada Paper 74-30, v. 2, p. 51-106.

Kent, D.V. and Opdyke, N.D., 1978, Paleomagnetism of the Devonian Catskill Redbeds: evidence for motion of the coastal New England – Canadian Maritime region relative to North America: Jour. Geophys. Research, v. 83, p. 4441-4450.

Keppie, J.D., 1977, Tectonics of southern Nova Scotia: Nova Scotia Dept. Mines Paper 77-1, 34 p.

_____, 1978, Browns Mountain Group, Antigonish Highlands, Nova Scotia – preliminary reassessment: Geol. Soc. America, Abstracts with Programs, v. 10, no. 2, p. 50.

_____, 1979a, Precambrian tectonics of Nova Scotia: Geol. Soc. America, Abstracts with Programs, v. 11, no. 1, p. 19.

_____, 1979b, Geological map of Nova Scotia, Scale 1:500,000: Nova Scotia Dept. Mines and Energy, Halifax, N.S.

_____, 1979c, Structural map of Nova Scotia, scale 1:1,000,000: Nova Scotia Dept. Mines and Energy, Halifax, N.S.

Keppie, J.D. and Dostal, J., 1980, Paleozoic volcanic rocks of Nova Scotia: in Proceedings "The Caledonides in the USA": IGCP Project 27, Caledonide Orogen, Dept. Geological Sciences, Virginia Polytechnic Institute and State University, Memoir No. 2, p. 249-256.

Keppie, J.D. and Muecke, G.K., 1979, Metamorphic map of Nova Scotia, Scale 1:1,000,000: Nova Scotia Dept. Mines and Energy, Halifax, Nova Scotia.

King, L.H., Hyndman, R.D. and Keen, C.E., 1975, Geological development of the continental margin of Atlantic Canada: Geosci. Canada, v. 2, p. 26-35.

Lane, T.E., 1975, Stratigraphy of the White Rock Formation: in Harris, I.M., ed., Ancient Sediments of Nova Scotia: Soc. Econ. Paleontol. Mineral., Eastern Section Guidebook, 1975, Field Trip, p. 43-62.

Lefort, J.P. and Haworth, R.J., 1978, Geophysical study of basement fractures on the western European and eastern Canadian shelves: transatlantic correlations and late Hercynian movements: Canadian Jour. Earth Sci., v. 15, p. 397-404.

Murphy, J.B., Keppie, J.D. and Hynes, A., 1979, Geology of the northern Antigonish Highlands: Nova Scotia Dept. Mines, Rept. of Activities 1978, Report 79-1, p. 105-108.

Patel, I.M., 1973, Sedimentology of the Ratcliffe Brook Formation (Lower Cambrian) in southeastern New Brunswick: Geol. Soc. America, Abstracts with Programs, v. 5, p. 206-207.

Rast, N., O'Brien, B.H. and Wardle, R.J., 1976, Relationships between Precambrian and Lower Palaeozoic rocks of the "Avalon Platform" in New Brunswick, the northeast Appalachians, and the British Isles: Tectonophysics, v. 30, p. 315-338.

Reynolds, P.H., Kublick, E.E. and Muecke, G.K., 1973, Potassium-Argon dating of slates from the Meguma Group, Nova Scotia. Canadian Jour. Earth Sciences, v. 10, p. 1059-1067.

Reynolds, P.H. and Muecke, G.K., 1978, Age studies on slate: Applicability of the $^{40}Ar/^{39}Ar$ step wise outgassing method: Earth and Planetary Sci. Letters, v. 40, p. 111-118.

Ruitenberg, A.A., Fyffe, L.R., McCutcheon, S.R., St. Peter, C.R., d'Arinki, R.R. and Venugopal, D.V., 1977, Evolution of Pre-Carboniferous tectonstratigraphic zones in the New Brunswick Appalachians: Geosci. Canada, v. 4, p. 171-181.

Sheridan, R.E. and Drake, C.L., 1968, Seaworth extension of the Canadian Appalachians: Canadian Jour. Earth Sci., v. 5, p. 337-374.

Smith, P.K., 1978, Geology of the Giant Lake area, southeastern Cape Breton Island, Nova Scotia: Nova Scotia Dept. Mines Paper 78-3, 30 p.

Strong, D.F., Dickson, W.L. and Pickerill, R.K., 1979, Chemistry and prehnite-pumpellyite facies metamorphism of calc-alkaline Carboniferous volcanic rocks of southeastern New Brunswick: Canadian Jour. Earth Sci., v. 16, p. 1071-1085.

Van der Voo, R., French, A.N. and French, R.B., 1979, A paleomagnetic pole position from the folded Upper Devonian Catskill red beds, and its tectonic implications: Geology, v. 7, p. 345-348.

Wait, J.H., 1959, Geology of the Mabou area, Cape Breton Island, Nova Scotia, Canada: B.Sc. Thesis, Bates College, Lewiston, Maine.

Watkins, R. and Boucot, A.J., 1975, Evolution of Silurian Brachiopod communities along the southeastern coast of Acadia: Geol. Soc. America Bull., v. 86, p. 243-254.

Webb, G.W., 1969, Paleozoic wrench faults in Canadian Appalachians: in Kay, M., ed., North Atlantic geology and continental drift: American Assoc. Petroleum Geol. Memoir 12, p. 754-786.

Weeks, L.J., 1954, Southeast Cape Breton Island, Nova Scotia: Geol. Survey Canada Memoir 277.

Williams, H., 1978, Tectonic-lithofacies map of the Appalachian Orogen: Memorial University of Newfoundland, Map No. 1.

—————————, 1979, Appalachian Orogen in Canada: Canadian Jour. Earth Sci., v. 16, p. 792-807.

Manuscript Received October 13, 1979
Revised Manuscript Received October 3, 1980

The Geological Association of Canada Special Papers

Orders and requests for information should be sent to:
Geological Association of Canada Publications,
Business and Economic Service, Ltd.,
111 Peter Street, Suite 509,
Toronto, Ontario M5V 2H1.